A Journey Through Water: A Scientific Exploration of The Most Anomalous Liquid on Earth

Authored by

Jestin Baby Mandumpal
Department of Chemistry & Chemical Engineering
Khazar University
Baku
Azerbaijan

A Journey Through Water: A Scientific Exploration of The Most Anomalous Liquid on Earth

Author: Jestin Baby Mandumpal

eISBN (Online): 978-1-68108-423-7

ISBN (Print): 978-1-68108-424-4

© 2017, Bentham eBooks imprint.

Published by Bentham Science Publishers – Sharjah, UAE.
All Rights Reserved.

First published in 2017.

advertisements or ideas contained in the Work.

Limitation of Liability:

In no event will Bentham Science Publishers, its staff, editors and/or authors, be liable for any damages, including, without limitation, special, incidental and/or consequential damages and/or damages for lost data and/or profits arising out of (whether directly or indirectly) the use or inability to use the Work. The entire liability of Bentham Science Publishers shall be limited to the amount actually paid by you for the Work.

General:

1. Any dispute or claim arising out of or in connection with this License Agreement or the Work (including non-contractual disputes or claims) will be governed by and construed in accordance with the laws of the U.A.E. as applied in the Emirate of Dubai. Each party agrees that the courts of the Emirate of Dubai shall have exclusive jurisdiction to settle any dispute or claim arising out of or in connection with this License Agreement or the Work (including non-contractual disputes or claims).

2. Your rights under this License Agreement will automatically terminate without notice and without the need for a court order if at any point you breach any terms of this License Agreement. In no event will any delay or failure by Bentham Science Publishers in enforcing your compliance with this License Agreement constitute a waiver of any of its rights.

3. You acknowledge that you have read this License Agreement, and agree to be bound by its terms and conditions. To the extent that any other terms and conditions presented on any website of Bentham Science Publishers conflict with, or are inconsistent with, the terms and conditions set out in this License Agreement, you acknowledge that the terms and conditions set out in this License Agreement shall prevail.

Bentham Science Publishers Ltd.
Executive Suite Y - 2
PO Box 7917, Saif Zone
Sharjah, U.A.E.
Email: subscriptions@benthamscience.org

BENTHAM SCIENCE

CONTENTS

FOREWORD

On the surface of the Earth planet, water is everywhere: very visible in oceans, rivers, rain or clouds, less visible in rocks or even in the body of living organisms. Being almost the only easily accessible natural liquid, many processes depend on its properties. No life would be possible without the presence of water in its three states. Moreover, its properties play a major role in seasonal cycles and in weather stability. From the largest scales of ocean currents till the molecular scale of biological reactions the singularity of water is evident. As a consequence, water is at the core of research in a large variety of scientific disciplines, as it is nicely presented in the first chapter of this book.

One may ask why, more than any other substance, is water so important and subject of a huge literature, given the apparent simplicity of both the chemical composition and shape of the molecules. The answer cannot be simple because, indeed, complexity comes from different origins, such as the local tetrahedral arrangement or the intermolecular hydrogen bonds and their fast dynamics.

The great merit of the presentation of J. B. Mandumpal is the gradual introduction of the main concepts of the physics of water, with a pedagogical approach based on experimental results and on computer simulations. The text encompasses essentially all the properties of water from the better known till the still controversial models of supercooled water and glass transition.

Given the amount of data available and the enormous number of research papers (more than 400 publication every day!), the content of the book constitutes a remarkable review of the state-of-the-art in water physics, which will be source of information and inspiration for graduate students, scientists and engineers.

José Teixeira
Laboratoire Léon Brillouin (CEA/CNRS)
CEA Saclay 91191
Gif-sur-Yvette Cedex
France

PREFACE

Water is one of the most puzzling substances on earth despite its relatively small size and simple molecular formula. Importance of water in our life need not mention to the scientific community any further as numerous theoretical and experimental investigations have already been performed on water, and staggering volume of research work has been appeared for last several decades in hundreds of scientific journals and books. The second half of twentieth century witnessed a sudden expansion of scientific repertoire due to the refinement of the existing experimental equipment with better resolution and the introduction of computer simulations into basic and applied sciences. It has been widely accepted by now that both experiments and simulations are not independent subjects; rather they are mutually complimentary disciplines; this warrants much more concerted effort in future for better understanding of complex systems like water. The proposed book, titled, **"A JOURNEY THROUGH WATER: A scientific exploration of the most anomalous liquid on earth"**, is an attempt to provide the reader an account of computational and experimental investigations on water.

This work is expected to serve the reader as a useful secondary source of information, with appropriate references to the primary sources, research articles and reviews by pioneers in the field. In addition to the anticipated readers of the book (physicists and chemists), scientists and engineers who indulge in water–based investigations, for example cryobiologists and chemical engineers, can also make use of this work for sharpening their understanding on water. The contents of this book are presented in such a manner that a person with minimal understanding of physics and chemistry can comprehend most of them without much laborious effort. An important feature of this book is the way by which the introductory chapter has been presented, contrary to the traditional approaches: I venture into providing a wider outlook to water from a socio–economic, political and technological perspectives. This, I hope, will make the scientific community much more aware of the importance of their research, and prompt to design their aim according to social needs as well. The first four chapters serve as a platform for the subsequent five core chapters (5-9). Chapters from 2 to 4 are devoted to cater the needs of people who do not have fundamental understanding of various simulation and experimental methods as well as theories that have been developed over the years for explaining the properties of water and liquids in general.

Conceived in the beautiful city of the South African coast, Capetown, I proceeded to write this book part by part in several countries including the Republic of Maldives where I was later appointed as a Teaching Service Officer (TSO) under the Ministry of Education of the Republic, and Baku, where I am living now. Completion of this book was a long haul, and took more than three years to the present form after several reorganisations of the chapters. Even the title of the book has been revised several times: initially I thought of focussing mainly on computational investigations of water, but had to change my mind since such a move would become futile attempt given the fact that neither experiments nor computer simulations are standalone as I mentioned before, at least in the case of water. I thought to include clathrates suggested by one of the reviewers, but had to abandon this idea due to the never ending task of completion and also non−familiarity of the topic, but it will definitely be included in my future assignments!

This book is dedicated to my departed father Baby Mandumpal, who has been very inspirational in my life, and other family members including my mother Filomina, wife Priyanka and our little daughter Joann. I would like to thank Professor Ricardo L. Mancera at Curtin University for introducing me to this marvellous topic and for the stimulating discussions during my PhD research. I owe much to him for the training I obtained during my stay in Perth in the art of scientific presentations and more importantly structured academic writing. I would like to mention my gratitude to Professor Hamlet Izaxanli, the president of Khazar University, and Professor Hassan Niknafs, rector of Khazar University for inviting me to the wonderful city, Baku. I confer gratitude to my friends Rev. Dr. Paul Kattookaran (Co-ordinator, **Art-i** (Indian Christian Artists' Forum)) for motivating me towards fulfilling this task, Dr. Rajesh Komban (Research Scientist at Center for Applied Nanotechnology (CAN) Hamburg, Germany) and Dr. Thiruvancheril Gopakumar (Assistant Professor at Indian Institute of Technology, Kanpur) for providing numerous manuscripts for the completion of this book. Without these vital supports, this book could not have been materialised. I thank Professor José Teixeira (Laboratoire Léon Brillouin, France) for his willingness for writing an appropriate "Foreword" to this treatise. Finally I thank Bentham Science publishers for inviting me to write this book and in particular their acquisition editors Ms. Dur−e−Shahnaz Shafi and Ms. Humaira Hashmi for reminding me the deadline constantly. In fact, there are more people that I could mention here, I tender my apology for not including all of them here.

As one of the prominent theoretical physicists of our era Stephen Hawking in the preface of his seminal book "A brief history of time − from big bang to black holes" mentioned, the more one includes complex mathematical equations in a book, the more readers can be deterred from it. I support for the notion of books written in plain language, especially when they are aimed at general audience, for quick understanding of the subject. As a result, I tried to minimize the number of mathematical equations as far as I can, without losing the rigour of the subject. I have tried my level best to provide the already available research work (most of which are written in academic language with complex physics and mathematics) that has been carried out hitherto on water as much as I can. Nevertheless, the expanding volume of research on water year by year makes this work an endless task. I would therefore encourage readers for making constructive criticisms on the content of this work. This book's shortcomings, in terms of its contents and style, weigh heavily upon me. I can only say that this book does not serve to account of everything we know about water. Nevertheless, I venture to hope that the readers will enjoy a short journey through this book!

Jestin Baby Mandumpal
Baku
Azerbaijan

A Journey Through Water: A Scientific Exploration of The Most Anomalous Liquid on Earth

A Journey Through Water: A Scientific Exploration of The Most Anomalous Liquid on Earth

CHAPTER 1

Water, The Centre of Life

Abstract: The ever increasing demand for clean water has prompted the world to consider water scarcity in a serious way. Some regions in the world are already at the brink of war over the ownership of major water resources, and it is feared that the situations may become worse. The marginalised people living in the impoverished regions of the world are struggling to obtain clean water, non–availability of which puts their life in utmost misery. Despite the fact that technological innovation provides some solution to this matter, water's growing demand surpasses what technology can offer. A joint approach unifying various facets of human life is necessary to overcome the issue, and hence they are discussed in detail. It must be appreciated that several organisations including the U. N., representing all nations around the globe, is taking proactive steps to curb this problem by setting up various committees to study the matter in depth and taking appropriate measures to decentralise the resources to all. With the development of robust computer simulation methods, and water models, it is now possible to study water at microscopic level. Together with state–of–the–art experimental techniques the properties of water can be unravelled further. This is expected to have tremendous impact upon improving the quality of water refinement process since most of them are fundamentally of a chemical nature.

Keywords: Disinfection, Filtration, IWMI, MENA, Peptide, Reverse osmosis, Salinity, Solar pasteurisation, Speciation, Water crisis, Water logging, WHO.

INTRODUCTION

Our earth, a blue water planet when observed from space, contains approximately 75% of water, but the vast majority of it (a whopping 97%) is salty and too concentrated to be useful for most of the habitats. This means that the sustainability of life heavily depends on the remaining 3% of water on earth (fresh water). The need for pure water, in particular, creeps through all spheres of life has become foci of our attention: from political summits to economic and scientific conferences, as evident from the emergence of specialist academic journals, particularly aimed at discussing different perspectives of water and life [1]. Ever increasing demand for this substance in quality (in its pure state) and in quantity is a challenge for the world in the coming years. Since the global human population is skyrocketing and proportional increase in natural water resources does not seem to be realistic, finding an overarching solution for this problem is a daunting task. The explosion in population also means proportional rise in water–consuming industries, both leading to a reduction in per capita water availability [2]. Other detrimental effects such as climate change, over–exploitation of natural resources and environmental degradation are also associated with them. It has been pointed out by the experts that the demographic explosion generates much more water scarcity than the environmental hazards such as climate change [3]. The studies conducted by Lazarova *et al.* underline this observation: with the environment forecast for the coming 80 years, they demonstrate that the effect of Climate Change does not necessarily have negative influence around the world at the same level [4]. This can be explained by the fact that Climate Change does not reduce amount of rain received on earth, but it only alters volume and timings of river flow, causing damages only at some places. Hence it is very evident that only with proper water management involving the following core principles, namely development of new water technologies, inter basin water transfer, efficient irrigation systems and incurring appropriate charges for water, these issues can be resolved [4]. Before going to a deep analysis of various aspects of water, we need to define "water crisis".

Water Crisis

Water crisis is the shortage of water for internal and external consumption, which occurs due to growing imbalance between supply restricted by stagnated natural resources and demand increased by growing number of consumers. This is a much oversimplified statement because under this definition only human being's needs (internal consumption (for *e.g.* clean drinking water) and external consumption (for *e.g.* irrigation, industry operations and power generation)) are included at the expense of the basic rights of other organisms, inclusion of which magnifies the issue than it is appeared now.

The following facts describe in a nutshell the gravity of water crisis. There are 345 million people living in Africa without proper water access. 3.4 million people (equal to the population of Los Angeles city in the United States) die every year due to water related diseases such as diarrhoea, a potential threat that killed a child in the continent in every 6 seconds between 1980–1990. Every year almost 60 million people migrate to the major cities in the world, overwhelming majority of which live in slums and do not have proper water access [5]. Another issue is the gender discrimination (against women) existing in many parts of the world related to transporting water: to collect water from long distance (usually miles away from their living places) falls upon women's shoulders. One of the World Health Organisation (**WHO**) reports indicates that women and young girls work approximately over 150 million hours a year for just bringing drinking water for their household activities. These distressing facts have been summed up by a report by the International Water Management Institute (**IWMI**), set up for overcoming water crisis in the developing world, according to which approximately 40% percent of people are living in the developing world affected by water shortage [6]. In Fig. (**1.1**), some representative pictures of water crisis have been shown.

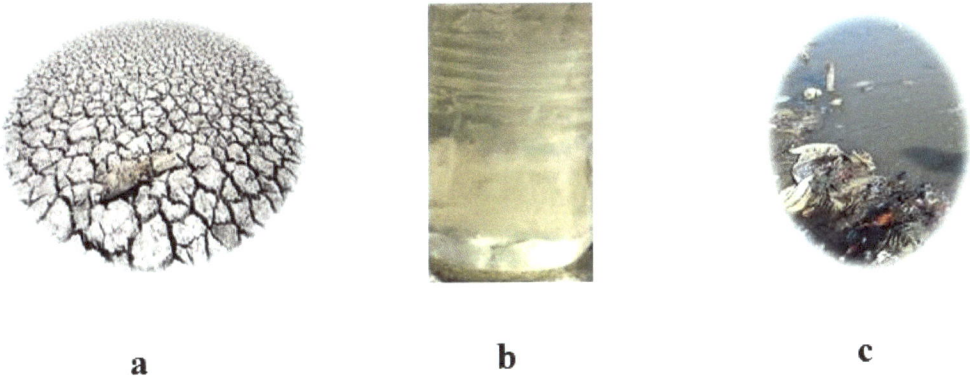

a   **b**   **c**

Fig. (1.1). Snapshots of the water crisis. (**a**) Shrinking volume of primary water sources due to human encroachment and climate change. (**b**) A sample of impure, undrinkable water containing pathogens spreading diseases. (**c**) A large chunk of population has only got access to dirty water.

Having rummaged through the introduction section, an intelligent reader may come forward with an immediate solution to the water crisis by stressing upon strong birth control policy. There may have some truth in it, but we must also consider the fact that the population growth is also contributed by better life care in many parts of the world. Average life expectancies all over the world, with exception to war–torn regions, have risen due to the appropriate intervention of the local governments and non–governmental organisations. This evidently rules out birth control as one and only viable option in order to overcome issues like water shortage. Since the crisis curbs the living population, a concerted effort is required to find overarching solution to this cataclysmic issue within the coherent framework of political, economic, scientific and technological aspects. In the following sections, I would like to investigate various grass root issues that are associated with the problem of water scarcity.

POLITICS OF WATER

Countless problems do exist between and within countries in sharing world's fresh water resources [7]. Several countries are already in political turmoil in various parts of the world in particular the Middle East and North Africa [6] and claims over water is one of the principal reasons for this situation. This region (also known as MENA), accommodating 5% of the world population but has only 0.9% of the world water resources, can be very vulnerable to conflicts, if the rate of population grows at the current alarming rate. It is predicted that within next fifteen years the population in the zone would reach twice as that was in 2000, indicating that the non–availability of clean water will aggravate accordingly [3, 8].

In countries including Saudi Arabia, United Arab Emirates (UAE), Oman, Qatar, Kuwait, Jordan, Israel, Bahrain & Palestinian territories in the Middle East, all conventional water resources have been completely utilised [4]. Tensions of similar nature are growing at the confluence of Turkey, Syria and Iraq over water in Tigris River. Several academicians also point out to the role of water in another notorious conflict that is boiling the Middle East for over more than sixty years: tensions between Israel and Palestine over the ownership of mountain aquifers between these two countries [9]. Gaza is the most vulnerable place to the scarcity of water in this region or even in the whole world due to several reasons: it does not have independent natural water resources that cater the need of its citizens which forces them to rely on Israel for highly expensive water transportation [10]. The bitter irony is that these two countries spend billions of dollars for buying or developing defence equipment that always had lethargic impact upon the ordinary citizens of both countries. In North Africa, where most of the people are under the poverty line, countries such as Egypt, Sudan, Somalia and Ethiopia are already at loggerheads over building up of new dams in the river Nile. Without any doubt, MENA can be considered as an extreme stress region as per Water Resource Index (WRI): a region is water–stressed if region has water availability of less than 17,000 m^3/person/year (water threshold) [4, 11]. In addition to MENA countries, countries surrounding Mediterranean, Eastern and Southern Africa, South West Asia are either under high stress or extreme stress with water threshold 1000 and 500 m^3 per person per year respectively [4, 11].

It must also be noted that akin to economic disparity, there emerges another disquieting difference in the form of water availability in the region. Iraq, Sudan and Mauritania are well ahead in terms of per capita of water availability; whereas countries like Kuwait, Djibouti and Palestine read the lowest [3]. It is feared that the gap could widen in the coming years. Situation is not so promising in other parts of the world either: in the European Union (EU), several countries including Germany, Belgium and The Netherlands have utilised more than 50% of their natural water resources, strongly suggesting that relying on alternative methods for safe and clean water is becoming mandatory.

Not all is that bleak as per the latest global developments as several local and international agencies are actively engaged in several target setting initiatives. Some of such notable endeavours is Clean Water Act (CWA) introduced in the United States in 1972 [12], UNESCO's (United Nations Educational Scientific and Cultural Organisations) Dublin principles [13] and more recent Europe's Water Framework Directive (WFD) [14].

One of the aims of CWA was to put a strong restriction on waste disposal into the natural water resources in the United States, which have polluted ever since the industrialisation and urbanisation started in 19th century. This target was primarily met by constructing numerous wastewater treatments plants across the United States [15].

The Dublin Accord was of broader perspective, and made four key resolutions, as summarised below, to meet water quality at acceptable standards:

1. Clean water is very fundamental for sustaining life on earth but is a depleting resource.
2. A concerted effort is required for better water management incorporating users and policy makers at all levels.
3. The role of women is so crucial in water management.
4. Since water is a public property, its social and economic value cannot be undermined.

An expansive version of Dublin principles (Dublin Accord) with more pragmatic approaches can be found in [2]. The principal objective of Water Framework Directive (WFD) was to centralise the initiatives taken by the member states of European Union (EU), ensuring the water quality equal across the continent [16]. As a result, the map of Europe was redrawn by hydrological parameters instead of political or administrative considerations in order to meet the water requirement.

Responding positively to the UNESCO's propositions, several countries have reached in mutually acceptable settlements. One of such initiatives is known as Inter Basin Transfer, transfer of water from one geographically distinct river catchment to other locations [17]. The Senqu−Vaal transfer, between South Africa and Lesotho, and South−North water project in China are some of the glittering examples for solving water crisis by water management. The estimates show that around 43% water withdrawal in North America was Inter Basin Transfer, suggesting that many water−related conflicts can be alleviated by properly executed plans [17]. Senegal's recent progress in resolving water issues demonstrates that empowerment of women can tackle water issues to a great extent [18]. In Asia Pacific, "The Living Murray", the famous river restoration project is Australia's initiative to bring back river Murray to its former glory. The project encompasses various disciplines such as ecological modelling and empirical research for ensuring high rate of success in the endeavour [19]. Australia has also reached two bilateral agreements with Japan and China to ensure the preservation of natural habitat: JAMBA (Japan Australia Migratory Bird Agreement) and China-Australia Migratory Bird Agreement (CAMBA) [19].

SOCIAL AND ECONOMIC IMPACTS OF WATER

The economic disparities among various classes of society and nations also play a key role in the water crisis. The statistics below sums up the precarious economic situation of unprivileged people: people living in slums (informal settlements) often have to pay five or ten times higher than the privileged living in wealthier regions of the same city. Since the urban water is more affected than water in rural areas due to the discharge of municipal and industrial waste, this situation is likely to worsen. Most of the people who can't have access to clean drinking water live on very paltry amount of money (1−2 dollars per day). It is important to note that nearly 95% of the problems related to the scarcity of clean water are reported from the developing world [20].

The developed world too is not spared from this growing menace albeit the reasons are different. Increasing demand for agriculture, and various thermal and hydroelectric projects has eroded natural water resources beyond recovery in the Western World. Researchers have found that over 65% of ground water is consumed for agriculture, which is non−recoverable, followed by industries and hydroelectric projects [21]. The required volume of water has been skyrocketed in many cities after Second World War due to the huge influx of people from urban areas, which led to the emergence of private companies commercialising water. Often tinted with corruption this move resulted in violent protests in many countries, including Argentina and Philippines [1].

An important economic pact has been signed to placate growing tensions between various countries under the auspice of academicians from the Middle East and North America [10]. The pact, The Harvard Middle East Water Project (HMEWP) aims to promote peace and harmony among the warring countries in the Middle East by implementing a protocol that is based on effective distribution of water among the countries and setting up infrastructure for posterity [10]. The major finding of the project was that the monetary value of water claimed by Israel, Palestine and Jordan is "surprisingly low" such that an amicable settlement can be achieved to ease the strained relationship between these countries [10]. The current water deals between countries, for instance between Israel and Jordan, were monitored in the project. It found that the current water transaction between these two countries is very negligible compared to actual need, and suggested increasing the transaction fivefold to meet the requirement by 2020 [10]. The report recommends for constructing better water conveyance facilities across the region, facilitating an increase in transporting of water 10–15 times higher than that of the current capacity. The project, in addition, proposes the development of various national water pipeline systems, for example connecting Jordanian water transporting facility to Nablus region of Palestine, and water pipelines in Gaza to Israeli National Water Carrier [10]. Nevertheless, the success of this proposal heavily depends upon the political climate of this region: as Israel and Palestine fight over their sovereignty on yearly basis, the economic viability of such plans cannot be guaranteed.

External consumption of water (not necessarily in its pure form) also serves as basis for our existence: for example the water we use for generating electricity or agricultural purposes, shortage of which clearly push us into two more perilous crises pertaining to energy and food. Though alternative modes of generation of electricity replaced the traditional hydro power electricity in many parts of the world, 20% of the global electricity is still obtained from this way, notably in China and India where 50% of the electricity is produced from hydro electrical projects [22]. In some countries, hydropower is the principal source accounting for more 90% of electricity generated underpinning the importance of water among the essential commodities that sustain human life. Statistics predicts that the amount of water required for agricultural purposes will see a dramatic leap from merely 1021 million tons in 1993 to 1634 million tons by 2020 [22a].

Is there an effective economic model for water management? Hydro–economic modelling, a post war discipline initiated in the United States and Israel is a powerful mathematical model in which hydrologic engineering is integrated with economic nature of demands and costs of water [23]. Unlike most of the management models proposed over the years, it amalgamates various perspectives of water management: geographical, economic, environmental and technological, and offers an optimum solution to the water related issues. Compared to the traditional approaches which relied either on economic aspects or technological aspects, hydro–economic modelling is demand related in nature considering the dynamic nature of demographic landscape, agricultural and industrial water requirements which can cope with any surge in demands [23]. Hydro–dynamic modelling has been employed for various purposes such as water supply, engineering infrastructure, capacity expansion, ground and surface water management, water monetizing, trans–boundary management, managing weather fluctuations, floods and improving water quality. Countries in the developed world including the United States, Australia, France, Germany and Spain and in the developing world such as India, Iran and Palestine are greatly benefitted by this state–of–the–art tool [23].

WATER AND ENVIRONMENT

Natural resources of water depend first and foremost upon hydrological cycle, but activities and priorities of the world reshape this natural landscape of water in a negative way. Hydrological cycle (precipitation–absorption–evaporation) is nature's delicate way for providing clean water to its inhabitants. Solar energy distils approximately 40% of water in earth's atmosphere, and out of these 14% is from land and the rest from oceans. But a higher proportion, 24% of the water, returns onto earth as precipitation falls on land which is stored in lakes, streams, ice caps, glaciers, soil moisture and ground water [7]. Around one fourth of the total fresh water on earth is store as ground water than any other forms, making it the most important source of fresh water across our globe. A huge volume of water, approximately 7×10^{15} m^3, in aquifers underneath the earth's surface has been stored from annual hydrological cycles occurring for millions of years [7].

Increasing demand for water has led to over exploitation of the ground water mostly by ground water mining, for example in countries like the United States, which results in two environmental problems in several parts of the world: increasing salinity of water and water logging [7, 24]. Other anthropogenic activities such as building dams over rivers, creation of artificial lakes, altering the natural course of water ways and discharging treated water to water bodies too have adverse effect on quality of water. Industrialisation has made very serious fingerprints on the purity of water such as change in temperature (thermal pollution); turbidity; presence of organic or inorganic chemicals such as heavy metals, oxygen depleting substances, nutrients; and alteration of pH [15, 25].

Thermal pollution is primarily caused by nuclear and thermal power generating industries. The chemical waste ejected by these industries reduce the capacity of water to absorb oxygen, and hence diminishes the purity of natural clean water reserves as well as systematically terminate the existence of aquatic species such as fish [15]. Growing concerns over the pollution (including water impairment) caused by these energy sources, accounting for over 20% of greenhouse gas emission in the world, prompted technologists to search for other sustainable means for power generation, for example Ocean Thermal Energy Conversion (OTEC), economically viable for smaller and poor developing nations like Ghana, Somalia, Cuba, Bahamas, Fiji and Maldives crippled by water scarcity [26].

Aquatic life is also severely perturbed by the presence of materials such as metals (arsenic, lead, mercury), halogenated aromatics such as DDT, solvents like benzene and toluene, nitrosoamines, nitrates and phosphates (the list is endless). These materials often cause cancerous tumours on the skins of fish and birth defects on aquatic birds [15]. Traces of toxic materials like poly chlorinated biphenyls and mercury have been found in several species such as Weddell seals in Rose sea in Antarctica and Cree Indians of Canadian High Arctic respectively, putting them at the brink of extinction [27]. Dissolved oxygen in water is very fundamental for the survival of many aquatic species. Lack of sufficient oxygen dissolved in water (minimum 4 milli grams per litre) puts their existence into perilous situation. Industrialisation has played a non−negligible role in the discharge of various oxygen depleting substances including food processing waste, pulp from news paper industries and animal waste into fresh water sources [15].

The non−expansion of natural resources upon the rising demands can be clearly noticeable in shrinking capability of water aquifers [2]. There exists a huge imbalance between average annual recharge and average annual use in several aquifers across the globe: Saharan basin, Saq, Tenerife and Ogallala are some among them [2]. In most of these aquifers, the amount of water withdrawal is three or four times higher than annual water recharge, suggesting that these aquifers are on the brink of extinction. The major environmental threat to safe, clean and fresh water is the presence of pathogens such as a variety of helminthes, protozoa, fungi, bacteria, rickettsiae, viruses and prions in water which can cause disease in humans, animals or plants, threatening the very existence of human beings in many parts of the world, in particular Asia and Africa [20]. Contaminated water is responsible for many contagious diseases such as hepatitis gastroenteritis, meningitis, fever, rash, conjunctivitis which can be spread at amazing pace due to the presence of human enteric viruses in water [28]. The following Fig. (**1.2**) outlines outbreak of waterborne diseases by direct and indirect ways. In addition, behavioural abnormalities, cancer, genetic mutations, physiological malfunctions are also common in communities devoid of access to clean water [15].

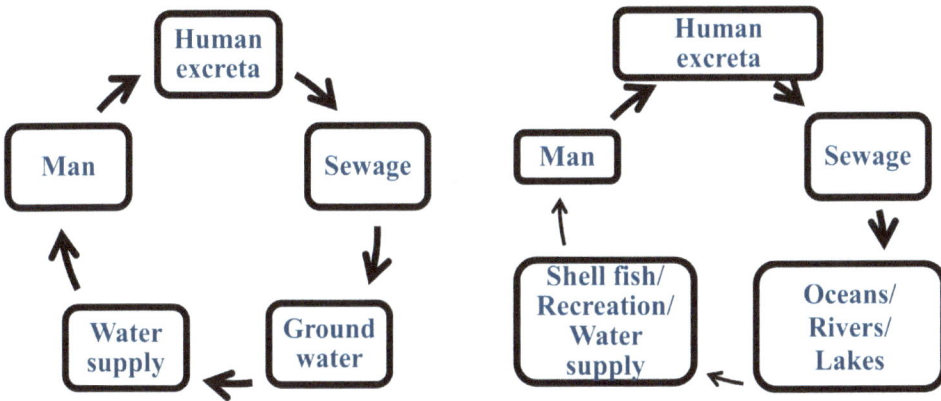

Fig. (1.2). Different routes of enteric virus transmission. (**a**) Direct virus transmission to human body occurs from ground water through water supply. (**b**) Virus enters into human body indirectly from oceans, rivers and lakes polluted by sewage water.

One of the classical examples of deteriorating quality of water upon the natural habitat is Georgia's (in the United States) decreasing availability of natural edible species such as oysters [29]. Oyster reef, built by accretion of oysters over the years, is central to the preservation of the land against shore line erosion, to maintenance of water quality by filtering pollutants and excessive nutrients, and to provide breeding environment for other species such as shrimps and blue crabs [29]. Impaired water does not stifle the growth of all species on earth but promotes pervasive growth of certain species called Invasive species which endanger the very existence of natural habitat like oyster reef in Georgia as studies indicate [29]. An invasive species is a plant or animal that foreign to a specific location, has a tendency to spread to become the dominant species in the location, suppressing the growth of native species. They are characterised by high breeding rate, broad diets, wide environmental tolerance, longevity, and being gregarious [29]. Notable invasive species include Australian tube worm, Island Apple Snail, Charrua Mussel, Green Mussel, Green Porcelain Crab, Alligator weed, Water Hyacinth, Hydrilla, Marsh Dewflower, Giant Salvinia, Red Lionfish, and Titan Acorn Barnacle. Some of these are found universally, whereas others grow at specific geographical areas. Some invasive species in particular those grow in water known as Aquatic Invasive Species pose threat to its purity since they impair water quality by altering pH and lowering oxygen levels.

None can deny the benefits of modernisation achieved through industrialisation, but a key question remains is "how to reap the benefits of industrialisation without costing the sanctity of nature". One may argue that industrialisation cannot be unavoidable, but there must be proper regulations and their implementations must be put in place strictly. Even CWA was only a partial success due to lack of propositions to cut the several other water contaminating activities that are more individualistic in nature such as farming and timber harvesting [15].

WATER FROM BIOLOGICAL PERSPECTIVE

The role of water in biological activities of human beings and other living creatures is not by any means negligible. Water is the most abundant chemical in our body, and its presence is pivotal in many functions in animals and plants including nutrient transportation, biochemical reactions, reproduction and fertilisation. Transportation of fluids is very fundamental for the survival of organisms with which essential nutrients are carried through. In plants, vital molecules such as sugars, amino acids and hormones are transported in the water medium. A huge proportion of water is contained in blood and cytoplasm, major transporting fluids in animals. Complete solubility of molecules such as glucose, amino acids and certain minerals in water is essential for metabolism, by which new cells are generated, and energy is produced. High stability of water in the normal temperature domain and excellent solvation properties make water indispensable reaction medium for various enzymatic reactions in our body, for example photosynthesis. Our reproduction system is primarily based on water, without which male and female gametes cannot move freely rendering the fertilisation process impossible. Water also acts as a comfortable cushion for the development of embryo for the whole gestational period.

Water is absorbed into human body from the food intake; it is then consumed for the biological functions inside organs such as lungs and kidneys and goes out *via* perspiration and urination. Smooth repetition of this cycle (physiological water cycle) warrants the quality of water very high in order to safeguard one's body from serious pathological conditions. The water circulating inside participates in variety of reactions including redox, condensation and hydrolysis [6]. Water's role in photosynthesis is critical and well known: in one of the steps of ATP synthesis, water releases oxygen that combines with glucose and Adinosine Di Phosphate (ADP) to form Adinosine Tri Phosphate (ATP), the store house of chemical energy, with water being one of the by–products of this reaction itself.

Water's role is much more versatile, inside the cell, in interacting delicately with wide range of biomolecules. Numerous volume of research work is published every year on these interactions in relevance to various cellular activities, yet researchers have not been successful in developing a coherent understanding of structural, dynamic and morphological properties of water in this biochemical context. The activity of biological macromolecules such as proteins and nucleic acids (DNA and RNA) is negligible without water. One of the notable properties water is that the macromolecular interactions in cells are mediated by water through hydrogen bonding (an electrostatic attraction when a hydrogen atom bonded to highly electronegative atoms such as nitrogen, oxygen and fluorine in a molecule experiences attraction towards highly electronegative atom of a neighbouring molecule), hydrophobic association (tendency of nonpolar substances to aggregate themselves in the presence of water) and hydrophobic hydration (the orientation of water molecules around macromolecules resulting in a formation of microscopic cage of water molecules around them) [30].

Water indulges in its characteristic hydrogen bonding network composed of vast number of water molecules especially in low temperature and normal temperature domains. Albeit reduced in terms of coordination number, water retains the network with large proteins, crucial in mitigating protein–water dynamics [31].

Consequence of hydrogen bonding is extended to another peculiar phenomenon that occurs in confined water trapped between large macromolecules, proton hopping, by which proton is passed through water *via* rearrangement of hydrogen bond network [32]. Fig. (**1.3**) demonstrates water in the cellular environment.

Fig. (1.3). Schematic diagram of water in cellular environment. Water (shown in blue) sequestered between large macromolecules participates in numerous activities in cell, the smallest part of an animal or plant that is able to function independently. The water molecules trapped between the macromolecules behave differently with respect to the bulk water.

Hydrophobic association is so critical in protein folding, a process by which proteins fold into a more compact form (three dimensional structure) by reconfiguring its nonpolar residues to the core and polar residues to the surface to carry out vital functions in every cells. It also plays a central role in aggregation of lipid bilayers. Lipids contain two different molecular segments that are different in their solvation characteristics: polar units and non−polar units. Hydrophobic interactions keep the non−polar segment away from water, whereas interactions between water and polar units are strengthened by hydrogen bonding. These two concomitant interactions result in layering of non−polar (hydrophobic) segments in lipids. Hydrophobic hydration on the other hand has been found to enhance the catalytic activity of proteins, known as functional tuning [33].

Other amazing properties of water have also been found out from experimental and computer simulations. Water facilitates protein dynamics, the motion of peptide bonds (the chemical bond that connect various amino acids in a macromolecule) by maintaining the optimum balance between rigidity and mobility of protein structure [30]. Water has an exceptional ability to speed up biochemical reactions (its catalytic activity), making use of which it can act as both proton acceptor and proton donor fulfilling several roles including splitting ring closures, hydrolysis of peptide bonds, eliminating chemical groups from substrates and proton transfer between adjacent protein surfaces [34, 35]. Recent theoretical and experimental investigations indicate water's amazing morphological transitions within the close proximity of large macromolecules, which are otherwise observed only at very low temperatures [36]. These morphological transitions in the presence of large macromolecules enable water to suppress its crystallisation temperature further below. This clearly explains the natural ability of several species living in extreme conditions such as polar bear to overcome very extreme cold weather [30]. The binding of smaller molecules (ligands) from adjacent protein surface to protein receptors either by hydrophobic association or hydrogen bonding is known as ligand binding.

Water's role in ligand binding is also a subject of immense debates [30]. One school argues that during ligand binding water molecules are completely replaced by other molecules, while the other advocates that the water molecules confined between the protein surface and the approaching ligand act as flexible adhesive facilitating ligand–protein interactions. Both arguments can be plausible, as accounted for the natural cryopreservation conferred by sugars [37]. Disaccharides, a class of sugars, interact with the biological membranes in two ways: sugars replace water molecules, and offer protective shells around the membranes by indulging in hydrogen bonds. This stabilisation mechanism is known as Water Replacement Hypothesis; whilst the presence of sugars increases the degree of hydration known as Water Entrapment Hypothesis [37]. Water present between two adjacent proteins is reported to have a role in generating different protein conformations, observed in cytochrome c2 redox protein and haemoglobin (the oxygen carrier in blood), which facilitates improved functionalities of protein substrates [38].

The properties of water are so unique in organising biological functions in our body. Contamination of water due to the presence of salts and other organic compounds limits the multi−functional role of water. Every person living on earth has right to have access to clean drinking water, non−availability of which can lead to perennial diseases that continue to pose a constant threat to our health, which in some cases may even lead to loss of life.

WATER FROM TECHNOLOGICAL PERSPECTIVE

Development of robust technologies for producing clean water can mitigate the impact of water scarcity to a greater extent. These techniques are based on two concepts: water reuse and water retrieval from an impure water source. Affluent countries such as Belgium, England and Germany have implemented successful water reuse technology to face growing demand for water in their cities [4]. I venture into explain various classes of these techniques in this section, some of which have been in use for over centuries and others are in developmental phase. The most applicability of these methods critically depends upon three factors: the geographical peculiarity, the size of targeted communities and economic viability. For example, implementing a method suitable for a large city in a small town can be considered as economic mismanagement. Similarly, techniques suitable for cold conditions of Finland might be inappropriate for hot and humid Indian conditions.

Water purifying techniques can fundamentally be divided into two: treatments in which chemicals are used for (*e.g.* osmosis), and treatments that are based on physical methods (for *e.g.* treatments driven by solar energy). The fundamental protocols for water cleaning techniques by traditional chemical methods are based on disinfection, decontamination and re−use and reclamation [20]. Indiscriminate use of these methods that are widely employed for water purification can result in extremely lethal outcomes as the waste generated from these processes cannot be bio−degraded resulting in environmental catastrophe. Moreover, the present chemical methods employed for water purification are limited in their usefulness since they can only effectively act at very low concentrations; this makes the cleaning process for large volume of water very expensive [20]. Though chemical treatments are widely in use for the purification of water from the contaminants, detection of impurities can sometimes be impossible due to speciation, the process by which the distribution of an element amongst defined chemical species in a system (for example, the generation of pentavalent Arsenic (As(V)) from trivalent Arsenic As(III) or *vice versa*) [20].

Various physical water purification methods have been evolved since the time immemorial, the first being natural filtration which make use of the natural ability of soil to sieve off the contaminants [39]. The filtered water is stored in sources like natural aquifers. A variant of natural filtration is River Bank Filtration (RBF), popular in industrialised countries like Germany, by which water in the river bank is collected in wells connected to aquifers by conduits. Another popular variety of natural filtration is Slow Sand Filtration, by which water is filtered through a mixture of sand and bio film containing tiny organisms which purify water passing through it [39].

Membrane Filtration (MF) is a highly sophisticated technique, used in majority of desalination plants. In this process, only water can pass through a membrane (a sieve), blocking contaminants such as bacteria, viruses or chemicals. Major types of membrane filtering techniques fall into two categories: pressurised systems and gravity fed systems [39]. Pressurised systems, as the name indicates, are pressure–induced such that the membranes, serving as filters, can act as more sensitive in detecting and removing the contaminants. Gravity fed systems are used to separate large suspended impurities from water, and hence they can be used prior to the other sophisticated techniques.

Simplistic in design and at the same time very much energy efficient solar distillation technique draws energy form sun (solar energy). Water is vaporised using solar energy and collected afresh for household use. Due to its low cost and universal availability of solar energy without any break, solar distillation process is growing in popularity, in particular in the developing world. In order to limit the energy wastage, a modified form of solar distillation is employed, Solar Pasteurisation. In this technique, water to be purified is heated only to 65°C (Pasteurisation Temperature), the temperature above which pathogens such as Hepatitis A virus is perished. Many of these techniques have been implemented in developed and developing world alike: in the United States, Mexico, India and Kenya, at varying success rates [39].

A wide array of techniques combining aforementioned chemical and physical methods is being developed to increase the efficacy of water purifying technology for combating growing challenge of water scarcity. Lead DNA sensor with micro−nanofluidic device, highly effective in trapping contaminants such as lead ions, is worth to mention here. Major component of this instrument is a combination of a sensor which detects the contaminant, and sweeper which has high affinity for the contaminant cation (Pb^{2+} ion) removes the contaminant from the impaired water [20]. Another state of the art technique is membrane bioreactor treatment system, which is a common platform of the various pressurised membrane filtration techniques: Reverse Osmosis (RO), Nano Filtration (NF) and Ultrafiltration (UF) and Micro Filtration (MF), cumulative application of which is proved to be the most successful approach among the various desalination techniques available today [20]. In Reverse Osmosis (R.O), pressure is applied to overcome the influx of water molecules towards the other side of the membrane (filter) where the solute (contaminant) is present through the barrier (with pore size 0.0001 micron) known as osmosis. In Fig. (**1.4**) how various techniques can be combined for producing clean water is demonstrated.

The filters employed in Ultra Filtration technique have the pore size 0.01 micron, with which large particles and some viruses can be filtered off. On the other hand, the filters employed in Nano Filtration (UF) is smaller, 0.001 micron with which smaller particles can be filtered off including most of the viruses, many organic matters and ions. However, membrane fouling, occurring due to the absorption of various contaminants commonly known as organic foulants (for example sugars and proteins) present in impaired water leading to reduction in the pore size, can limit the effectiveness of the water filtration process [20]. Development of fouling resistant and economically viable filters is therefore needed to ensure the continuing use of Membrane Filtration (MF), and membranes made up of Comb copolymers are proved to be highly resistant to organic fouling [40].

Despite aforementioned progresses in the development of robust water purification technologies, a large chunk of the population in the world are still without the access of these innovations, and hence live without clean water. One of the reasons for this precarious condition is the great imbalance existing between the affluent and poor countries in terms of availability of technical expertise. Organisations working around the globe must take note of this fact as well while planning for "Clean Water for all" to alleviate water scarcity.

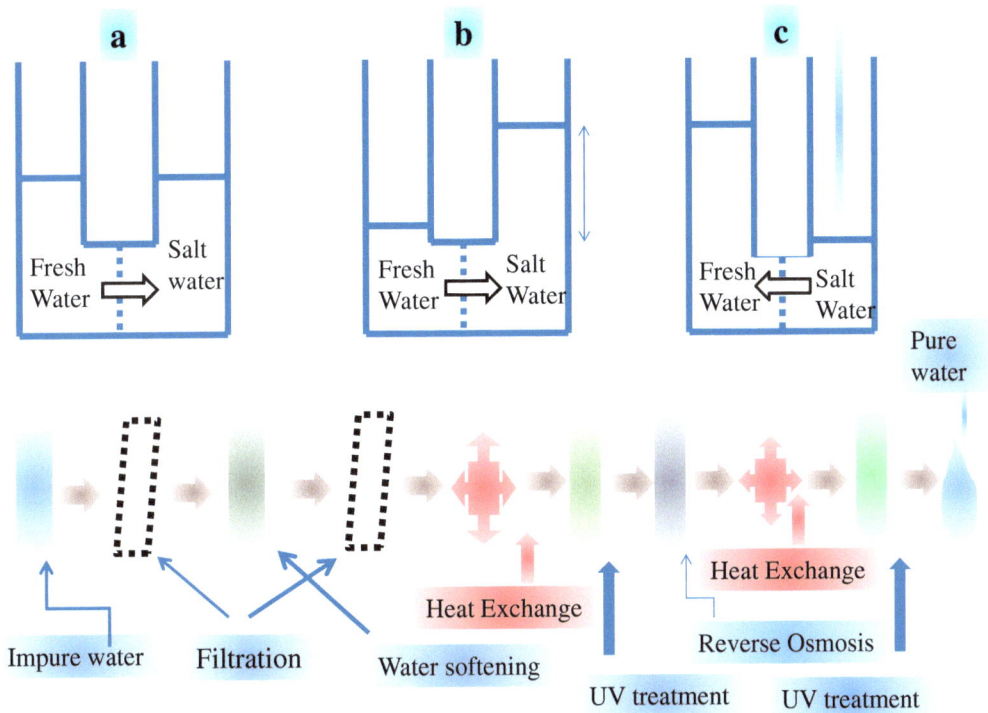

Fig. (1.4). Schematic diagram of Water Filtration Process. In upper panels (**a-c**), the process of Reverse Osmosis (RO) is shown. (**a**) During osmosis, water flows towards the side where concentration of solution is higher (higher number of solute molecules). (**b**) Continuous flow of pure water towards compartment containing salt water results in osmotic pressure. (**c**) To counteract the osmotic pressure, an opposite pressure must be exerted (shown with thick black arrow; note the direction of it), forcing water molecules in the compartment containing saltwater towards side with lower solute concentration (fresh water). (Bottom) Impure water is subjected to various filtration processes and chemical treatments like softening (for reducing water hardness) as shown by the arrows in clock wise direction. Pure water is collected after subsequent Reverse Osmosis (RO) along with proper thermal and ultra violet (UV) treatment. The idea has been adapted from www.membrane-solutions.com and www.safewater.org.

COMPUTER MODELLING

In the final section of this chapter, I would like to correlate computer modelling with the core theme of the book, "water". Based on the extent of applicability, computer modelling is divided into two: macroscopic (macro meaning large) and microscopic (micro indicating small). Giving a clear–cut definitions of these two terms is not straight forward, as they differ across various disciplines. However, one can say that macroscopic modelling deals with the matter in very large dimension (for example, Hydrodynamic Modelling mentioned earlier) and with microscopic modelling, one can investigate matter at atomistic or molecular level.

Macroscopic modelling can be applied for variety of purposes, from Weather Forecasting to Share Value Prediction. It is common to have featured weather forecast sessions by all major broadcasters, which make use of large scale computer modelling of ocean current, wind direction and other associated effects. Macroscopic modelling is very central to effective Water Resource Management (WRM) based on principles of statistics, and it is one of the growing research areas recognising the importance of smooth distribution of water across the world [41]. In space science, researchers extensively employ computer modelling to explore the space, and to predict the location of distant planets. Several countries now spend billions of dollars to launch satellites whose tracking is mainly achieved by the state–of–the–art computer software. In finance, it is being used for predicting stock market values and for obtaining economic projection for various companies and countries. In social science, computer models are widely used by experts and political parties to project demographic data bases for arriving at certain factors that shape people's verdict, for example, in elections at national and local levels.

On the contrary, microscopic modelling is purely a science and engineering discipline, in which the maximum limit of time and special scales are seconds and milli meters (mm) respectively, and is performed to investigate various phenomena (bond vibrations, lipid diffusion, protein folding & membrane fusion) occurring at time scales ranging from femto seconds (10^{-15} seconds) to seconds [42]. The following Fig. (**1.5**) shows how time scales and spatial scales are inextricably linked in simulating various physical and chemical phenomena.

With the development of modern computer simulation software, it has become a routine task to perform simulations of large biological matter, which seemed impossible half a century ago [43]. The simulations can now be performed with short span of time (probably by one or two weeks) thanks to the development of multiprocessor computers that can perform billions of calculations per second and efficient algorithmic approaches such as parallelism. Computer simulation has been emerged as a powerful tool to verify numerous theories and hypotheses based on experiments regarding structure and function of matter, and sometimes altered the already existing views. This facilitated unlocking the mysteries of biological matter by structure determination, tuning molecular properties and tracking movements of large macromolecules [43].

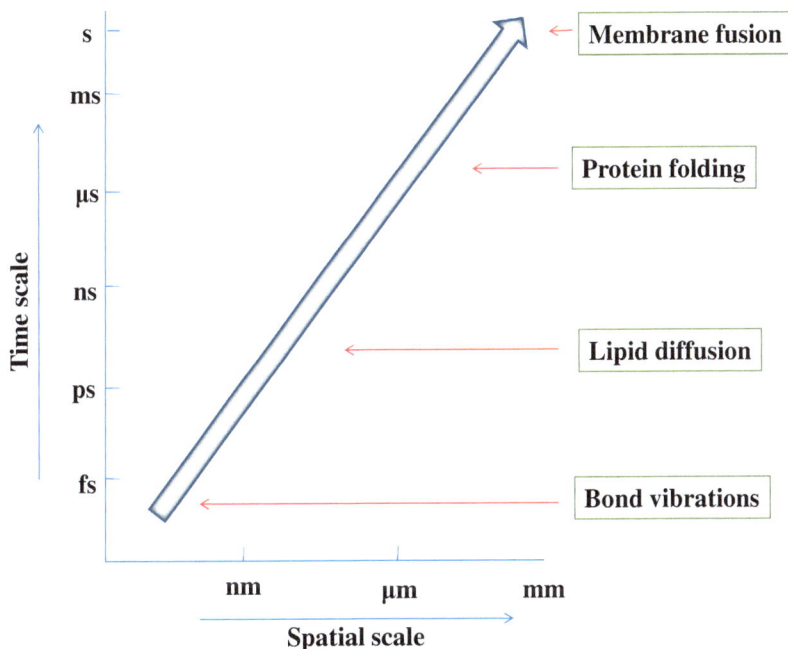

Fig. (1.5). The relationship between time and spatial scales in the investigation of various physical, chemical and biological phenomena. Spatial scales (in the order of nano meters, micro meters and milli meters) shown in the x axis denotes the size of materials simulated, whereas the time scales shown in the y axis (in the order of femto second (fs), pico second (ps), nano second (ns), micro second (μs), milli second (ms) and second (s) indicates the time required for estimating various processes at a reasonable accuracy. The idea has been adapted from [42].

Rudimentary knowledge of physical and chemical nature of water has led researchers towards constructing robust computational models, which enabled extraction of amazing properties of this puzzling liquid. One of the principal achievements of computer modelling of water is thermodynamic contour map, inaccessible experimentally due to practical constraints, by which stability (in terms of thermodynamic variables such as pressure and temperature) of various forms of water can be known. This has a profound impact upon the modification of existing technologies or development of new methodologies in water purification.

The flow chart shown below summarises how advancement in multifarious disciplines are connected to each other, and how they can contribute to solving the cataclysmic water crisis. As shown in the Fig. (**1.6**), water resource management plays a pivotal role in resolving or mitigating the chronic issue of water crisis. Efficient water management schemes can be modelled considering societal, economic and political parameters. On the other hand, robust water technologies have to be developed in order to meet the growing demand for pure water, and successful implementation of these technologies heavily depends upon information generated from experimental and computer simulations on water itself.

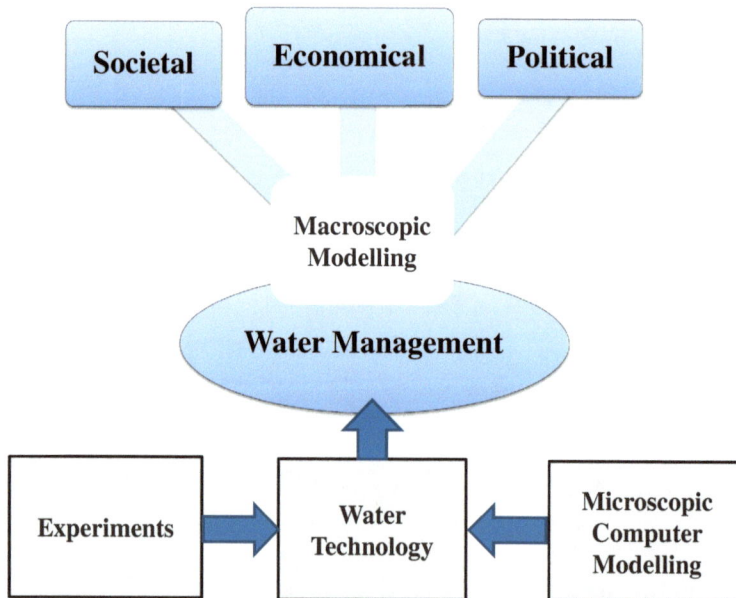

Fig. (1.6). Towards efficient water management schemes. The water crisis can be overcome by appropriate water management initiatives, as shown in the diagram, greatly influenced by socio-economic-political-technological developments around the world. Advancement of the state–of–the–art water technologies driven by microscopic level computer simulations and experiments ensure effective water management.

STRUCTURE OF THE BOOK

Apart from this chapter, the book contains nine more chapters. The second chapter begins with a comparison of liquid state with two other prominent states, solid and gas. A brief discussion on intermolecular forces is followed then, before viewing water as a normal as well as a supercooled liquid. The chapter ends with a detail discussion on the qualitative aspects of glassy water. Though it is digressing from the core theme of the book, I feel that the readers will greatly be benefitted by experimental approaches, explained in chapter 3, that are generally employed for investigations on water. In this chapter, diffraction and spectroscopic techniques are reviewed bit detail since these are two of the most widely used techniques for the microanalysis of water. Chapter 4 provides the basic aspects of computer simulations for the readers who are not familiar with computer modelling. In this chapter, fundamentals of atomistic and molecular dynamics simulations are discussed. The chapter concludes with a discussion about water models (Force fields) which are the input structures for molecular simulations. Chapters 2 through 4 are intended to bridge the gap between theorists and experimentalists. Chapter 5 addresses investigations of ambient water, in particular its structural, thermo dynamic and kinetic properties. I tried my level best to cover most of the important published work in this chapter. In the following chapter, the properties of water in sub−zero temperatures, popularly known as the supercooled and glassy water, are discussed. Chapter 7 summarises all important computational findings so far on ice, the low temperature allotrope of H_2O. Chapter 8 attempts to gather available literature of water beyond its boiling point, in particular Super Critical Water (SCW). I think more efforts are required to unravel the properties of water at elevated temperatures, and therefore anyone who wants to pursue this area will greatly be benefited by a good number of important references given at the end of the chapter. Chapter 9 depicts all well−known anomalous properties of water. The objective of the final chapter is to summarise what have been discussed in the chapters 1 through 9, and a brief account of how molecular scientists should align their research interests with global initiatives to tackle water related issues.

CONFLICT OF INTEREST

The author confirms that he has no conflict of interest to declare for this publication.

ACKNOWLEDGEMENTS

Declared none.

CHAPTER 2

A Snapshot of Liquid State

Abstract: Liquid is one the three principal states of matter and its properties are known to be intermediate between gaseous and solid phases. Several types of intermolecular forces, categorised into long range and short range, play important roles in defining liquid structure. Long range forces are of three types, namely electrostatic, induction and dispersion, whilst the short range forces are of quantum chemical nature, due to exchange of electrons. A wide range of materials, including elements, oxides, mixtures of salts and dilute acids, are known to form glasses, which are non–crystalline, amorphous matter. Methods such as lattice theories have been devised long time back to understand the structure of liquids. Several other theories have been put forward as well in order to explain complex behaviour of liquids at lower temperature such as formation of highly viscous glassy materials. Most notable theoretical propositions include Adam – Gibbs theory, Mode Coupling Theory and Energy Landscape theory. Inherent Structure (IS) analysis is a powerful tool to identify the fundamental structures of the system under investigation, and to obtain a pictorial characterisation of the energetics between strong and fragile glasses. Relaxation times exhibit two distinct kinetics, alpha and beta relaxations, which can be properly explained by Mode Coupling Theory. Aqueous solutions of sugars and alkali salts such as lithium chloride are known to be good glass formers, which require only low cooling rates in order to form glasses, bypassing crystallisation.

Keywords: Atactic polymer, Expansion coefficient, Fragile, Freezing, Glass transition, Hole theory, Inherent structures, Kauzmann's paradox, Lindemann ratio, London forces, Metastable, Strong liquids.

INTRODUCTION

The three principal states of matter are solid, liquid and gas, and properties of these states differ in many aspects including molecular arrangement & shape,

Jestin Baby Mandumpal

speed of movement, energy and forces of attraction. Major differences in properties among these three states are summarised in Table **2.1**.

Table 2.1. The distinction between the three principal states of matter.

Properties	Solids	Liquids	Gases
Distance between particles	very close	less close	far
Molecular arrangement	regular	irregular	irregular
Shape	well defined	shape of the container	no shape at all
Volume	fixed	Fixed	not fixed
Speed of movement	the slowest	Faster	the fastest
Forces of attraction	the strongest	weaker	the weakest

The principal distinction among these three forms of matter cannot be more accurately explained than by Thermodynamics, in particular entropy, a measure of how available energy in the system is distributed in its constituent particles. When entropy increases, the ways by which the energy can be distributed (degree of disorder) increases. One can clearly see that solids are of high order, followed by liquids and then gases (highest disorder and therefore highest entropy). This is also reflected in higher densities of solids and liquids (collectively known as condensed phases) over gases.

Higher densities of liquids over gases have big impact on the nature of forces that binds molecules in liquid, implying that they are very closely packed compared to gases, and the distance between two molecules in liquid state is simply the molecular diameter. This evidently points in to the significant role of intermolecular forces in binding molecules in liquid phase. Conversely, the distance between two gas molecules are approximately ten times than that of diameters, limiting the role of such forces in gases [44]. Closer packing arrangement (therefore lack of ample free space between molecules) in liquids also implies that they have lower compressibility than gases. When compared to tightly–packed crystalline solids, liquids and amorphous solids do not have long–range order. However, they possess short–range order, as shown by Fig. (**2.1**). Due to the lack of "perfect" order, a vast number of holes or voids can be observed in the micro structure of liquids, suggesting that liquid occupies larger volume than a solid in general.

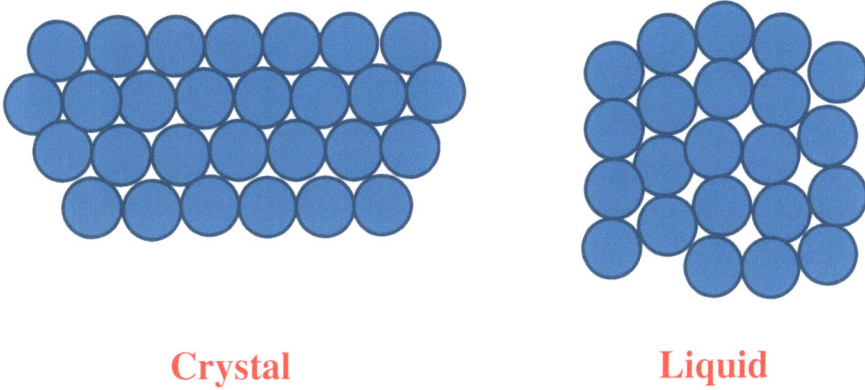

Fig. (2.1). Molecular view of crystal and liquid. The molecules in crystal (shown in the left) have been locked in regular pattern, compared to liquids (shown on right) which possess random arrangement.

Amorphous solids are highly viscous liquids with most of their features akin to liquids. The structural difference between a perfect crystalline and an amorphous phase cannot be better understood than using the tiling model as shown in Fig. (**2.2**).

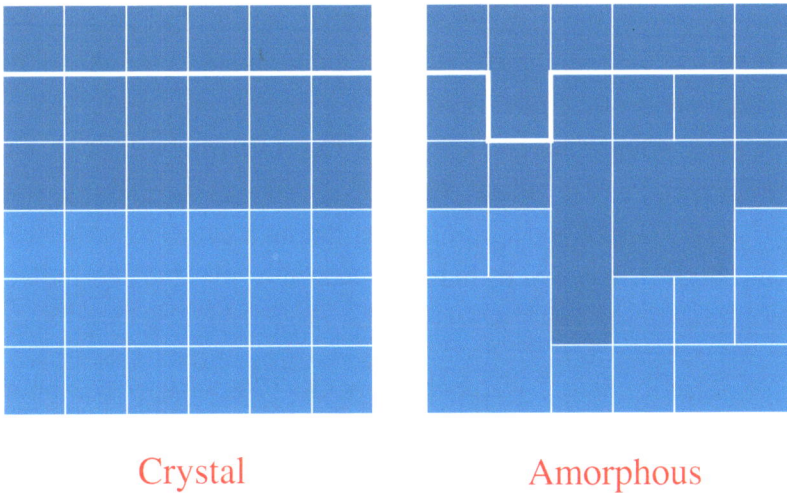

Fig. (2.2). Crystalline and amorphous phases. On the left shown is crystal, represented by arrays of regular periodic squares, and on the right shown is amorphous phase constituted by irregular squares.

As evident from Fig. (**2.2**), in the crystalline structure, the small squares are of equal size and well−ordered with all the lattice points are filled by particles. On the other hand, amorphous phase is a collection of squares of different sizes, and not all lattice points are occupied by particles.

The dependence of volume with respect to temperature and pressure is manifested in two notable thermodynamic quantities: coefficient of thermal expansion (α), the measure of variation of volume of a material with respect to temperature at constant pressure, & coefficient of compressibility (κ), the measure of variation in volume with respect to pressure. It has to be noted that the value of coefficient of thermal expansion is more or less same for all gases, whereas a material has distinct values of α & κ (albeit smaller in magnitude than gases) in liquid and solid phases. A notable difference between solids/liquids and gases can also be made from compressibility. Volume of solids or liquids decreases linearly with pressure, in gases, on the contrary, volume is inversely proportional to pressure (Readers are requested to refer [44] for procuring a better understanding of Thermodynamics.)

Potential energy dominates in solids and liquids, whereas the motion in gas molecules is dictated only by kinetic energy. The prime difference in the motion of particles in solid and liquid phases is that in solids the motion is purely vibrational character; whereas molecules in liquid phase, in which any two molecules are unlikely to stay together for any significant time, have two principal ways of movements: some molecules are able to move freely through the phase (similar to molecules in gas phase yet not so freely), while the remaining molecules are held by neighbouring molecules like in a cage (in particular at low temperature and high density) and therefore can only vibrate like molecules in solids.

How would one describe liquid water theoretically? Perhaps, one cannot find a better description of this enigmatic liquid than Eyring *et al.* have given to it (concerning the fact that several distinct features that distinguish it from other liquids): the simplest of "non−simple liquids" [45]. The main purpose of this chapter is to provide theoretical concepts that are central to the investigation of liquid and supercooled water much more readable to the non−experts in the field. The concepts developed for liquid matter in general are found to be useful in unravelling several mysteries of water, in particular in low temperature region. Firstly, I provide a brief introduction to the various types of intermolecular forces that dominate the internal structure of liquid matter, which is followed by a discussion on the generic theoretical treatment of liquids, and then, the concept of metastability of supercooled state is discussed. The chapter is concluded with a discussion on glassy physics and theories developed to explain glass transition. The latter sections of this chapter may be find useful in comprehension of chapter 6.

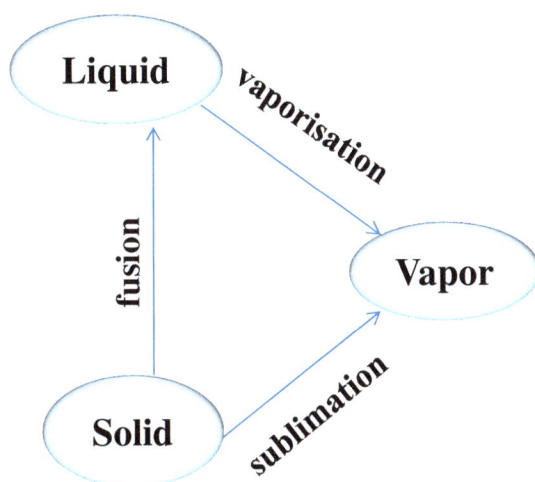

Fig. (2.3). Interconversion of three phases. The energy required to convert solid to vapour phase is equal to the sum of energy required to convert solid to liquid and liquid to vapour.

INTERMOLECULAR FORCES

The three aforementioned forms of matter are inter–convertible. The energy that is supplied for converting liquid to vapour phase is known as heat of vaporisation. The energy cost to dislocate the arrangement in solids is known as heat of fusion. A greater amount of energy, known as heat of sublimation, is required into convert a solid to its gaseous phase. Succinctly speaking, the heat of sublimation is equal to the sum of heat of vapourisation and heat of fusion, as summarised in the schematic diagram, (Fig. **2.3**). Heats of vaporisation, fusion and sublimation are greatly influenced by intermolecular forces.

Intermolecular forces are of two types: long range, in which the energy of interaction follows power series in $\frac{1}{R}$ (R stands for distance of separation), and short range, in which energy shows an exponential decrease with distance. Long range forces are further classified into three types: electrostatic, induction, and dispersion. Electrostatic effects are originated from the static charge distribution between two molecules (the attraction between electronegative elements (for example fluorine (F)) and less electronegative elements (*e.g.*, sodium (Na)). Fig. (**2.4**) shown below demonstrates the interaction pattern within a polar molecule such as hydrogen chloride (HCl).

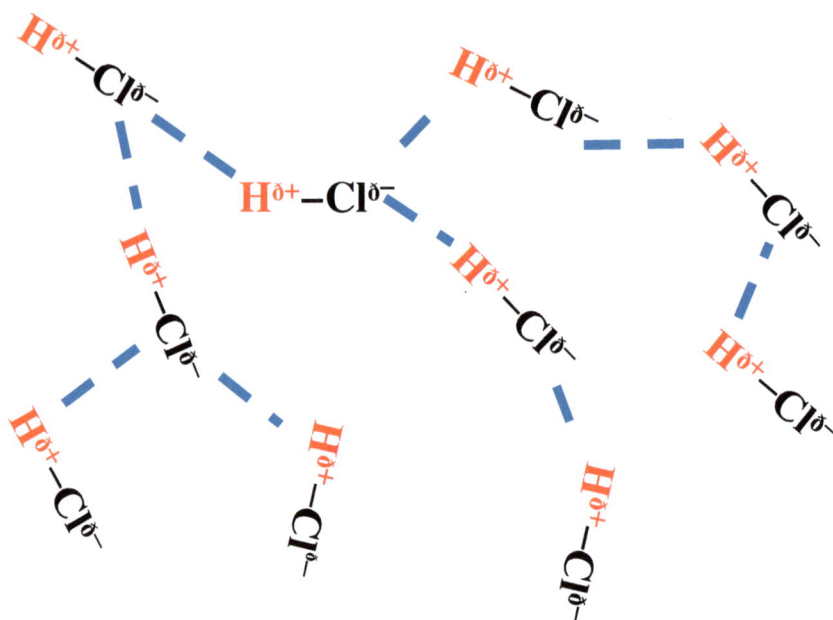

Fig. (2.4). Permanent dipole – permanent dipole interactions existing within polar molecules such as hydrogen chloride (HCl). The broken red line indicates the interactions existing between the opposite partial charges of neighbouring molecules.

A rudimentary way to treat electrostatic interactions is to use well-known Coulomb's expression, according to which the charges separated by a fixed distance is proportional to the product of charges over the distance of separation [46]. Using Taylor series approach one can expand this equation to a series of infinite order to include more complex electric moment (multipole) terms, known as multipole expansion [46, 47]. The simplest multipole moment is total charge of a molecule, which is the sum of all charges of constituent particles including all electrons and of course nuclei. The subsequent multipole moments are called by their ranks: dipole corresponds to rank 1, quadrupole corresponds to rank 2 and so forth. Interactions between multipoles are strictly distant dependent: for example, dipole−dipole, dipole−quadrupole, quadrupole−quadrupole interactions are inversely proportional to the third, fourth and fifth power of distance of separation (R) respectively. In a dipole, two opposite charges are aligned; in a quadrupole two pairs of positive and negative charges are arranged in a two dimensional array; and eight charges (four positive and four negative) are arranged in a three dimensional array in an octopole, as shown in Fig. (**2.5**).

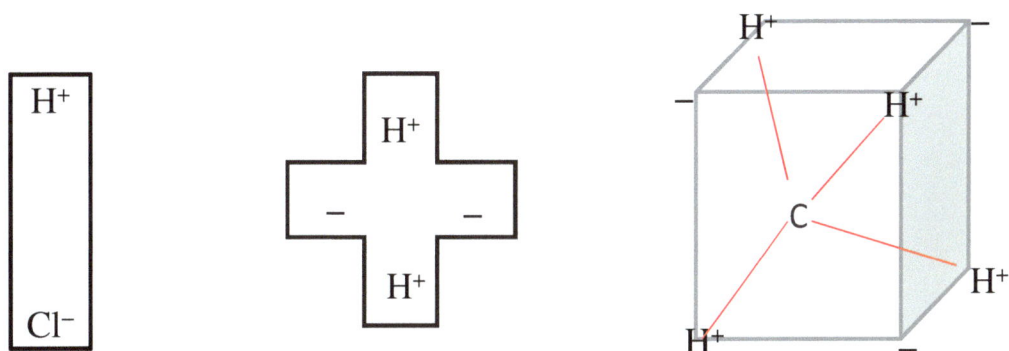

Fig. (2.5). The charge distribution in dipole, quadrupole and octopole. The total number of charges on multipole depends upon the rank 'n' of multipole moment (2^n−pole moment). Dipole, quadrupole and octopole have n equal to 1, 2 and 3 respectively.

Distortion of a molecule due to the electric field generated by the neighbouring molecules gives rise to another long−range effect, induction forces. Consider a molecule (molecule 1) with a permanent dipole, whose charges are separated by a finite distance, oriented in a particular direction. When it is approached by another molecule, a dipole moment is induced on latter. The electric field of the first molecule is proportional to induced dipole moment, and the proportionality constant is known as polarizability, often represented by the symbol alpha (α). Upon the application of electric field on an atom in a molecule, the atom is polarised (charge distribution is distorted), spurring on higher order multipole moments in the atom, which is extended to the whole molecule.

In general, one can say that induction forces result from the distortion in a particular molecule due to the electric field generated by its neighbouring molecules. Unlike electrostatic interactions, induction forces are non−additive, dependent upon the orientation of surrounding molecules, as shown in the following schematic diagram, (Fig. **2.6**). As one can see from the diagram (please see case B), induction energy is accumulated when the molecules with their electric fields aligned in the same direction, whereas it is annulled when the field aligned in the opposite directions (case C).

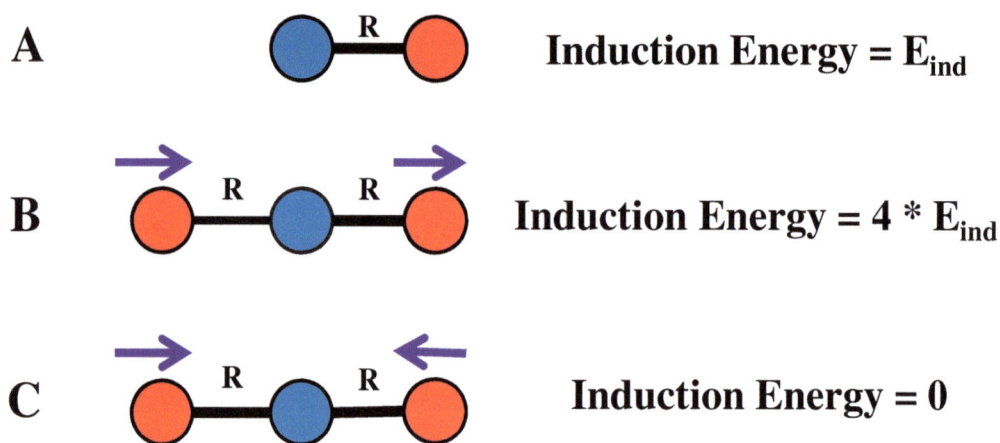

A   **Induction Energy = E_{ind}**

B   **Induction Energy = 4 * E_{ind}**

C   **Induction Energy = 0**

Fig. (2.6). Induction energy in different molecular environments. (**A**) Induction energy is produced due to the effect of a single neighbour molecule (shown in red) on the molecule shown in blue. (**B**) Induction energy is cumulated (four times than the first case) by the presence of two molecules whose electric fields align in the same directions(shown by the violet arrows). (**C**) When the electric fields of the two neighbouring molecules are aligned in the opposite direction, induction energy is annulled.

Like electrostatic interactions, induction effects in a molecule can be computed by variety of mathematical techniques including multipole expansion, perturbation theory, Applequist model and distributed polarizability approaches [47].

Dispersion forces, also known as London forces, are generated by constantly fluctuating charge distributions about the nucleus as electrons move. As a result, an instantaneous dipole (not permanent) within an atom is created. This can induce dipole in neighbouring molecules, and motions of electrons in two neighbouring molecules are correlated as schematically shown in Fig. (**2.7**).

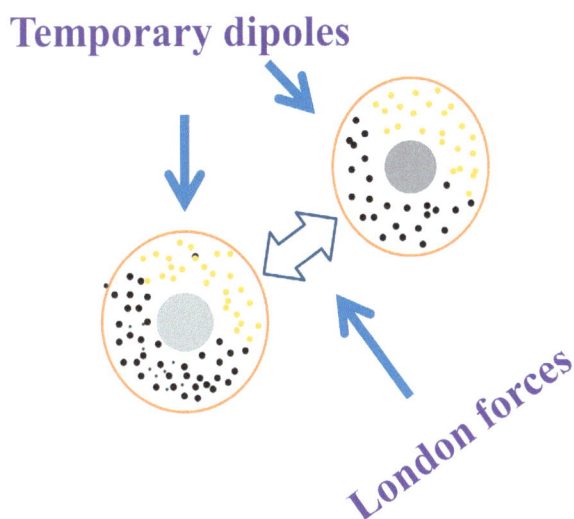

Fig. (2.7). Generation of London forces between two molecules. Uneven distribution of electrons due to the motion of electrons around the nucleus in atoms results in charge separation (dipoles, as indicated by black (positive) and orange (negative) doted regions in the figure).

The interesting fact is that once a dipole is created in an atom, it is induced on neighbouring atoms, and is spread across throughout the molecule as in the case of induction forces. One can say that as the number of electrons in the atom increases, London forces get stronger as exemplified by increase in boiling point upon increase in the molecular weight of hydrocarbons. It must be however noted that this dipole is not created permanently, rather it is formed momentarily (instantaneous). Dispersion energy can be modelled by various methods. A rudimentary way of dealing with dispersion forces is to employ Drude model, in which electrons (treated as cloud) are bound to the nucleus, like a harmonic oscillator. For other complex mathematical treatments see [47].

Hydrogen bonding is a typical intermolecular interaction that is very fundamental to the understanding of water. An appropriate treatment of this intriguing intermolecular interaction has been given in chapter 5, therefore I don't venture into providing a detail account of it here. The aforementioned long range intermolecular forces, instantaneous dipole – induced dipole (the weakest), dipole

– dipole & hydrogen bonding (the strongest), can act simultaneously or individually, depending upon the type of molecules, whether polar or non–polar. One has to note that the term 'long range' is originated because of the survival of these three types of interactions at large separation, and these forces can also act at very short distances [47].

Short range effects include exchange–repulsion, exchange–induction, exchange–dispersion, and charge transfer. Exchange–repulsion term comprises two terms: attractive and repulsive, and its origin lies in the overlapping of molecular wave function making possible of exchange of electrons, which makes way for the free movement of electrons. As a consequence, the precise location of the electrons remains uncertain. At the same time, electrons with same spin experience repulsion between them as per Pauli's anti symmetric requirement that electrons with same spin cannot occupy same space. Exchange–induction and exchange–dispersion are quantum mechanical counterparts of induction and dispersion effects discussed previously. Charge transfer effects originate from the electronic transition between lower energy ground state and higher energy excited states in an electron acceptor – electron donor complex. Fundamental long range and short range intermolecular forces hitherto discussed in this chapter are summarised in Table **2.2**.

Table 2.2. **Various contributions to Intermolecular forces.**

Long Range	Short Range
electrostatic	exchange–repulsion
induction	exchange–induction
dispersion	exchange–dispersion
	charge transfer

The accurate estimation of intermolecular forces is very vital for understanding liquid structure as it is inextricably linked to their structural as well as thermodynamic properties.

THEORIES OF LIQUID STRUCTURE

In the introduction of this chapter, we saw that the order in liquid lies in between solids (the most ordered of the three principal states) and gases (the least ordered of the three). This intermediate random movement of particles in liquids warrants a statistical treatment: by means of a distribution function, with full set of which one can describe the structure of matter very accurately. This statistical approach has two principal advantages: the distribution function can be emulated by both experiments and computer simulations, as we will see in the coming chapters, and immediate characterisation of matter is possible from it, which is distinct for gases, liquids and solids and therefore is a powerful visual tool for characterising various states of matter [48]. For perfect gas, the distribution function is a straight line, as its density is same throughout the gas sample. The peaks in the distribution plot become more evident, as the density of gas increases. An excellent comparison of the distribution functions of gases, liquids and solids can be found in [48].

Having realised the usefulness of the distribution function, it is important to obtain its functional form, which can be achieved by an approach of Integral Equations. In this approach, based on a pair distribution, correlation between every particle–pair in the sample can be mapped. This method has been modified several times, resulting in the following approaches: Yvon–Born–Green equation, Kirkwood equation, Ornstein–Zernike (OZ) equation, Percus–Yevick (PY) equation, Hyper–Netted Chain equation, and Mean Spherical approach. For an excellent comparison of these methods, the readers are required to refer [48]. Integral equations can be used to estimate compressibility, pressure and energy. With these three quantities, other thermodynamic quantities can be calculated [49]. In addition, several perturbation approaches such as Barker–Henderson and Weeks–Chandler–Andersen are also in use. The fundamental basis of the perturbation methods is the refinement of the interactions with addition of incremental effects.

Besides these theoretical approaches explained above, several physical models have also been developed in order to describe liquid structure, including Lattice Theory, Significant Liquid Structure (SLS) Theory, and Scaled Particle Theory [48, 49]. According to Lattice Theory, liquid and solid particles occupy the lattice points of a regular structure and neighbouring molecules are placed in a lattice around the central molecule (random close packing). Depending on the ratio of number of particles to number of cells wherein they are accommodated, two further approaches are emerged within the scope of Lattice theory: Cell model and Hole model [48]. In Cell models, this ratio is one, whereas in Hole model, number of cells (vacancies in the lattice) is always larger than the number of particles. Significant Liquid Structure Theory (SLST), developed by Eyring and co−workers, stress the structural peculiarity of liquids as having intermediate structural features between solids and gas [45]. The third approach for modelling liquid structure is by means of computer simulations, which I mention exclusively in chapter 4.

METASTABLE WATER

The state of supercooling, reported first by Fahrenheit in 1724, is one of the intriguing characteristics many liquids including water exhibit. Water as any other matter can exist in three principal forms: solid, liquid and gas. Under normal conditions, it exists as liquid. Interestingly upon decreasing temperature below their freezing point, materials like water can exist as liquid and this state is known as supercooled.

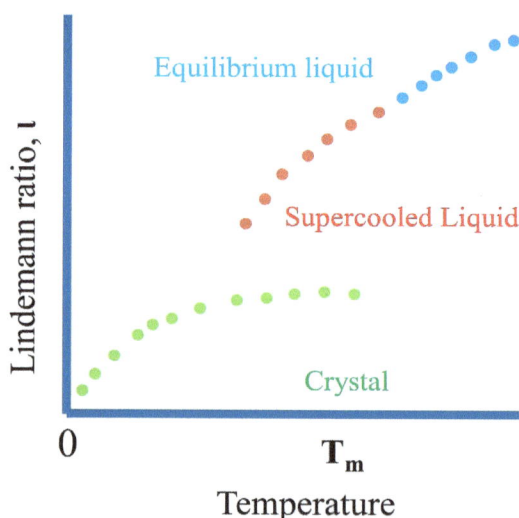

Fig. (2.8). Lindemann ratio of liquids, supercooled liquids and crystals across a wide temperature range. Lindemann ratio is the highest for liquids (shown in blue), followed by supercooled liquids (shown in orange). Both have higher ratio than crystals (shown in green). T_m is the melting temperature, at which the ratio for liquids is three times higher than that of crystals.

Lindemann(l) ratio, which measures root−mean−square particle displacement of particles over nearest neighbour spacing, provides a useful means to differentiate the crystal and liquid forms of a material, as shown above in Fig. (**2.8**) [50]. According to this criterion, liquid has larger Lindemann ratio than the solids as shown in the figure. Both phases clearly show increase upon hike in temperature. Supercooled state, the continuation of the liquid phase, indicates a moderate value. Lindemann ratio for solids can be computed by both computational and experimental methods (X−ray and neutron diffraction), and for liquids no experimental device has been set up for finding Lindemann ratio yet [50]. The characteristic differences in the magnitude of Lindemann ratio for various phases of H_2O have been shown in Fig. (**2.8**).

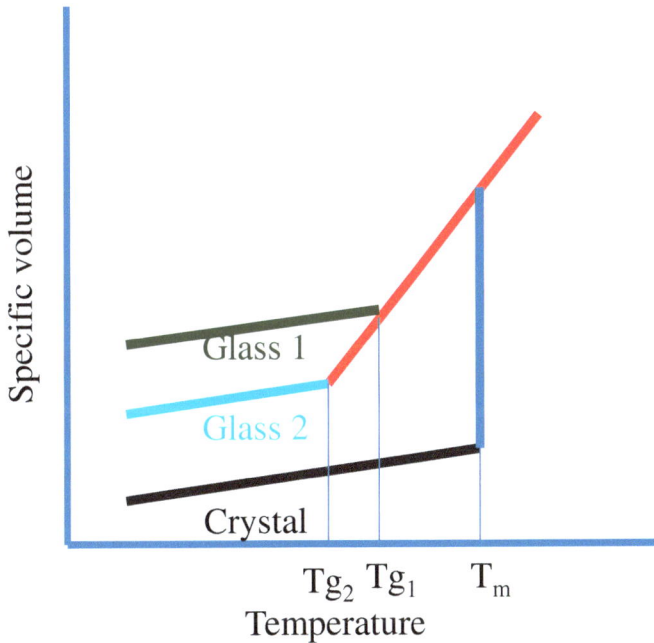

Fig. (2.9). Specific volume function of temperature. Glass 1 and glass 2 are obtained from the sample but with two different cooling rates. Sample 1 is cooled at higher cooling rate than sample 2. If the cooling rate dips below the threshold cooling rate, crystallisation is unavoidable.

Although more natural is the tendency to undergo crystallisation, which is thermodynamically favourable due to the release of heat (leading to higher entropy), some liquids such as water can remain in supercooled state under certain conditions if they are cooled at a threshold rate. If the sample is cooled at a slower rate, as shown in diagram (Fig. **2.9**), it will end up in crystalline phase [51].

Supercooled state is characterised by liquid like properties, and hence is said to be in metastable state with respect to the more stable crystalline phase. However, the transition from supercooled state to crystalline state is obstructed by an impenetrable high free energy wall, characterised by a huge gradient in potential energy. The free energy imbalance is due to the fact that supercooled states pass through several intermediate states with higher free energy than the liquid or crystalline state as shown in Fig. (**2.10**) [52]. Difference in free energy generates free energy barrier, precluding the transformation from metastable to stable states. One must note that the free energy barrier between liquid and crystal states is higher in higher temperatures and smaller in low temperature domain.

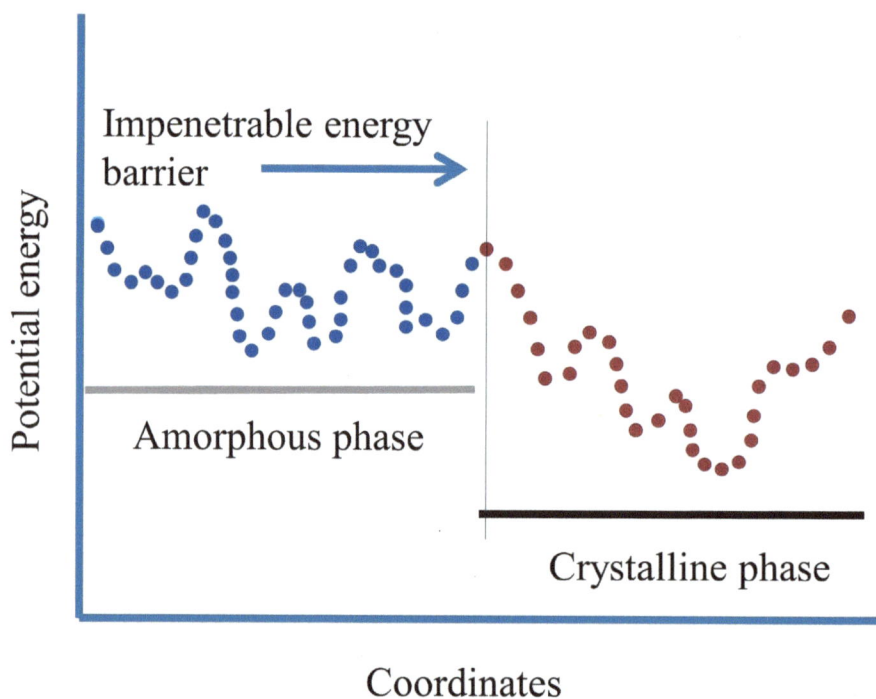

Fig. (**2.10**). Amorphous state *versus* crystalline state. In the left (shown in blue dotted line) is amorphous state, which is separated from the crystalline state shown in red dotted line by high energy barrier.

The concept of metastability cannot be applied in some exceptional liquids. For example, supercooled o–terphenyl can remain in supercooled state without yielding to crystallisation under laboratory conditions; and crystallisation never occurs in the case of some atactic polymers, a class of polymers with random molecular arrangement [53].

The Transition from Supercooled Liquid to Glassy State

In this sub section, I venture into providing a general account of the transition from supercooled state to glassy state. Chapter 6 is devoted exclusively to the supercooled and glassy waters and the transition between them.

The supercooled liquid can attain glassy state (disordered solid state) at glass transition temperature, T_g, a state characterised by high viscosity around thirteen orders of magnitude [54], and this temperature is about two–thirds of the freezing temperature, T_m [52a]. It is estimated that the difference between melting temperature, T_m, and the glass transition temperature, T_g, is approximately 30% of the melting temperature of the material [55]. The glass transition temperature, T_g, is the characteristic temperature which separates a liquid from its glassy form. However, the transition cannot be said to occur at a precise temperature, rather in a wide range of temperatures [52a]. This is primarily due to inter molecular interactions: the temperature width of the glass transition varies from as low as 50K to as high as 1500K, depending on the material [56]. In other words, one can correlate the glass transition temperatures with inter molecular strength.

A large class of materials are found to form glasses including elements (P, S & Se), oxides (SiO_2, GeO_2, B_2O_3, P_2O_5, As_2O_3, Sb_2O_3), sulfides (As_2S_3), halides (BeF_2, $ZnCl_2$), mixtures of salts (potassium and calcium nitrates ($KNO_3/Ca(NO_3)_2$) & potassium and magnesium carbonates ($K_2CO_3/MgCO_3$), aqueous solutions of acids (dilute H_2SO_4), organic compounds (glycerol, methanol, ethanol), polymer compounds (poly styrene, poly vinyl chloride), and metal alloys (Ni/Nb, Cu/Zn) [57].

Organic compounds can also attain glassy state. Highly noteworthy is that glass transition temperatures (T_g) of isomers, (for instance, aromatic compounds), can have large difference with each other depending upon the location of substituents attached to the aromatic ring [58]. It is important to note that T_g (glass transition temperature) cannot be considered as a fundamental property because its value can be changed depending upon the experimental conditions [59]. This attainment of glassy state is technically known as vitrification, with many potential applications including cryoprotection of biological organs & organic materials [54].

When supercooled liquids become glassy or vitrified, thermodynamic quantities such as viscosity, enthalpy, dielectric relaxation times, thermal expansivity, and heat capacities show dramatic variations. Table **2.3** shown below summarises the values of these two quantities for several materials at liquid and glassy state. It has to be noted that the values of the specific heat capacity (C_p) and coefficient of expansion (α) in the glassy state in general are close to those of crystalline state, with some exceptions. In the following table, a comparison has been made, and one can find that both specific heat and coefficient of expansion show a marked decrease at glassy state.

Table 2.3. Specific heat and coefficient of expansions of some materials liquid and glassy phases. The values can be found in [52a] and references therein.

Material	T_g (K)	Specific Heat Cal/g		Coefficient of Expansion x 10^4	
		(Liquid)	(Glass)	(Liquid)	Glass
Ethanol	90-96	0.417	0.27	11.2	--
B_2O_3	470-530	0.436	0.30	6.1	0.5
Rubber	200	0.39	0.27	6.0	2.0

Attempts to provide a theoretical explanation on the drop in these two quantities, C_p and α, at the onset of glass transition started as early as the middle of twentieth century, one of such is highly intuitive Hole theory, which was introduced to the reader in the last section [52a]. According to this theory, liquids are treated as quasi crystals in which some lattice points are occupied and others are not. The vacant lattice points are known as holes. When temperature is raised more holes are created as a result of transformation of liquid particles to vapour, and more energy is supplied which is contributed to heat capacity and thermal expansion coefficient. As the temperature decreases, the reverse process occurs, *i.e.* more number of holes is occupied, but this process requires rearrangement of molecules about new positions. At lower temperatures, the relaxation time (response to mechanical, electrical, thermal, and optical or any other physical perturbations) is so large that the energy contributions of the molecular rearrangement to the heat capacity or expansion coefficient are negligible, leading to a notable decrease in these physical quantities at glass transition temperatures.

It might be confusing for some readers when we talk about heat capacity: which heat capacity must be taken for account: heat capacity at constant pressure (C_p) or heat capacity at constant volume (C_v)? The heat capacity referred to the experiments related to liquids and solids is generally C_p unless otherwise stated. Unlike for gases, the difference between the two heat capacities is subtle for liquids because all the heat withdrawn from the surroundings is utilised to increase the spacing between the molecules. Here we must remember that their thermal expansion coefficients are negligible [44]. On the other hand, measuring heat capacity at constant volume is cumbersome due to the fact that the pressure change occurring during the experiments may cause an explosion of the container in which the solid or the liquid is kept. This is due to very small compressibility of liquids.

The variation in responses to various aforementioned physical properties, in particular relaxation times and viscosity, in the vitrified state results in two kinds of liquids (glasses): strong and fragile, which we discuss in the following section.

Strong and Fragile Liquids

The intimate relationship between glass transition temperature, T_g, and relaxation times were briefly mentioned before. Glass transition temperature is the temperature below which the molecular motions are too slow to establish the equilibrium between various phases [59]. In other words, it can be said that the long relaxation times at glass transition cause delay in transformation from one phase to the other, for example from liquid to the solid phase. Strong glasses follow Arrhenius behaviour, in which relaxation time is directly proportional to Arrhenius factor. A high school chemistry student is very familiar with Arrhenius factor, connecting activation energy and Boltzmann factor to absolute temperature, while studying chemical kinetics. On the other hand, fragile glasses follow non−Arrhenius behaviour following Vogel−Fulcher−Tamman (VFT) equation. Materials such as silicon dioxide (SiO_2) follow Arrhenius behaviour and therefore are strong liquids. Some liquids such as o−terphenyl exhibit non− Arrhenius nature and therefore are classified as fragile liquids [54]. Fig. (**2.11**) shows the mathematical description of strong and fragile liquids.

Fragility   Temperature below T_g Activation Energy

$$\tau = exp^{\frac{DT_0}{T-T_0}} \qquad \tau = exp^{\frac{A}{kT}}$$

Fragile liquid   Strong liquid

Fig. (2.11). Mathematical description of fragile and strong liquids. (Left) Expression for liquids showing fragile character - it does show exponential behaviour for relaxation times, but does not follow exact exponential (involving activation energy and Boltzmann's constant) like in strong liquids. Rather, a new term, D (fragility) must be introduced in the expression for fragile liquids.

The contrasting behaviour of strong and fragile liquids can be clearly understood from a viscosity or relaxation times *versus* glass transition temperature (T_g) scaled temperature plot. In the plot (Fig. **2.12**), the glassy behaviour of three different liquids is shown: SiO_2 as mentioned earlier demonstrates a strong linear relationship between temperature and viscosity, in contrast to o−terphenyl. Liquids such as glycerol exhibit an intermediate behaviour [60]. The interesting fact is that although liquids such as o−terphenyl exhibit non−Arrhenius behaviour in the supercooled regime, it does exhibit Arrhenius dependence at higher temperature, above its melting temperature, T_m [61]. This would mean that if o−terphenyl too had followed the Arrhenius behaviour in the supercooled region, the observed glass transition temperature would have been very far below of its estimated glass transition temperature, 243 K. The extrapolation of physical properties towards the supercooling regime results in thermodynamic catastrophe including negative configurational entropy, as noted by researchers earlier [52a, 59]. The classification of liquids as strong and fragile can also be explained on the basis of coordination number. In the strong liquids, such as SiO_2, the coordination number is fixed, whereas a variable coordination number is observed in the case of fragile liquids [62].

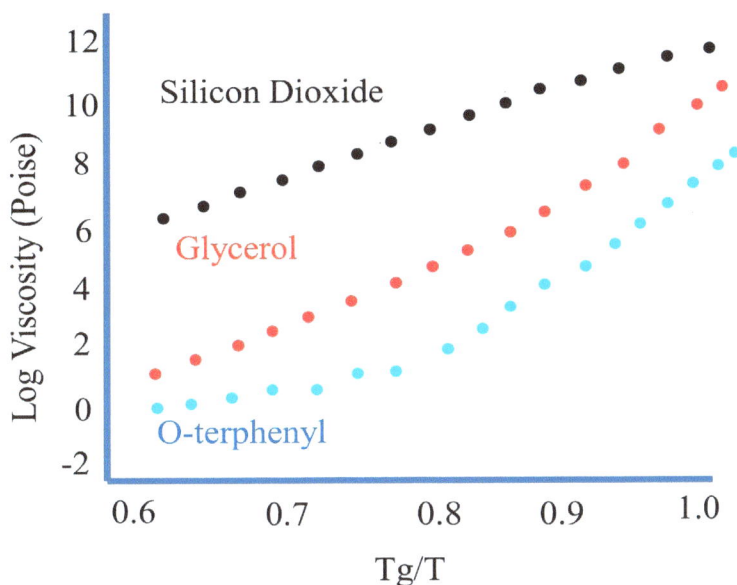

Fig. (2.12). Viscosity *versus* T_g scaled temperature. The glass transition temperature approaches from left to right when T_g/T equal to one. Silicon dioxide shows a regular linear relationship across the temperature domain. The fragile liquid o-terphenyl demonstrates non−Arrhenius behaviour and can be classified as a fragile liquid. Glycerol shows an intermediate trend. The idea has been adapted from [60].

Substituting A (Fig. **2.11**) in the expression for strong liquids by the difference between absolute temperature and the glass transition temperature $(T-T_g)$, one can easily construct similar plots akin to Fig. (**2.12**). In strong liquids this difference is independent of temperature. Even from temperature above the melting temperature, T_m, one can observe a steady increase in activation energy upon a dip in temperature [55]. On the contrary, a sudden jump in the activation energy can be observed in the case of fragile liquids, as one can observe the behaviour of o−terphenyl (Fig. **2.12**).

Entropic Considerations

The strong –fragile classification of liquids can also be explained basis of entropy (S) as well. Liquids are more disordered than crystals and therefore have higher entropy. The entropic difference, alternatively called excess entropy, between these two states of any material obviously becomes zero when the crystals are formed from liquid phase. The temperature at which the entropic difference between liquid and solid becomes zero is known as Kauzmann's temperature, T_K. From the ΔS Vs $\frac{T}{Tm}$ plot, one can extrapolate the temperature at which excess entropy becomes zero [63]. For strong liquids such as silicon dioxide (SiO_2) or boron trioxide (B_2O_3) the difference is close to absolute zero, 0 Kelvin, as shown in (Fig. **2.13**). However, for fragile liquids such as lactic acid, the excess entropy would vanish at a temperature, which is $\frac{2}{3}$ of melting temperature (T_m). The zero "excess entropy" would be reverted to excess entropy if a kinetic phenomenon such as glass transition (T_g) occurs before the liquid is converted to crystals [63]. The sluggish motion of particles prevents the formation of much ordered crystals then, averting the entropy loss. The following figure demonstrates the characteristic difference between two liquids: strong and fragile, in terms of entropy.

The problem nevertheless remains as a big paradox when liquid remains as supercooled beyond Kauzmann's temperature as it would mean that the excess entropy (entropy difference between the solid and liquid phases) becomes 'negative'. This ambiguity is known as Kauzmann's paradox, eponymous of Walter Kauzmann who noted this situation first [52a, 63].

Theories of Low Temperature Liquids

The notion of Kauzmann's paradox however has been challenged, attributing it to the result of inaccurate measurements [64]. At the same time, various models have been suggested to accurately describe the properties of liquids at lower temperatures, and to explain Kauzmann's paradox. Free volume model, which is inextricably linked to Lattice models discussed previously in this chapter, is one of the earliest models to account for the characteristics of supercooled liquids, based on two assumptions: first, a supercooled liquid exists in two distinct phases, liquid and solid, and second, if a molecule has a 'free volume' around it, and it can diffuse through the whole liquid without any activation energy (please see [48]).

I would like to discuss only three important theories that are historical landmarks in glass physics in considerable detail in the remaining part of this section: Adam–Gibbs theory, Mode–Coupling Theory (MCT), and Energy Landscape model.

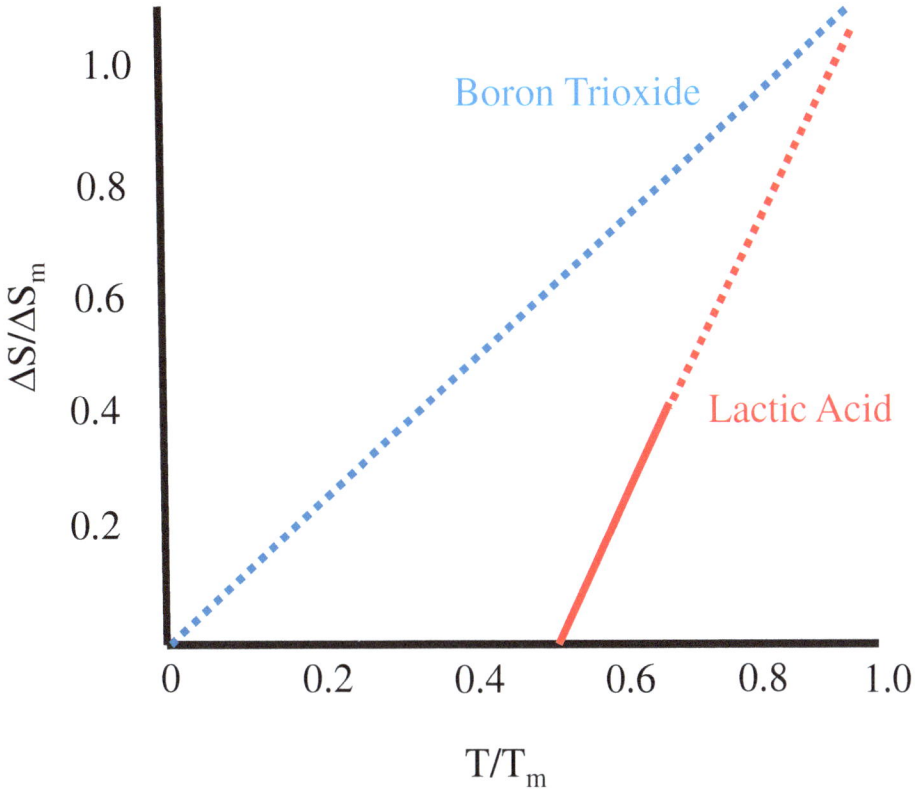

Fig. (2.13). Entropic effect in strong and fragile liquids. The relationship between scaled entropy and temperature is demonstrated in the figure. For strong liquids like boron trioxide (B_2O_3), the excess entropy attains zero only near the absolute zero. On the contrary, fragile liquids such as lactic acid, the surplus entropy attains zero well before the absolute temperature, if the glass transition does not occur. The idea has been adapted from [52a].

Adam–Gibbs Theory

The significance of Adam–Gibbs theory is that it provides a solution to the paradoxical negative entropy (Kauzmann's paradox) below glass transition temperature. Adam–Gibbs model is a molecular kinetic theory which correlates temperature dependence of relaxation times of glass forming liquids to cooperative motion of molecules [59]. This gives rise to a cooperative rearranging region, the smallest region that can take new configuration without requiring configurational change outside its boundary [59]. This is strictly temperature dependent, as shown by Adam and Gibbs: at lower temperatures, there can only be one of such regions. On the contrary, there can be several cooperatively rearranging regions at higher temperatures. The cooperative regions are related to a new thermodynamic variable, called configurational entropy, S_{conf}, which is inversely proportional to relaxation times. Glass transition occurs when the relaxation times diverge (approaching infinity), leading to zero configurational entropy.

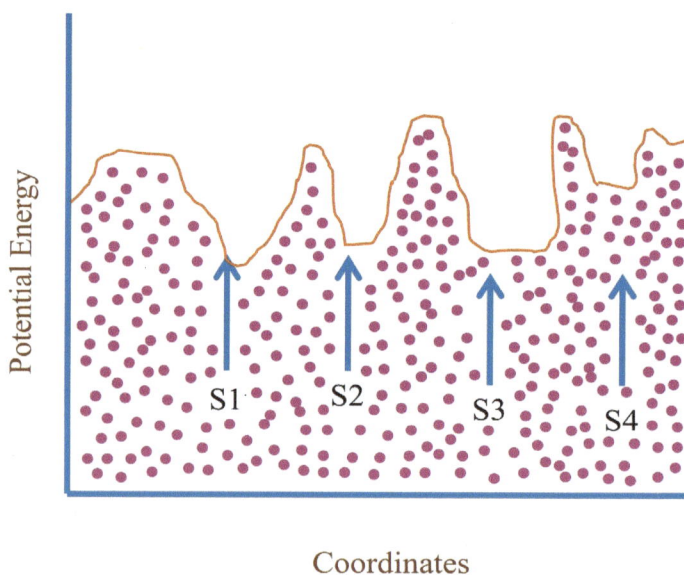

Fig. (2.14). Schematic illustration of concept of Inherent Structures. Inherent structures (indicated in the figure from S1 through S4), also known as fundamental structures or metabasins, are sequentially visited by system during specified time interval.

Energy Landscape Theory

Energy Landscape Theory (ELT) provides a topological view of the nature of glass forming liquids at molecular level, by connecting their thermodynamic and kinetic properties. Energy Landscape Theory (ELT) introduces the notion of Inherent Structures (IS), which are basic structures joined together to create a continuous composite potential, as schematically shown in Fig. (**2.14**).

According to Energy Landscape Theory, two distinct internal motions occurring in liquids dictate their dynamics at low temperatures, allowing one to assign two entropic contributions to the configurational entropy: first, vibrational motions localised in every basins and second, hopping between the basins leading into the structural relaxation [65]. In conjunction with Normal Mode Analysis (NMA), Inherent Structure (IS) analysis is very powerful tool to distinguish various physical processes in liquids such as hydrogen bond rearrangement in water [66]. With Energy Landscape Theory (ELT), topological characterisation of strong and fragile glasses can be achieved, as shown in the Fig. (**2.15**). It can be seen that in the case of strong glass forming liquids there involves only a single basin. By contrast, there can be several widely separated traps present in the topology of fragile glasses, which are accounted for a wide range of relaxation times [57].

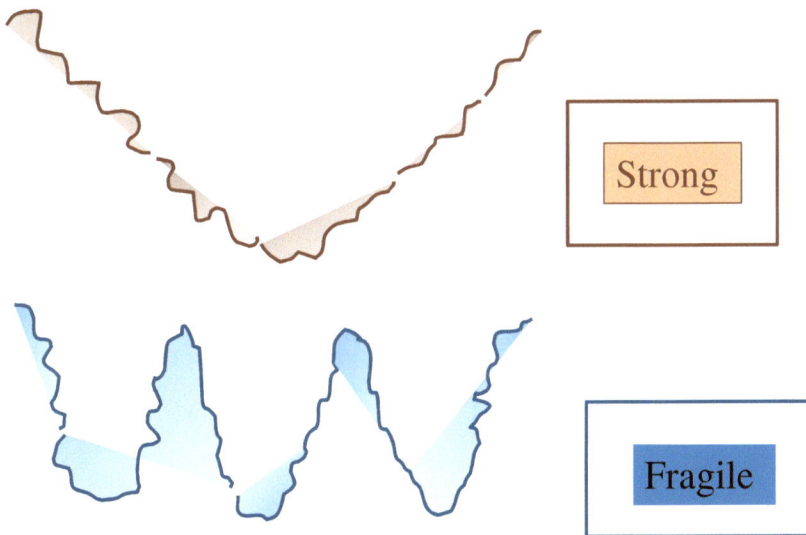

Fig. (2.15). Illustration of Energy Landscape Theory (ELT). Schematic illustration of the difference between strong (shown in blue) and fragile (shown in red) glass formers, according to Energy Landscape Theory. In the strong glasses there exists only single basin. On the contrary, there exist numerous basins separated by large energy barrier in the case of fragile glasses.

Mode Coupling Theory (MCT)

Mode Coupling Theory (MCT) considers glass transition as a transition occurring at a critical temperature (T_c), from ergodic (higher temperature than T_c) to non–ergodic (lower temperature than T_c), estimated to be 1.2 times that of actual glass transition temperatures, T_g. Ergodicity refers to the ability of particles to diffuse freely and hence to be delocalised. Within the context of relaxation bifurcation phenomenon (alpha and beta relaxations), the glass transition can be explained by Mode Coupling Theory (MCT).

According to Mode Coupling Theory (MCT), the intermediate scattering function, (a measure of density correlations) of a liquid when supercooled clearly exhibits dual relaxation mechanisms, known as α relaxation (slower) and β relaxation (faster) (also known as primary and secondary relaxations respectively), as shown in Fig. (**2.16**) [67]. From the previously discussed Energy Landscape Theory (ELT), we learn that the faster β relaxation is concomitant with the inter basin (which are separated by small energy barrier) transfer. In normal temperature domain, however, this dual relaxation behaviour is dormant, and the function decays very fast. The drawback of classical MCT is that it does not quantitatively predict the behaviour of supercooled liquids at around respective glass transition temperatures [57]. Modifications of MCT theory, which offer better accurate results, are nevertheless widely applied to investigate glassy systems.

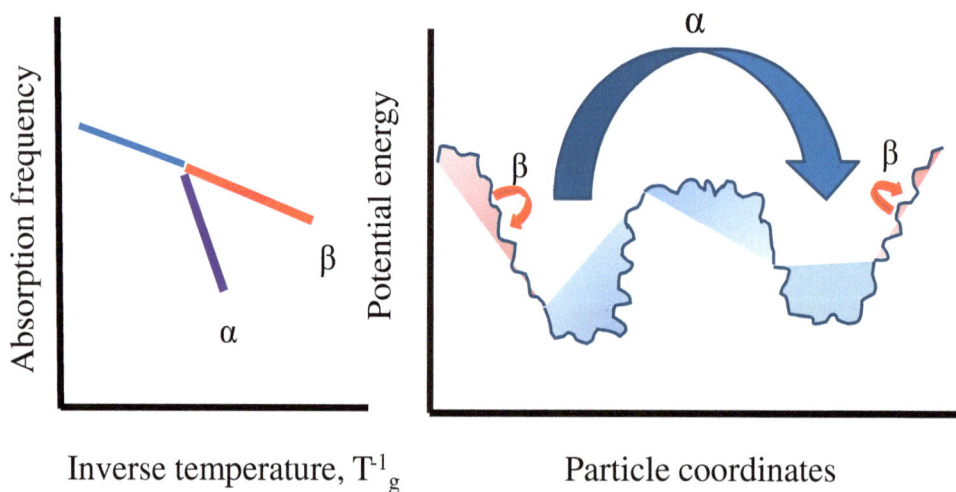

Fig. (2.16). Temperature dependence of absorption frequencies of glass forming liquids (left). Upon lowering temperatures below T_g, the relaxation is split into alpha and beta relaxation process (right). The alpha relaxation is associated with transfer from one metabasin to the other shown by the thick blue curved arrow, whereas beta relaxation is identified with elementary relaxations between neighbouring basins.

In addition to these three models, several other theoretical models including kinetically constrained model, Trap model, Frustration model, spin–glass model, and Two Order Parameter (TOP) model have been developed in order to explain the behaviour of materials at low temperature [67].

Good Glass Formers

The definition of strong and fragile glass formers and their theoretical interpretations were given in earlier sections. However, one direct question remains from a practical point of view. Which is a good glass former? According to the criterion given by Angell, the cooling rate at which the liquid be cooled must be low to avoid the formation of ice [56]. Aqueous solutions of multivalent cationic salts at a critical concentration – aqueous solutions of lithium salts such as LiCl – H_2O are prime examples for good glass formers, as per the above definition. Other type of good glass formers are aqueous solutions of sugars such as trehalose – water system. However, the validity of this classification is strongly debated as both cooling at faster and slower rates can result in crystallisation [37].

CONCLUDING REMARKS

The primary aim of this chapter is to provide the reader a brief overview of the theoretical treatment of liquids and glasses. A good grasp of various types of intermolecular forces that hold molecules intact is essential to understand the complex character of liquids such as water. These interactions are represented by complex mathematical expressions, which can accurately be computed by high performance computers. This renders a realistic estimation of the properties of matter, of liquids in particular, from computer simulations. With enhanced computational facilities, current crude approximations employed to account for the complex intermolecular interactions are expected to be replaced by "accurate" mathematical constructs. We must also note that intermolecular potentials are central for obtaining thermodynamic as well as structural quantities (such as radial distribution functions). Several theories take account for the complex behaviour of liquids in low temperature range. Most notable of them is Energy Landscape Theory (ELT), very effective tool to characterise the strong and fragile glasses. On the other hand, Adam–Gibbs formalism provides a benchmark framework to investigate the thermodynamic aspects of glass transition. More development in this direction is required for further understanding of water in low temperatures, given the fact that more amorphous and solid phases of water are being unearthed. In the next two chapters, experimental as well as theoretical tools for understanding liquids are explained, to provide the reader necessary background for the core chapters following them.

CONFLICT OF INTEREST

The author confirms that he has no conflict of interest to declare for this publication.

ACKNOWLEDGEMENTS

Declared none.

CHAPTER 3

Experimental Tools for Microanalysis of Water

Abstract: A wide range of experimental techniques has been developed and applied for investigating matter at high resolution. Scattering experiments are considered as powerful tools for structure elucidation of liquids including normal water and supercooled water. Employing techniques such as Differential Scanning Calorimetry (DSC), one can record the temperature of phase changes, the glass transition temperature. Quasi Elastic Neutron Scattering (QENS) spectral analysis suggests distinct relaxation behaviour of diffusive motions of water molecules. Nuclear Magnetic Spectroscopy (NMR) is very useful tool in elucidating molecular structures of systems including liquids and aqueous solutions. By Compton Scattering and NMR techniques, estimation of average number of hydrogen bonds has been achieved to a considerable level of accuracy. Extensive studies have been made on water clusters using a sophisticated spectroscopic technique namely Far Infra−Red Vibration−Rotation−Tunneling (VRT) spectroscopy. Optical Kerr Spectroscopy has been employed to investigate the relaxation process at femtosecond and picosecond levels. Properties such as compressibility and diffusion coefficient have been experimentally measured by simple capillary tube techniques. Electron microscopic techniques have become invaluable tools to obtain high resolution of molecular structure materials. Electron microscopic techniques equipped with better resolution can yield further information regarding the microstructure of materials including liquids.

Keywords: Bragg, Compton scattering, Density, Diffraction, DSC, Infra−Red, NMR, QENS, SANS, SEM, Spectroscope, TEM, X−ray, Zeeman.

INTRODUCTION

Obtaining evidence to a claim has paramount importance in science, and experimentation is an integral part of this process. The evidence can be used to further development of theories, thereby increasing our understanding of the world we live in.

Due to immense growth in instrumentation, which modern science relies heavily on, information can be gathered automatically in areas that are beyond human sense perception [68]. An investigator first finds a research theme, and then carries out his/her investigations after procuring a sound theoretical and practical understanding in relevant experimental procedures. The design of experiments depends on the level of information required. Collection and subsequent analysis of data is followed, and the propositions of solutions to the investigating theme are ensued, before communicating his findings to the scientific community.

The experimental cycle just mentioned now is very helpful to arrive at developing models, by which generalization of data can be achieved, and, more importantly, to derive laws which can be used to make predictions [68]. A considerable understanding of both experimental and theoretical procedures is also required for accurate handling of the data, which evidently points into the fact that both experimentalists and theorists should gain a solid understanding of each other's working tools.

The purpose of this chapter is to provide the reader a general idea about some experimental techniques. These techniques are constantly applied for the investigation of various phases of H_2O. However, several challenges are encountered during experimental investigations on liquids such as water in extreme conditions, due to the high probability of crystallisation as temperature drops [69]. By contrast, performing experiments in normal temperature domain, precisely between melting point of water (ice) 273 K, and boiling point of water, 373 K is rather "straight forward". The chapter will aid a normal reader, who is not familiar with them, in understanding of several following chapters.

Scattering and spectroscopic techniques are the two major tools that are widely employed for understanding water structure. These include X–ray and neutron scattering experiments, Infra–Red (IR), and Nuclear Magnetic Resonance (NMR). In addition, a thermometric technique Differential Scanning Calorimetry, abbreviated as DSC, is widely exploited for investigating phase transitions occurring in materials. In addition to these experimental techniques, a reference has been made to capillary technique, one of the oldest experimental methods for investigating water for quite long time.

SCATTERING EXPERIMENTS

In scattering experiments, a sample (liquid or crystal) is subjected to radiation of a particular wavelength. The radiation is scattered by the sample, the intensity of which is determined as a function of scattering angle. Applying suitable mathematical treatments such as Fourier transformation on the resulting data, properties such as Radial Distribution Function (RDF) can be calculated. RDF is the measure of the probability of finding an atom with respect to another atom. Other quantities such as coordination number and nearest–neighbour distance can also be calculated from scattering experiments. Various crystalline ice forms differ in oxygen–oxygen distance and the angle between three neighbouring oxygen atoms (please refer the chapter 7 on ice), and hence differentiated among themselves in terms of aforementioned physical quantities. This explains why scattering experiment serves as an important tool for identifying various crystalline forms of ice. A simple sketch of scattering experiment protocol (known as Laue method) is shown in Fig. (**3.1**).

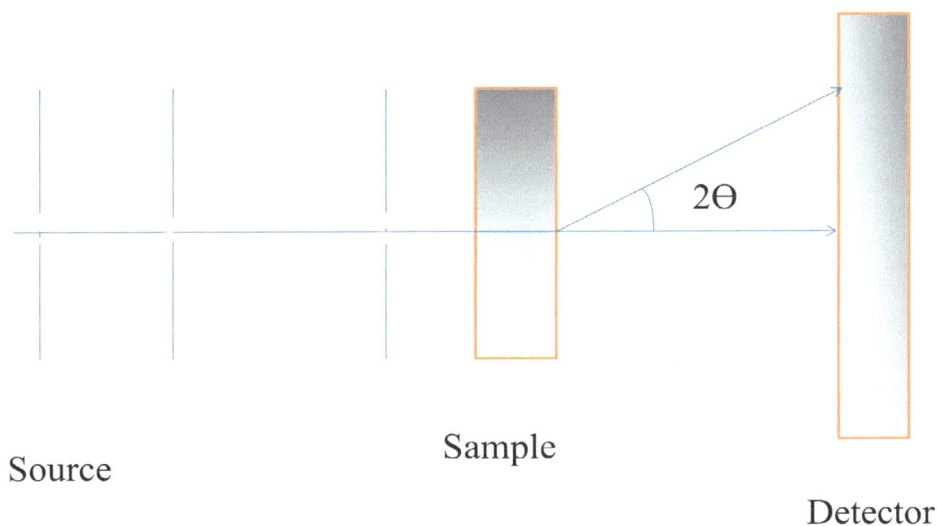

Fig. (3.1). The schematic diagram of scattering experiments. A sample is irradiated with some type of radiation (produced by a source), and the resulting intensity is measured as a function of angle between the incoming beam and scattered beam.

As beam (X–ray, neutron, electron or light) passes through the sample (crystal), it is reflected from every possible plane of atoms in the crystal since they are oriented in multitude of planes. One would expect that the emerging pattern spreads over all angles, but this is not the case. According to the fundamental law of scattering, known as Bragg's rule, the ratio of the product of the spacing between two adjacent planes and the sine of angle the incident rays makes with planes to the wavelength yields only whole numbers. The emergent beam appears only at certain particular angles, due to the presence of numerous parallel planes in the crystal. Laue method went through several improvisations, aimed at overcoming the complexities of mounting sample on a precise axis. Powder method, also known as Debye and Scherrer method, was a major breakthrough in scattering experiments. Many shortcomings of previous methods, most notable of them being the limitation of the number of crystals that can be mounted in alignment with the incident beam, were overcome by Powder method. In Powder method, several crystals (in powder form) can be mounted simultaneously [44]. If a set of planes obeys the fundamental Bragg's rule, a spot is produced on the detector. Rotation of these set of planes generates a cone, which finally yields a line pattern. A consequence of the rotation of the planes is that even wide deflection can be recorded in powder photograph. From the distance between lines recorded in the photograph and dimensions of camera, inter–planar spacing can be found out using Bragg's equation. Here we make use of diffraction, which is the bending of light when it passes around the edge of an object. The dimensions of unit cell can be found out from the spacing of lines in the spectra and intensities of diffraction pattern.

I have just explained briefly how the structural elucidation of crystals can be achieved using scattering experiments. In liquids, a similar protocol is followed with slight changes: instead of line patterns, what we have is broad band of reflected radiation from which Radial Distribution Functions (RDF) can be derived, after some cumbersome computational efforts [44] (Interested readers can refer several articles for procuring a much detail understanding on these tedious computational procedures [70]). It is important to note that several theories on liquids, which were discussed in the previous chapter, are based on distribution functions, highlighting their importance in elucidating liquid structure.

In contrast to crystals, the reflected radiation is represented as broad bands in liquids as was just mentioned. Scattering experiments on crystals provide information regarding cell dimensions and arrangement of atoms. On the other hand, in liquids which possess only short range order, we can only obtain an average estimate of number of particles around the atoms/molecules we are concerned *via* radial distribution functions. The following schematic diagram (Fig. **3.2**) illustrates the concept of radial distribution function and coordination number in liquids.

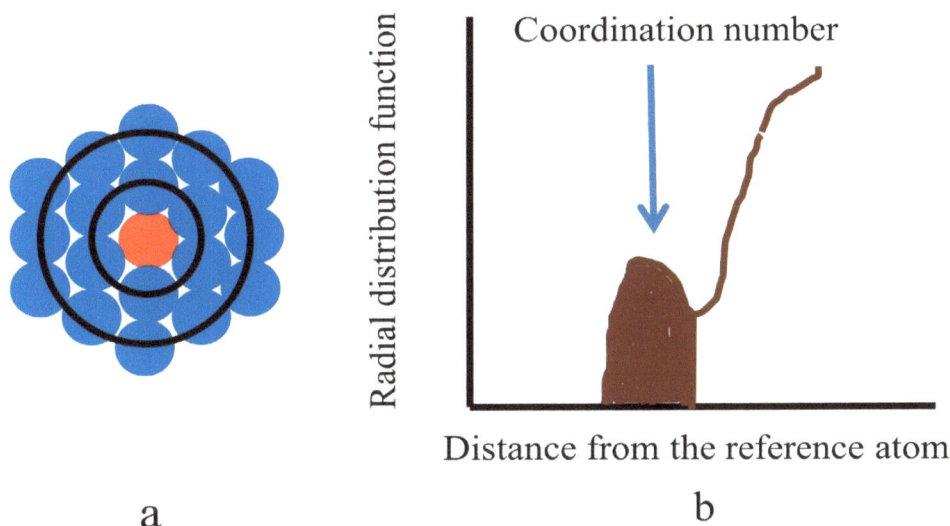

Fig. (3.2). Radial distribution analysis in liquids. In figure (**a**) shown in the concept of radial distribution function. Neighbouring molecules can surround in shells (shown in black circles) a given reference atom shown in red. In figure (**b**), radial distribution function is shown. The peaks observed in the figure correspond to the shells around the reference atom. The shaded region in figure b gives an estimate of coordination number.

X–ray Scattering

In X–ray scattering, X–rays are scattered by a sea of electrons in atoms of a sample. In materials, the atomic density is not alike and therefore some planes scatter the incoming rays better than others resulting in high intense reflective beam. Small Angle X–ray Scattering (SAXS) has been employed to investigate the microstructure of supercooled water and aqueous solutions [71]. Numerous X–ray diffraction investigations have also been made in order to investigate the structural patterns of ice, the low temperature solid form of water [72].

Neutron Diffraction Experiments

The general principle of neutron scattering is similar to X−ray scattering, and the information obtained from these two techniques serve complimentary to each other. In neutron scattering experiments, neutrons are collided with the nuclei of atoms or molecules in the sample. The number of neutrons scattered is counted by a detector and this number is a function of angle of diffraction and neutron wavelength. The extract information about arrangement and interactions of the components of the sample can be obtained from this data. One of the fundamental advantages for neutron scattering methods over X−ray is that neutrons are sensitive to hydrogen atoms, and therefore it is much more efficient to track complex liquid structure in which intermolecular forces play an important role [70b]. Furthermore, neutrons can move large distances in the reaction chamber without being scattered or absorbed since they are not charged. Neutron diffraction data of water have been measured over a wide range of temperatures [73]. Small Angle Neutron Scattering (SANS) experiments have been employed to obtain density profile of water, by measuring temperature dependent neutron scattering density, as reported by Chen *et al.* [74]. In hydrogenous systems, such as water, isotope substitution is employed to obtain their structural features by substituting hydrogen by its isotope deuterium [73b]. By comparing with deuterium, the contributions of hydrogen into the structural features of water can effectively be identified.

Neutron scattering experiments have also been carried out for the structural investigation of other forms of water [75]. Oxygen−hydrogen vibrational profile of normal and supercritical water has been extensively investigated using light and neutron scattering experiments [76]. More importantly, this technique has been accepted as one of the confirmatory tools for the elucidation of ice structures [77]. Phonons, also called as energy packets, absorbed or emitted by neutrons when they are scattered by materials such as ice [72]. Alterations in momentum and energy during this process correspond to atomic movements of the material. In other words, the distribution of phonon frequency gives us the information regarding the various types of movements that occur within materials. Neutron scattering studies have thus become one of the important experimental tools for the structural investigations of ice.

The way the particles are collided with the sample is very fundamental in scattering experiments. There are two types of scattering processes: elastic and inelastic. The total kinetic energy of the colliding bodies is conserved in the former. In the inelastic scattering, a loss of kinetic energy comes about in the form of heat or sound. Further, one or both of the colliding particles may break up, if they are made of smaller constituents. In this case, some of the kinetic energy is consumed for breaking chemical bonds that hold the constituents together. Inelastic scattering is a fundamental scattering process in which the kinetic energy of an incident particle is not conserved (in contrast to elastic scattering). In inelastic scattering process, some of the energy of the incident particle is lost or increased. Both of these scattering processes, elastic and inelastic, have become popular for the investigation of water, as evident from numerous research articles published over the years.

Quasi Elastic Neutron Scattering (QENS) technique has been employed to investigate the dynamics of water at extreme conditions. Dianoux *et al.* have performed QENS experiments on supercooled water [78]. Diffusive motions of water molecules with two distinctive relaxation times have been identified with QENS spectra. Bellisent–Funel *et al.* have investigated the dynamics of supercritical water using this method [79]. Inelastic neutron scattering experiments have been widely employed to investigate the dynamics of glass, and the proton dynamics of supercritical water [80].

Electron Scattering

We all know that hydrogen bonding is one of the fundamental properties of condensed phase such as water. Techniques such as Compton Scattering, inelastic scattering of photons by electrons, can effectively be employed for enumeration of hydrogen bonds in liquids such as water [81]. In particular electron scattering experiments can be tuned for absolute measurement of number of electrons involved in the process of the formation of hydrogen bonds.

X−ray, Neutron and Electron Diffraction Methods

We have just reviewed how X−ray, neutron and electron scattering techniques are useful for investigating structural arrangement in a variety of materials. Although the experimental procedures are nearly identical, it is important to note that scattering of X−rays, neutrons and light are influenced by differences in electron density, scattering power of nuclei and refractive index respectively. This suggests that each scattering experiment provides information at various scales. We have already seen that the incident rays collide with the sample, and they are diffracted at a particular angle, which is proportional to the degree of resolution. Higher the diffracted angle is, lower the resolution we obtain. Thus, the scattering experiments can be categorised into two: Small Angle Scattering (SAS) & Wide Angle Scattering (WAS). Crystal structure at atomic scale (up to the order of 10^{-9} m) can be investigated using WAS experiments, while SAS is used to investigate samples for a much lower resolution (up to the order of 10^{-6} m).

Scattering techniques have been extensively used in order to identify the structural features of solids and liquids. As stated before, all forms of water including crystalline and amorphous ices and supercooled and superheated water have been investigated by the scattering techniques, pinpointing how useful these techniques are in structural identification of water [70b, 82 - 84]. Neutron and X−ray scattering experiments have been the two most widely used scattering techniques in the investigation of liquids. Light, with higher wavelength than X−rays and neutrons, has also been employed to investigate the dynamics of glass forming liquids [85].

The location of oxygen atoms can be identified from various diffraction techniques. However, the molecular structures we extract on water from these scattering techniques differ with respect to each other to some extent [72]. X−ray diffraction (one of the oldest experimental methods employed for elucidating liquid structure), from which cell dimension of a solid can be determined, and hence this technique is more useful in the structural elucidation and confirmation of various phases of ice. Neutron diffraction methods, by which the proton positions can be identified, have been widely employed to investigate the structure of materials including normal water and supercritical water [86]. Since the orientations of protons play a significant role in the formation of different solid phases of water, this method is equally effective in the investigation of ice. The crystallographic dislocations in ice can be identified with the X−ray diffraction technique. Molecular pair correlation functions obtained from these experiments can be used to identify the shortest intermolecular distances, characteristic of each material. Electron diffraction has been employed to investigate the solid phases of water at low temperatures, by depositing water vapour on to a colloidal film [72].

SPECTROSCOPIC TECHNIQUES

In this section, a brief overview of various spectroscopic techniques is provided. Spectroscopic techniques are based on the fact that the changes in frequencies are accompanied by the change in energy of the system during emission or absorption of light. Energy of the system decreases during transition from higher energy level to lower level, then radiation is emitted. The opposite process occurs when energy is absorbed. However, the transition from one state to the other is governed by selection rule derived from Quantum Mechanics. There are many energy levels that a system can have, but only limited number of transitions is allowed such that change in transition from a given state is restricted only to one state higher or lower.

Light is an electromagnetic wave (having electric and magnetic properties), emitted by the excited substances consists of radiation of number of frequencies (called spectrum), which is resolved by a spectroscope. Light emitted from a source can be split into radiations of various frequencies. These components with different frequencies are received at different locations of a detector. The following Table **3.1** shows various radiations, corresponding to various interatomic processes.

The shape of atomic spectrum depends upon the movements of electrons. The simplest atomic spectrum that we can have is of hydrogen, since it is the smallest element with one proton and electron. On the other hand, molecular spectroscopy is much more complex than its atomic counterpart due to the presence of numerous neutrons in a molecule. In this case, we have to deal with nuclear motions (along with movement of electrons) that are of three types: translational, rotational and vibrational. The selection rules discussed before are also applicable in molecular spectroscopy. In addition, we have to consider the dipole moment of the molecule to interpret appearances of lines in the spectrum. For example, vibrational spectrum of homonuclear molecules such as hydrogen (H_2), nitrogen (N_2) and oxygen (O_2) cannot be seen in the Infra-Red region, since these molecules do not possess a permanent dipole moment. On the contrary, the scenario is quite different in the case of heteronuclear molecules such as water (H_2O), which possess a permanent dipole moment. In water, partial positive charge rests on two hydrogen atoms, while oxygen atom carries partial negative charge. The fundamental difference between these two classes of molecules is that the emission or absorption accompanied by the change of state is not Infra–Red (IR) active. In other words, the frequencies correspond to the rotation and vibration in molecules that do not have permanent dipole moment do not fall in Infra–Red (IR) region. On the contrary, in molecules such as water, the magnitude of dipole moment changes as a result of vibration. The corresponding band appears in the Infra–Red (IR) region.

Table 3.1. Different radiations absorbed by atomic systems and the information gained from them.

Radiation	Process	Information Gained
X–rays	transition of inner electrons of an atom	elucidation of electronic structure
Ultraviolet and Visible	transition of valence electrons	elucidation of electronic structure
Infra–Red	changes pertaining to vibrational-rotational state of the molecule	distance between nuclei
Far Infra–Red and Microwave	changes in rotational state	distance between nuclei
Radio waves	changes in spin orientation of nucleus in magnetic field	structural elucidation through analysis of magnetic environment of spinning nucleus

It has to be noted that molecules do have neither pure rotational nor vibrational spectra. Vibrational and rotational energies of a molecule is additive, and we get a vibrational–rotational spectra. In simple molecules such as water, inter nuclear distance can be measured from the spacing of lines in the vibrational–rotational spectra. Three notable bands can differentiated from the spectra of water, two of which correspond to the symmetric motions of water, and the third corresponds to asymmetric stretching vibrations.

Infra–Red Spectroscopy (IR)

Infra–Red spectroscopy, as the name indicates depends on the Infra–Red region the electromagnetic spectrum, can be used to identify various substances including solids, liquids and gaseous substances. The recorder in the Infra–Red spectroscopic instrument records absorption, emission or reflectance of the sample when the infrared light passes through it.

Vibrational profile of materials such as ice can effectively be monitored even at extreme thermodynamic conditions. The vibration in ice is dictated by hydrogen bonds due to the movement of oxygen–hydrogen bonds (mainly in the form of vibration). Several pioneering experiments on water have been carried out using IR spectroscopy, such as the investigation of hydrogen bonds in supercritical water by Kalinichev *et al.* [86b]. Hydrogen bonding can effectively be monitored using Infra–Red spectroscopy since a marked decrease can be noted in the frequency of its stretching vibration [87].

IR techniques are also very useful for low temperature investigations of water and its other forms. Crystallisation of water molecules at low temperature can efficiently be monitored by Infra–Red (IR) technique [88]. IR can effectively be employed for the investigations on ice, as its spectroscopic profile (majority of its vibrations are weakly active) is quite distinct from perfectly ordered crystals. Infra–Red absorption is strongly active in ice, and it can be grouped into three: wave numbers below 1000 cm^{-1}, above 4000 cm^{-1}, in between 1000 & 4000 cm^{-1} [72]. The vibrations pertaining to the lowest wave number are due to the existence of intermolecular interactions in water and ice [72]. Infra–Red (IR) spectroscopy has been used for the investigation of supercooled water long time ago [89].

Saykally *et al.* have devised a variant of IR spectroscopy, namely Far Infra–Red Laser Vibrational Rotational Tunneling (VRT) spectroscopy to investigate the existence of clusters found in water, and they have made a considerable progress in identifying water networks in liquid state [90]. Using this technique, it is now possible to track the mechanism of the breaking and making of the hydrogen bonds [91]. The major application of Vibrational spectroscopy is in characterising cascade of atomic motions in a molecular environment. Not all vibrations in a molecule are equivalent in terms of energy. This means that each peak is associated with a particular movement of atoms, and therefore corresponds to unique energy levels.

Nuclear Magnetic Resonance (NMR) Spectroscopy

Since the release of first commercially available Nuclear Magnetic Resonance (NMR) spectrometer, NMR spectroscopy has become one of the popular and highly useful spectroscopic techniques. NMR exploits the absorption and emission properties of atomic nuclei in a magnetic field, and can detect the energy differences between the quantum levels of a magnetic dipole in a magnetic field. The energy of the magnetic dipole (for example a proton) and hence its frequency can be obtained by applying Zeeman Effect, splitting of a spectral line into number of components in the presence of strong magnetic field. One must however note that the energy absorbed is not attributed to the proton itself, rather the magnetic environment of the proton (for example, three different peaks in CH_3CH_2OH (ethyl alcohol) due to CH_3, CH_2 and OH protons), which can be measured *via* chemical shift (the difference in magnetic field with respect to a reference molecule).

Nuclear Magnetic Resonance (NMR) technique is a powerful tool in establishing the structure and properties of various compounds. It is also employed widely for the investigation of condensed systems in which hydrogen bonds plays a vital role [87]. Characteristic signals can be detected for various hydrogen bonding environments [92]. Like IR, the NMR technique has also been employed for the investigations of the properties of supercritical water [93]. A quantitative estimation of hydrogen bonds in various phases of water has been achieved by Nakahara *et al.* employing NMR technique. By measuring hydrogen chemical shift, they have managed to estimate the average number of hydrogen bonds in supercritical conditions [94].

The rudimentary setup for a NMR spectroscope contains a magnet, radiofrequency oscillator and bridge circuit, amplifier and chart recorder. A coil is wrapped around the testing sample, placed between the poles of a very strong electromagnet. The coil is connected to radio frequency oscillator, which sends signals to the sample. When the frequency of the oscillator reaches the Larmor frequency (the frequency corresponds to the transition between two spin states +1/2 and −1/2), the system can absorb radiation sent by the oscillator. The impedance (the ratio of voltage to current) of the radio−frequency circuit is altered due to absorption, which is measured by a bridge circuit. The signal is then amplified and charted in a recorder. In Fig. (**3.3**), basic Nuclear Magnetic Resonance (NMR) Spectroscopy set up is shown.

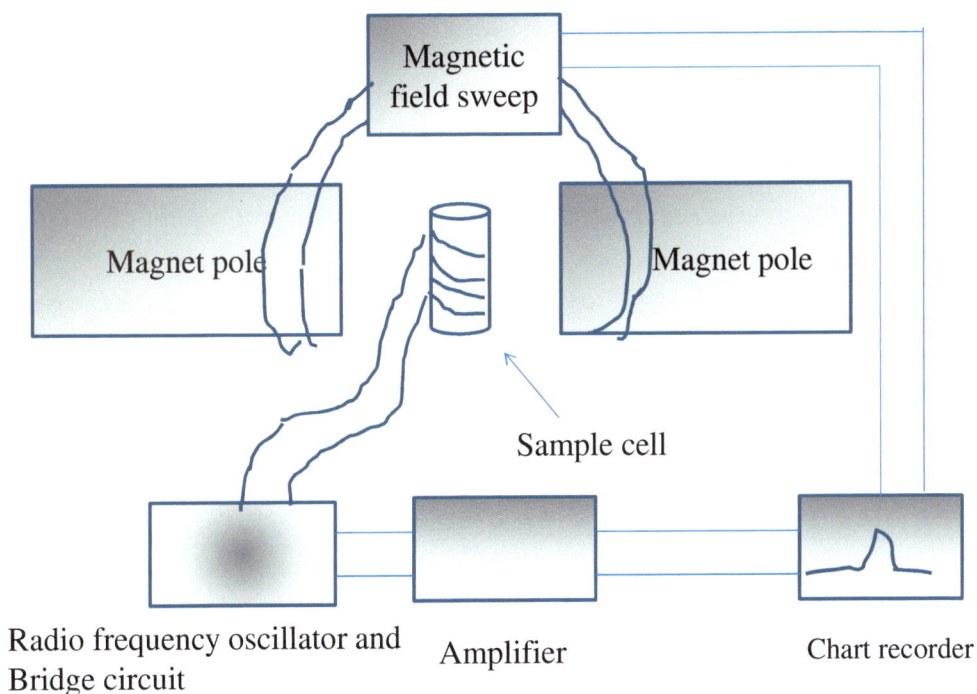

Magnetic field sweep

Magnet pole   Magnet pole

Sample cell

Radio frequency oscillator and Bridge circuit   Amplifier   Chart recorder

Fig. (3.3). Basic Nuclear Magnetic Resonance (NMR) set up is shown. Sample cell is placed between the two poles of a magnet connected to magnetic field sweep that is connected to the chart recorder. Sample cell is also connected to chart recorder through radio frequency oscillator and amplifier.

The strong magnetic field in the NMR spectroscope creates a net magnetisation in the material under investigation, which causes an excitation of protons to a higher energy state. However, this effect is temporary, as the protons will go back to the original lower energy state upon the removal of the magnetic field. This phenomenon is known as nuclear spin relaxation. This technique has been applied for the investigation of proton and deuteron NMR spin–lattice relaxation times of water as a function of pressure and temperature [95]. Pulsed –Gradient Spin–Echo NMR experiments have been performed to investigate the diffusion of supercooled water [96].

Raman Spectroscopy

In order to understand the fundamentals of Raman spectroscopy, first we must take note of another important phenomenon, namely Rayleigh scattering. Light is composed of electric and magnetic waves. A small fraction of light is scattered by molecules (sample), when they interact with an intense beam of light. As a result, the electron cloud in the molecule is deformed, under the influence of an applied electric field. The electron cloud deformation results in the generation of electric dipole, which oscillates with the applied electric field. It has to be noted that only a little amount of light is scattered and therefore high intensity beam must be used for the investigation. C.V. Raman noted that when incident light collides with molecules, energy is imparted to them, and this causes either an increase or decrease of the energy of scattered radiation. This effect is known as Raman effect. Fig. (**3.4**) illustrates the concepts of Rayleigh scattering and Raman effect.

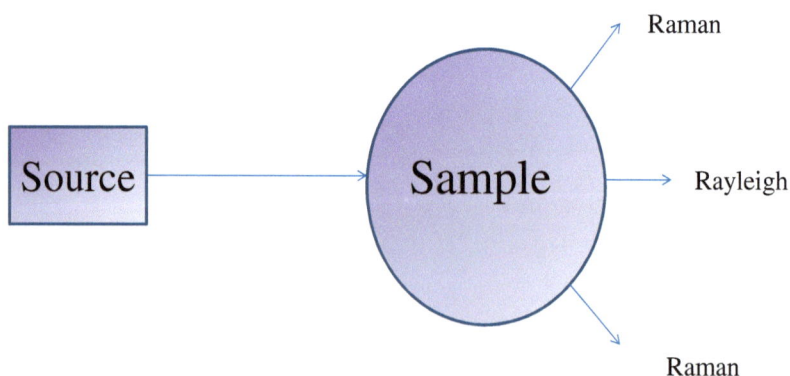

Fig. (3.4). Interaction of light with matter. A simple schematic diagram of the interaction of the light with matter is shown. In Rayleigh scattering the frequency of the incident radiation is equal to the frequency of the scattered radiation. Transference of energy between the incident radiation and the sample results in radiations with higher and lower frequencies, known as Raman effect.

It is clear from the above discussion that when light collides with matter, the resulting scattered radiation is split into the radiations of three distinct frequencies: one with frequency higher than the original incident light, which is formed when molecules transfer energy to the incoming radiation (Anti Stoke's line); when the sample absorbs energy from the incoming radiation, radiation of lower frequency than the incident radiation is produced (Stokes line); the third line corresponds to the radiation whose frequency matches with the incident radiation, (this is known as Rayleigh scattering). When the molecules absorb energy, the scattered radiation will have a reduction in the energy and hence frequency. As a result, the frequency of scattered radiation (Stokesline) shows a shift towards the red end of the visible spectrum. When the molecules release energy to the incoming radiation (the opposite process) the scattered radiation gains energy (by attaining higher frequency). This results in frequency of the scattered radiation being shifted towards the blue end of the spectrum. These effects are termed as red shift and blue shift respectively.

Water has been extensively studied by Raman Spectroscopy [70a, 97]. Structural features of water and some forms of ice have been identified using this spectroscopic technique [97c]. In addition, the relationship between hydrogen bonding and thermodynamic quantities such as density and pressure has been deduced by this experimental technique.

Dielectric Relaxation Spectrometry

Materials including liquids and aqueous solutions have been extensively investigated using Dielectric Spectrometry over wide range of temperatures [98]. The major focus of the Dielectric Relaxation Spectrometry has been identifying the dynamics of several glass formers including water [85, 99]. Using this technique, the structural dynamics of liquids can be investigated to pico second (ps) resolution such that the experimental findings can be compared with computer simulations [100].

Other Spectroscopic Techniques

Time Resolved Optical Kerr Spectroscopy is well suited to micro level analysis of liquids, at very short time intervals, from femtoseconds to picoseconds. This technique is very suitable to investigate the response of the sample to relaxation and vibration [101]. Intermolecular vibrations and structural relaxation of supercooled liquids have been studied by this method [102]. One advantage of this experiment is that the analysis can readily be compared with computer simulation techniques due to its high resolution (remember that in molecular dynamics simulations the time step is in the order of femto seconds, which will be discussed in the next chapter). Tremendous progress has been made in developing sophisticated spectroscopic experimental techniques in recent years, which includes the combined application of different spectroscopic techniques. For example, Time Of Flight Secondary Ion Mass Spectroscopy (TOF−SIMS) has been used in conjunction with Reflection Adsorption Infrared Spectroscopy (RAIRS) in order to investigate supercooled water [88].

ELECTRON MICROSCOPY

We have just seen that the interactions of light with matter can yield information about the structure of matter in a great detail. One of the drawbacks of these methods is the limitations in resolution imposed by their higher wavelength in the order of nanometer scale. When the light of varying frequencies is substituted by electrons having shorter wavelength of order of Angstroms, the resolution can be improved significantly, suggesting that it is possible to investigate matter more depth. In this subsection, two of the most promising microscopic techniques are discussed. These are Scanning Electron Microscopy (SEM) and Transmission Electron Microscopy (TEM).

Scanning Electron Microscopy

Scanning Electron Microscopy, abbreviated as SEM, is a high resolution tool in order to investigate the material composition with high resolution. In SEM, the image of sample is taken by electron scan by colliding it with the stream of electrons. By doing this, sample's topography and surface properties can be elucidated. SEM has already been employed to investigate the texture of ice [103]. One of the principal advantages of this technique is that it overcomes the limitations of the traditional light microscopes, by making use of the fact that electrons have shorter wavelength than the ordinary light, rendering better resolution.

The working principle of an electron microscope can be summarised as follows: electrons, emitted from the source (electron gun), are accelerated by anode and condensed by condenser lenses. The electron beam is then deflected by magnetic field produced by scan coils. Objective lens focuses the electron beam to very confined space, in the range of one to five nanometres. Interaction of electron with the atoms in the sample results in the emission of electrons and production of image. The following diagram (Fig. **3.5**) explains the experimental set up of the SEM.

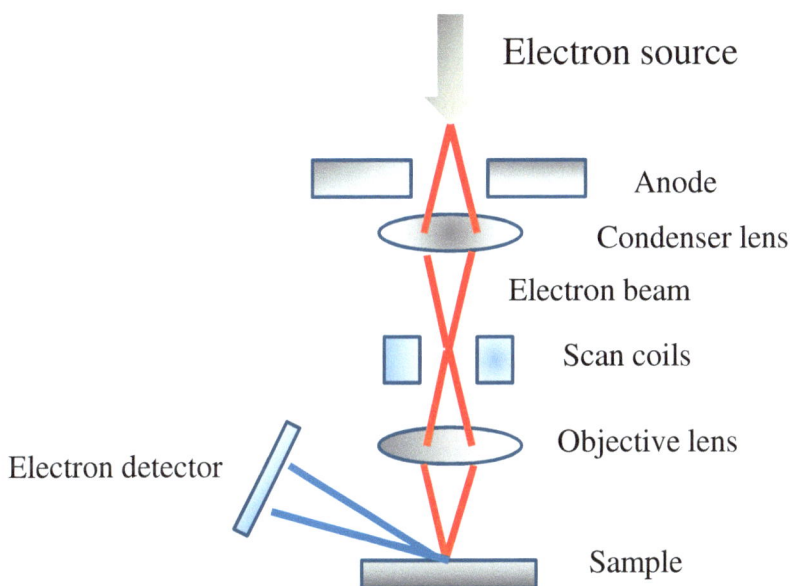

Fig. (3.5). A simple set up of Electron microscopy. Stream of electrons ejected from the source is passed through different types of lenses before interacting with the sample. The ejected electrons are captured in the electron detector placed over the sample.

Transmission Electron Microscopy (TEM)

Transmission Electron Microscopy (TEM) is another technique that is widely being used to obtain high resolution structure of materials. The working principle is same as that of SEM, but TEM provides at least 50 times better resolution than SEM. Another fundamental difference between SEM and TEM is the former is based on scattered electrons, while the latter is based on the transmittance of electrons. The important information that we get from SEM is the morphology of the sample, while with TEM, due to its superior resolution, we can explore deep into its internal composition. This suggests that TEM can be more useful in the investigation of ice structures due to its superior ability to capture their internal dynamics. One can clearly note the principal difference between SEM and TEM, and scattering and spectroscopic techniques explained earlier in this chapter. In electron microscopic techniques, a realistic picture can be obtained, while in the latter, the information we procure is abstract, requiring further manipulations to arrive at a meaningful conclusions. We can expect that more investigations using the state–of–the–art electron microscope techniques, although challenging under extreme thermodynamic conditions, will shed light into the internal structure of liquids such as water. Quite recently, Grigorieva *et al.* have applied this experimental technique to investigate the properties of water confined between two graphene layers [104].

DIFFERENTIAL SCANNING CALORIMETRY (DSC)

With Differential Scanning Calorimetry (DSC), one can monitor heat effects with respect to the changes in temperature. Since water undergoes transformations from one form to the other, by releasing or absorbing heat, this mode of investigation has special role in the physics and chemistry of water, in particular in low temperatures [72]. This technique has also widely been employed to measure the transition of various liquids to their glassy states [58]. Using DSC, heat capacity (C_p) can effectively be monitored [105]. One of the most remarkable experiments using DSC is the one conducted by Johari *et al.* in order to investigate glass–liquid transition in water [106]. The experimental set up for DSC is shown below in Fig. (**3.6**).

As can be seen in the diagram (Fig. **3.6**) the sample is taken in one pan, while the other is kept empty, taken as a reference, and both pans are placed on top of a heater. The computer turns on the heaters, and "directs" it to heat the two pans at a specific rate. The heating rate is kept exactly the same throughout the experiment by computer control. Since the reference pan is empty and the other pan has a sample in it, there is a difference in the heat absorbed by both pans. Having extra material means that it will take more heat to keep the temperatures of the sample pan increasing at the same rate as the reference pan. By noting how much more heat it has to put out is what we measure in a DSC experiment. Temperature and difference in heat between two heaters are noted. The change in heat capacity does not occur instantly; rather it takes place in a wide temperature range. The glass transition temperature is approximated by taking midpoint of the temperature range. Differential Thermal Analysis technique, closely related to DSC, has been employed to measure heat capacities of solutions [107].

Reference pan

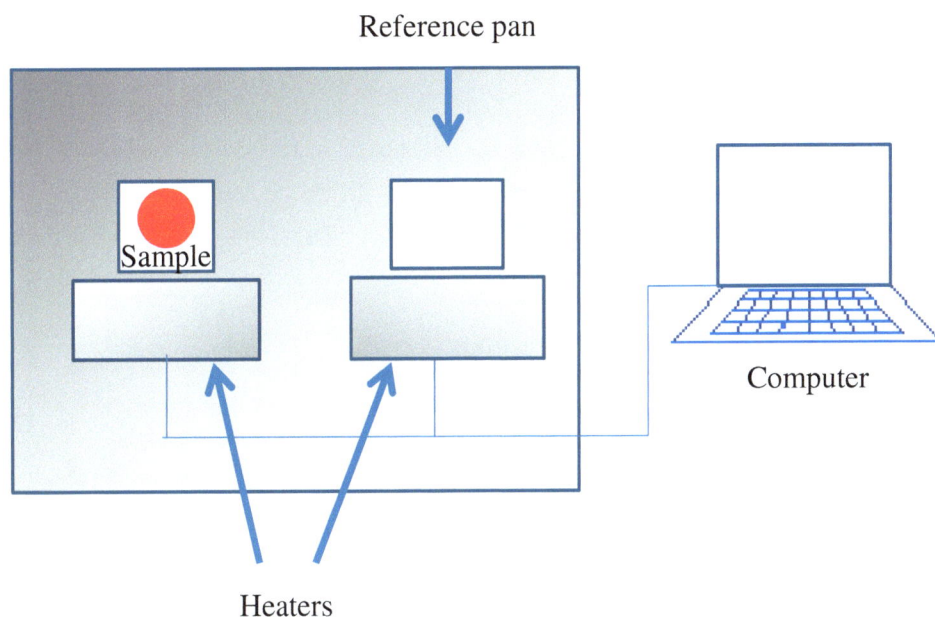

Fig. (3.6). Schematic diagram of a simple Differential Scanning Calorimetry (DSC). The equipment contains two pans, in one of which the sample is kept. These pans are heated during the experiment, and the heat flow to the sample is recorded using the computer connected to the calorimeter.

CAPILLARY TUBE TECHNIQUES

Capillary tube is one of the oldest methods and is still employed for the investigation of liquids. This is such a useful technique that it can be employed independently as well as in conjunction with other experimental methods including NMR for the calculation of wide ranges of physical quantities. Thermodynamic properties such as density and isothermal compressibility of supercooled water have been measured using capillary tube techniques, with a great level of accuracy [108]. Capillary technique has been used by Nakahara *et al.* for performing NMR investigations of supercritical water [94]. Viscosity of liquids can be measured by a technique known as capillary flow technique: this is achieved by noting the time taken for the liquid kept in a reservoir to flow [108b]. In this apparatus, as shown below (Fig. **3.7**), the thermodynamic variables (pressure and temperature) are controlled by pressure and thermostat tubes respectively. A thermostatic fluid, for example mixture of ethanol and methanol taken in equal proportions, is circulated through the tube for the temperature control.

When we rummage through experimental articles published over last several decades, we note that there is a visible dip in number of experiments exclusively based on capillary techniques. This is primarily due to vast technological developments that we have witnessed for the past three or four decades. Another reason is that nearly all of the macroscopic data on known materials have been unearthed, rendering traditional methods redundant.

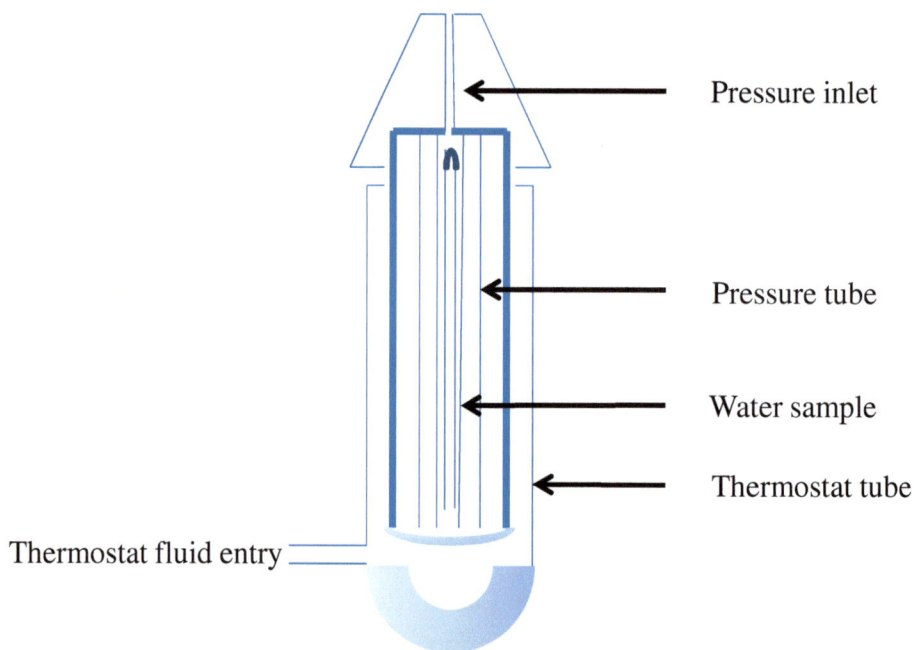

Fig. (3.7). Apparatus for measuring compressibility of supercooled water. The pressure and temperature of the sample is controlled by separate tubes.

GLASSY WATER & HYPERQUENCHING

Hyper quenching experiments play an important role in the investigation of H_2O–based materials at low temperatures. For example, hyperquenched glassy water is considered as a starting material in the investigation of the properties of cubic ice. This is due to the fact that the glassy water can be transformed to cubic ice under appropriate conditions. Loerting *et al.* have investigated the proton dynamics of cubic ice using hyperquenched glassy water [109]. The authors claim that the production cubic ice by this fashion yields better samples devoid of structural defects [109].

WATER IN CONFINEMENT

Water in confinement is a new way of conducting experiments on water in lower temperatures. Most of the earlier experiments on water were carried out in bulk medium, and this is not ideal scenario for performing experiments on water in lower temperature than its melting point due to the formation of ice crystals, the main hurdle for performing experiments at low temperature. As Angel notes, water-confinement experiments make use of the fact that water at the surface does not crystallise, thus making these methods effective alternatives than the traditional experimental routes [110].

CONCLUDING REMARKS

This chapter is exclusively devoted for summarising important experimental protocols hitherto employed for understanding water at microscopic level. Tons of experimental articles have been published so far, unearthing exceptional properties of water and ice. A vast number of these experiments belong to either diffraction or spectroscopic techniques. Diffraction methods have been the most popular choice for structural elucidation of ice, due to the fact that proton and oxygen positions can be identified accurately. Application of various classes of spectroscopic techniques has furthered our understanding of the structure and dynamics of water at various thermodynamic conditions. On the other hand, the state–of–the–art high resolution electron microscopic techniques can provide important structural information of water, which could become decisive in the determination of its internal structure. In addition, experimental techniques such as Differential Scanning Calorimetry (DSC) have also made a significant impact in exploring the characteristics of water in sub–zero temperatures, for example, in the investigation of phase changes. It is highly useful for the determination of glass transition temperature of water. Our understanding of the micro structure of water has been improved by successive applications of these sophisticated experimental methods. In the next chapter, I am going to describe computer simulations, which have already been emerged as a complementary tool to experimental methods discussed in this chapter.

CONFLICT OF INTEREST

The author confirms that he has no conflict of interest to declare for this publication.

ACKNOWLEDGEMENTS

Declared none.

CHAPTER 4

The Fundamentals of Molecular Simulations

Abstract: Computer modelling is a powerful enterprise for the investigation of matter at atomic and molecular levels, and has generally been accepted as a supplementary tool to traditional experimental methods. Its advantages over real experiments are primarily exemplified by its portability and cost effectiveness. Monte Carlo and Molecular Dynamics methods are two principal techniques that have gained a great level of popularity among various computer simulation methods. Numerous mathematical models, popularly known as Forcefields, have been developed in order to investigate water computationally. The application of computer simulation methods is limited by the choice of parameters that define the intra and inter molecular interactions within the framework of Forcefields. *Ab−initio* forcefields are expected to overcome the limitations of other types of water models. Concept of ensemble provides a theoretical basis for deriving physical properties by significantly reducing number of particles in a system. Mathematical devices such as Periodic Boundary Conditions (PBC) bypass the inconsistencies in simulations. Density Functional Theory (DFT) and Wave Function methods are two important classes of quantum chemical methods for investigating matter at electronic level. Born−Oppenheimer approximation provides a fruitful means for separating electronic and nuclear motions, which reduces the complexity of quantum calculations to a great extent. Hartree−Fock (HF) method is the most fundamental wave function procedure for calculating the energy of multi−electronic systems. On the contrary, Density Functional Theory (DFT) is based on the estimation of electron density, which can be validated by experimental means. Combined electronic and classical approaches are increasingly becoming popular in the scientific community.

Keywords: Basis sets, Boltzmann factor, Ensemble, Equilibration, Ergodic, Forcefield, Initialisation, Molecular dynamics, Periodic boundary, Production, Quantum mechanics, Shifting function, Switching function.

Jestin Baby Mandumpal

INTRODUCTION

Models are built based on our theoretical understanding of what is observed, and hence, they should serve as complementary tools to the experimental methods. If one wants to test the predictions based on the model created, performing experiments is not the only one option. Due to ever growing availability and capability of computers, it is now customary for researchers to apply various computer simulation protocols to extract physical and chemical properties of the systems of interest. One can represent a system by a model, and with simulation on can track the changes in the model by varying conditions.

Computer simulation is a virtual experiment, in which an abstract model of a system created artificially in a computer. In comparison to the laboratory experiments, it has several advantages: firstly, compared to an experimental set up, a simulation (computer experiment) kit is portable, *i.e.* without much hassle, a person can carry it and any associated instruments (for example printer or scanner) from place to place, and secondly, challenging physical conditions limit the applicability of certain real experiments. For example, performing experiments on liquid water in low temperatures is met with limited success due to its unstable nature with respect to its more stable phase, ice, but mimicking them on a computer (computer simulations) is rather straight forward. Selecting appropriate models, one must supply input structure to the computer software he/she employs, and can extract certain physical or chemical properties without much strenuous efforts. However, this does not imply that computational investigations are superior to experiments.

Despite their advantages, computer experiments too are not without limitations. One major drawback of the computer methods is lack of accuracy of the methods and models employed in the simulations. One can in fact say that models and methods are plenty but one requires a detailed awareness of models and various computational procedures. Certain methods or models might be very efficient for particular type of systems and inefficient for some other types. Therefore, one must be judicious in choosing them prior to running the simulations, and interpreting the results afterwards. On the ground of aforementioned advantages and disadvantages one can say that computational methods do not stand independently, rather they can supplement existing experimental investigation modes towards enhancing scientific understanding.

It becomes now clear that the role of computational chemists and physicists is to develop novel computational procedures, to perform molecular or atomistic simulations using these methods and to validate the findings when experimental results are available. Important modes of molecular investigations chiefly fall into two types: classical modelling grounded on laws of classical physics and quantum chemical modelling based on the revolutionary quantum mechanics founded in twentieth century. Dynamics of planetary objects (the best example being the motion of earth around the sun) can be explained by the laws of classical physics: that is they follow Newtonian Mechanics, the theoretical framework based on three Newtonian laws of motion. On the contrary, classical physics was proven to be inadequate for dealing with the motions of much smaller atomic particles that could only be described efficiently by a later theoretical development, called quantum mechanics. Important distinction between these two theoretical domains (classical and quantum) lies in the fact that the exact location of a particle (large particles) using classical physics can be estimated in advance without sacrificing accuracy, whereas quantum mechanics offers only a probable location for particles (small particles).

Over the years, more researchers are interested in computational methods as evident from the growing number of scientific articles appearing in national and international journals, and conferences proceedings. However, caution must be exercised in order to avoid pitfalls. For instance, a person without working knowledge of the basis sets will struggle to operate quantum chemical packages efficiently. On the other hand, the choice of appropriate forcefields is necessary to avoid embarrassing results in classical simulations. This demands a basic level of comprehension of computational methods.

The impetus for writing this chapter is to provide the reader a brief conceptual understanding of various computational methods that are being employed in material science. This chapter is organised as follows: firstly, I briefly outline the general simulation protocol. Then I discuss simulation techniques at molecular level (classical molecules); in this section concept of force field is developed and several aspects of water forcefields are discussed in a subsection. Although overwhelming majority of the theoretical research works on water that has been appeared in the scientific literature are based on classical simulations, with the advent of super computers, the volume of quantum chemical studies on water rises year by year. Keeping this in mind, I, therefore, also volunteer to provide a basic level of information on quantum mechanical tools that are widely employed in investigation on water. The quantum chemical methods discussed in this chapter are principally divided into two categories: wave function based, and electronic density based.

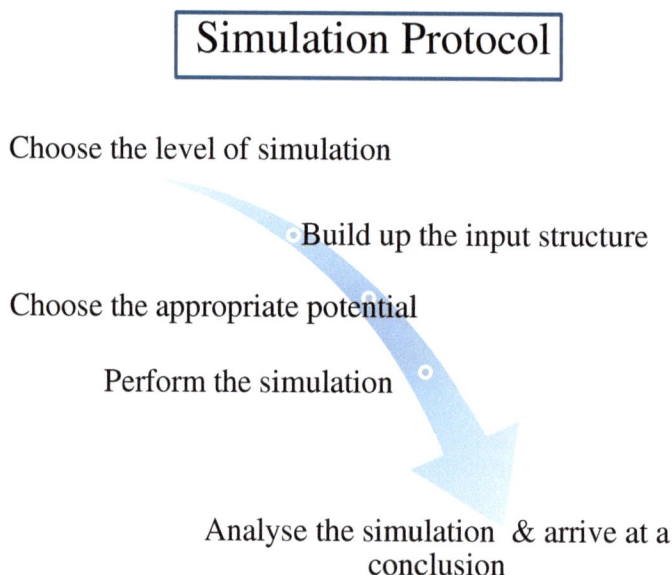

Simulation Protocol

Choose the level of simulation

Build up the input structure

Choose the appropriate potential

Perform the simulation

Analyse the simulation & arrive at a conclusion

Fig. (4.1). Protocol for atomic & molecular simulations. After fixing the systems for the simulations, potentials (for classical simulations) or approximations (in the case of quantum simulations) are chosen. Various physical quantities can be extracted from the trajectories, the output of simulations.

THE PROTOCOL OF MOLECULAR SIMULATIONS

First step in a computer experiment is choosing the level of theory by which the simulation must be carried out. For example, if one would like to extract very atomistic details of matter in order to trace complex chemical reactions, then high level quantum chemical calculations are mandatory. On the other hand, if one aims only for extracting molecular level understanding, classical molecular dynamics simulations can provide satisfactory information. In the classical simulation protocol, building initial structure is followed by choosing an appropriate forcefield, by which all information regarding the bonding and non–bonding interactions (including bond lengths, bond angles, dihedral angles and partial charges) are specified. In quantum chemical calculations, the analogous step is to choose the basis sets and/or functionals. The fact we must note is that the final outcome depends heavily on the forcefield/basis sets/functionals that we choose. It is not uncommon that the quantities extracted using various forcefields/basis sets/functionals show large variation with respect to each other. This means that unless we choose the suitable methods, the results can be meaningless. The output of simulation (called trajectory), can be analysed in order to obtain various quantities of the system of interest. The diagram (Fig. **4.1**) summarises principal steps in molecular/atomic simulations.

FUNDAMENTALS OF MOLECULAR SIMULATIONS

Molecular simulation, which provides information at molecular level of the system of interest, is categorised into two: deterministic and stochastic. The most common deterministic simulation protocol is Molecular Dynamics. Monte Carlo (a popular simulation technique) on the other hand is a stochastic method. Two important aspects common to both of these methods are ensembles and forcefields, which will be discussed next.

Ensembles

General purpose of molecular simulations is to derive properties of bulk sample from limited number of virtual particles. Finding time average of dynamics of system containing as large as 10^{23} particles is not feasible. By invoking the concept of ensembles (a collection of points in phase space obeying a particular set of thermodynamic conditions), this difficulty was overcome by Ergodic Hypothesis, which posits that time average be equal to ensemble average. There are three common ensembles in use: Canonical Ensemble, Microcanonical Ensemble, and Grand Canonical Ensemble. In Canonical Ensemble, number of particles, volume and temperature are held constant; number of particles, volume and energy are held constant in Microcanonical Ensemble, and in Grand Canonical Ensemble, chemical potential, volume and temperature are kept fixed.

One can obtain an estimate of other physical quantities, for example pressure, by taking average of the aforementioned fixed quantities.

Periodic Boundary Conditions

Representing bulk system with limited number of particles induces several artefacts such as uneven forces on particles and variable densities (due to the movement of particles during simulations), which can be avoided by replicating the box in three directions (x, y & z). This allows us to represent macroscopic behaviour of the system without resorting to the inclusion of millions of particles is known as Periodic Boundary Condition (PBC). The following diagram (Fig. **4.2**) illustrates this concept. Although it has been reported that systems with periodic lattice may not be suitable for simulating liquids due to the bias towards the formation of solid with lattice structure that matches with the chosen box shape, it is customary to simulate cubic systems containing $4M^3$ of molecules (M takes any integer from 1 onwards) [111].

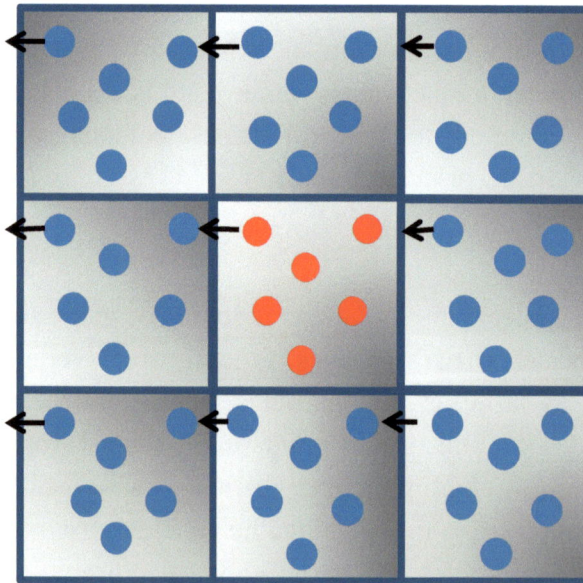

Fig. (4.2). Illustration of Periodic Boundary Condition (PBC) concept in molecular simulations. Identical boxes containing particles are joined together in order to overcome the inaccuracies during simulations. The image of particle moving outside (shown with an arrow) re−enter to the box avoiding the artificial fluctuations in density. With Periodic Boundary Conditions (PBC), the original box (with particles shown in red) is replicated in all three directions (x,y,z) resulting in a three dimensional construct containing 27 boxes.

Forcefields

We have discussed in depth various forces acting between atoms and molecules, commonly known as intermolecular forces, in chapter two. We can represent these interactions in a delicate mathematical form, and the conformational energy of a system can be estimated by appropriate potential functions. Forcefields contain various interaction terms including bonding and non–bonding terms. Various forms of energy, accumulated into a clumsy mathematical expression, contribute to the forcefield of a system. Therefore, I only provide a literal description of what it contains in the diagram (Fig. **4.3**) shown below.

Force Field =
$\sum$ bond stretching (bonded interaction) +
$\sum$ angle bending (bonded interaction)   +
$\sum$ proper and improper dihedral interaction
(bonded interaction) +
$\sum$ Van der Waals interactions
(non bonded interaction) +
$\sum$ Coloumb's interaction
(non bonded interaction)

Fig. (4.3). Components of a standard classical Forcefield. A classical Forcefield contains fundamentally five terms, three of which accounting for bonding interactions (in blue), and the remaining two accounting for non-bonded interactions (in red).

Expressions for more sophisticated treatments vary depending upon the molecular systems under investigations. As indicated above, mathematical expressions for a standard forcefield constitute bonded interactions and non–bonded interactions. Bonded interactions account for interactions due to connectivity (interactions between two, three and four particles or sites). A molecule can experience attraction or repulsion with its neighbouring molecules, which is altogether termed as intermolecular interactions, and within its parts (called intramolecular interaction). Bonded interactions and a part of non–bonded interactions (fourth term in the mathematical expression of the forcefield (Fig. **4.3**) and generally known as Van der Waals interactions) are straight forward to compute. In chapter two, we came across various forces contributing to the intermolecular forces and their role in estimating the properties of condensed matter. Hence it is vital to obtain accurate expressions for these binding forces in the condensed matter. One fact we should remember is that the mathematical model describing these interactions must be computationally tractable.

Van der Waals interactions account for short range repulsive and long range attractive interactions, and can be modelled by variety of ways such as Hard sphere, Lennard–Jones and Born–Mayer potentials. In Hard sphere atom model, each atom is considered as rigid hard sphere. If atoms overlap, the potential is taken as infinite. This model requires only atomic radius as parameters. Lennard–Jones model, one of the widespread models used in the contemporary computational chemistry, has both attractive and repulsive terms, which are represented by two variables epsilon (ε) and sigma (σ) as shown in Fig. (**4.4**).

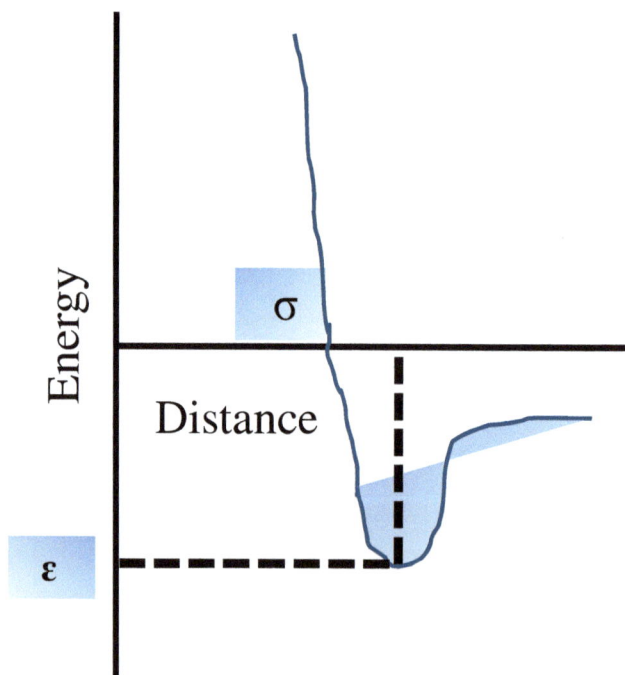

Fig. (4.4). Energy of interaction between two molecules. Epsilon and sigma represents the depth of energy well when energy of interaction between the molecules equal to zero, and the distance of separation at minimum energy respectively.

When used for multi atomic molecules, geometric and arithmetic mean for ε and σ are used instead, known as Berthelot and Lorentz rules respectively [47]. Popular forcefields developed over the years include AMBER (Assisted Model Building and Energy Refinement), CHARMM (Chemistry at Harvard Macromolecular Mechanics) and OPLS (Optimised Potentials for Liquid Simulations). Many other potential models have been devised as well, including Born Mayer potential in which repulsion between atoms is denoted by exponential dependence on distance. *ab−initio* derived potentials, meaning 'derived after solving from Schrödinger equation' (this will be discussed in a later part of this chapter), account for several contributions to inter molecular interactions such as electrostatic, induction, exchange−repulsion, dispersion and charge transfer.

Modelling Intermolecular Interactions of Atoms and Molecules

The computation of the remaining part of the non−bonded interactions, in particular long range electrostatic interactions (Coloumb) requires immense time and therefore further approximations are warranted. One of the approximations employed for calculating long−range interaction is the introduction of a simple cutoff scheme in the simulations. This is achieved by truncating the interaction beyond a certain distance, which, however, results in unphysical consequences during simulations because of discontinuous energies and forces [112]. A much improved way of handling long range forces is by Switching Function approach, by which the rendering of potential to zero smoothly over a distance range is achieved. In Shifting Function approach, which itself is a variant of Switching Function method, the lower bound of the distance range is made zero. The efficacies of all these methods are limited due to the artificial truncation of the potential.

Two of the later additions to the treatment of long range interactions are Ewald based methods, and Fast Multipole Methods (FMM), both of which splits the interactions into two: near and far field. These two methods revolutionised the treatment of long range interactions in the molecular simulations. In Ewald summation techniques, each point charge is represented as Gaussian functions, which is screened from each other. Outside of each Gaussians, the potential is counted as zero, thereby achieving a smooth truncation of potential, in stark contrast to Switching or Shifting function approaches [112]. In FMM methods, the simulation space are divided into smaller boxes, and interactions of each boxes are represented by multipoles centered in each boxes. Novel methods are being developed in order to reduce the computation time of non–bonded interactions, in particular electrostatic interactions, significantly, which takes approximately 90% of computational effort during a standard Molecular Dynamics run.

Molecular Dynamics

Particles in a simulating system are interacted through a potential defined by forcefield. From the given forcefield, one can easily calculate forces on each atom (particles), which are derivative of potential with respect to the coordinates of the particles. The main goal of Molecular Dynamics simulations is to predict the position of a given system in an advance time thorough tiny time intervals. Technically this is achieved thorough method of finite differences: if we know about position, velocity, acceleration and its time derivative at a time, t, the position of the system at another time (t+ δt) can be computed using Taylor expansion. There are many algorithms have been tailored to do this task, examples include Predictor–Corrector method, Verlet, Velocity Verlet method and so on. Every Molecular Dynamics (MD) package is driven by any of these methods. The schematic diagram (Fig. **4.5**) shows how a MD program works using Predictor–Corrector method. Using this algorithm, new positions, velocities and accelerations can be obtained from initial configuration. Then, forces and new accelerations are computed, and thereafter new positions, velocities and accelerations are corrected. This cycle is repeated for many steps, after which properties of interest are calculated.

In Molecular Dynamics (MD) simulations, Newton's equations are numerically integrated for all atoms, which requires evaluation of the atomic forces at each time step. The choice of time step (Δt) is very important for efficient progress of simulation: very small time step results in slow coverage of phase space and with large time steps, unrealistic collisions between particles occur, resulting in inconsistent simulations. The choice of time steps depends of the size of particles present and type of motions investigated. For example, a time step of 10^{-14} seconds may be appropriate if we are interested only in the translational motion of atoms. On the other hand, time step of either 10^{-15} or $5 * 10^{-16}$ is required for investigating the dynamics of molecules with flexible molecular models. The force evaluation is dominated by cumbersome non−bonded interaction computation.

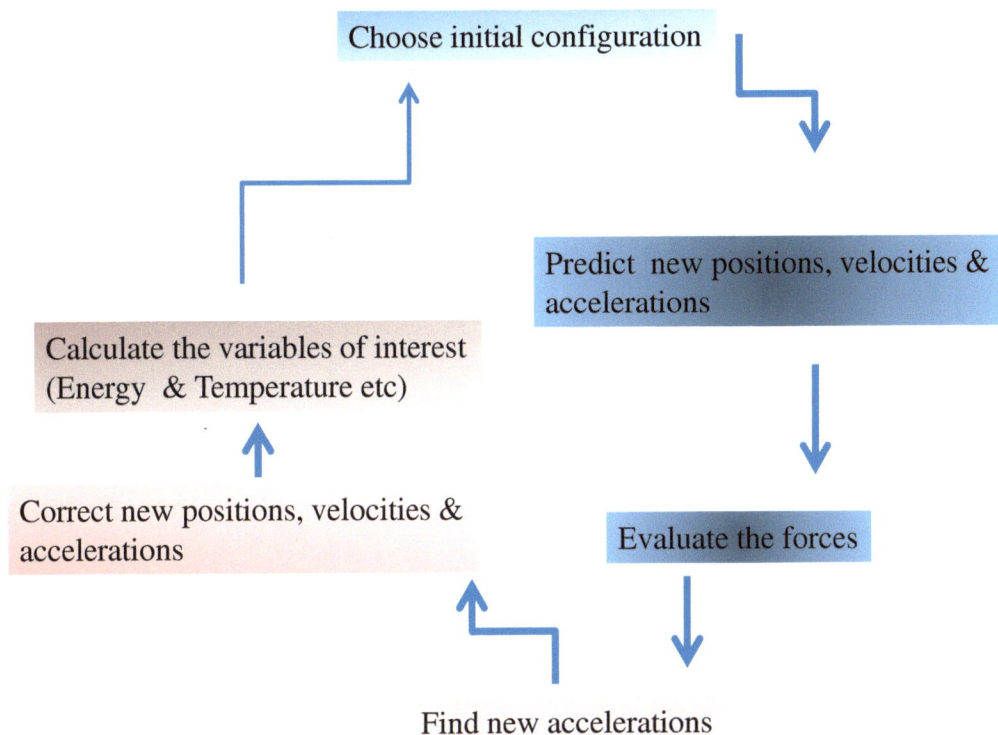

Choose initial configuration

Predict new positions, velocities & accelerations

Calculate the variables of interest (Energy & Temperature etc)

Correct new positions, velocities & accelerations

Evaluate the forces

Find new accelerations

Fig. (4.5). Molecular Dynamics algorithm using Predictor−Corrector method. This allows the computation of the mechanical quantities in an advance time.

Stochastic Simulations

In stochastic simulations, molecular configurations do not depend upon time, rather upon previous configurations [113]. This means that the chain of various configurations occur randomly. Metropolis Monte Carlo and Force−Biased Monte Carlo are the two prime examples of stochastic simulations. In Metropolis Monte Carlo simulation, often applied for studying liquids and solutions, firstly, initial positions of the particles are assigned; total potential energy of this configuration can be considered as the sum of isolated pairs in the system (known as pair additivity). Let us assume that the energy of this configuration be E_{old}. Second step involves in moving a particle in the system along an arbitrary distance to new position, which generates a new system configuration with energy E_{new}. We have then two possibilities: one case is that the energy of new configuration, E_{new}, becomes less than that of the old configuration, E_{old}; in this case, the move is accepted. If the opposite is the case (*i.e.* $E_{new} > E_{old}$), the move is accepted according to the condition that the probability of such move is equal to the Boltzmann factor. If we reject this move, then E_{new} is considered as E_{old} and the procedure is repeated. In order to obtain a reliable estimate, approximately a million moves are required, although there is no strict guide in this regard. The following diagram (Fig. **4.6**) demonstrates the flow chart of Monte Carlo simulations.

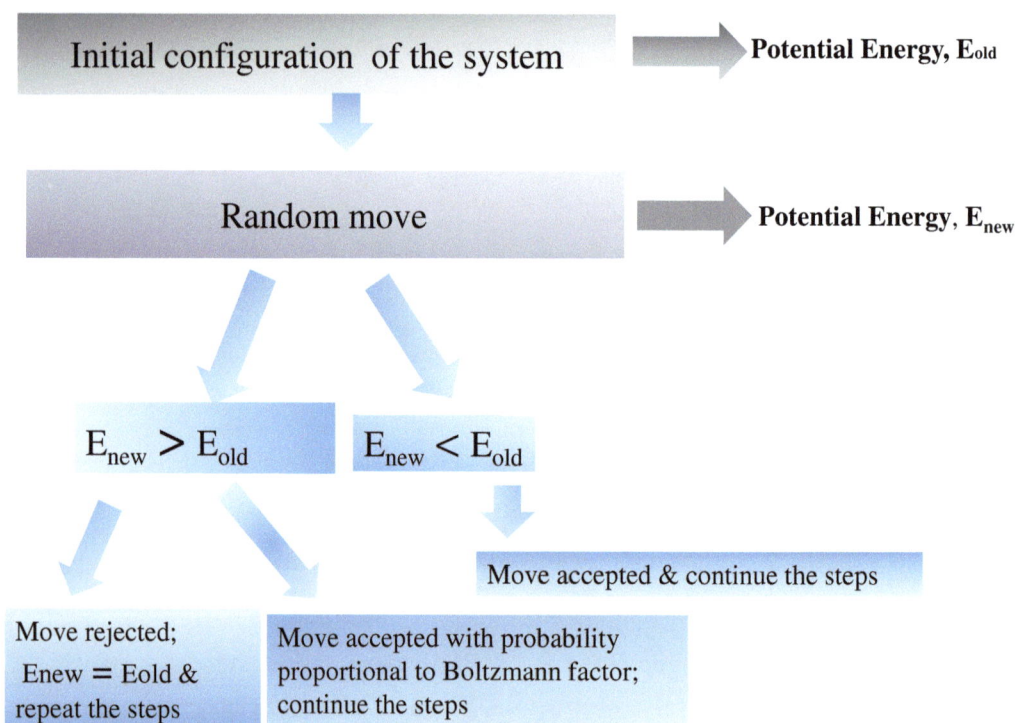

Fig. (4.6). Simplified algorithm of Monte Carlo Simulation. From the initial configuration, the direction of random move is chosen by energy criteria.

Variants of Monte Carlo simulations have also been employed to simulate systems such as hydrogen fluoride (HF), water (H_2O) and acetic acid (CH_3COOH), which are highly associated liquids due to strong intermolecular interactions [114]. Compared to Monte Carlo techniques, an explicit time variable associated with Molecular Dynamics formalism, and because of this extraction of dynamic properties such as time correlation functions and transport coefficients can be achieved by MD techniques.

Forcefields for the Simulation of Water

In the introductory section, a vague reference was given to classical and quantum simulation methods (here I refer to pure quantum chemical approaches only). Unlike quantum simulations, classical methods require models (forcefields) as one of the input for the simulations, and in the case of water, finding an accurate forcefield is one of the core issues we encounter while performing molecular simulations. This is due to the unpredictable nature of water. However, water is one of the most computationally studied liquids. More than 60 different water models have been proposed and employed in computer simulations so far [47]. These large number of water models, more than any other materials, were tailored to model and to simulate water for its different roles [47]. Over the years, there have been various attempts to accurately represent water in computer simulations. Guillot has given a historical list of water models that have been developed so far in his seminal review [115]. The conceptual model of water has even emerged well before the introduction of Molecular Dynamics methods. Simple electrostatic model with positive and negative charges near the oxygen and hydrogen atoms respectively was proposed by Bernal and Fowler in 1933.

Water models (forcefields) can be divided into three [46]: first class among these is simple interaction type models. In this framework, water molecule maintains very rigid geometry. Three to five point charges are used in the simple water models: for example, SPC, acronym for Simple Point Charge, one of the two most widely employed water models in classical simulations, uses three point charges. The other most widely used simple interaction type water models is variants of Transferable Interaction Potentials (TIPnP) (where n runs from 3 to 5, indicating number of point charges). These types are approximate water models such that they can speed up computer simulations to a considerable degree. For this reason, SPC and TIPnP models are widely employed in classical computer simulations. TIP4P model is appeared to be the most popular models among these, as evident by the fact that it has been re—parameterised several times (TIP4P/2005, TIP4P/Ice, TIP4PQ/2005), aiming to simulate different physical states of water including phases of ice [116]. It must be noted that quantum effects have been incorporated to TIP4PQ/2005, which will be crucial in investigating internal dynamics of ice [116]. Some of the variants of these two models are shown in the following diagram, (Fig. **4.7**).

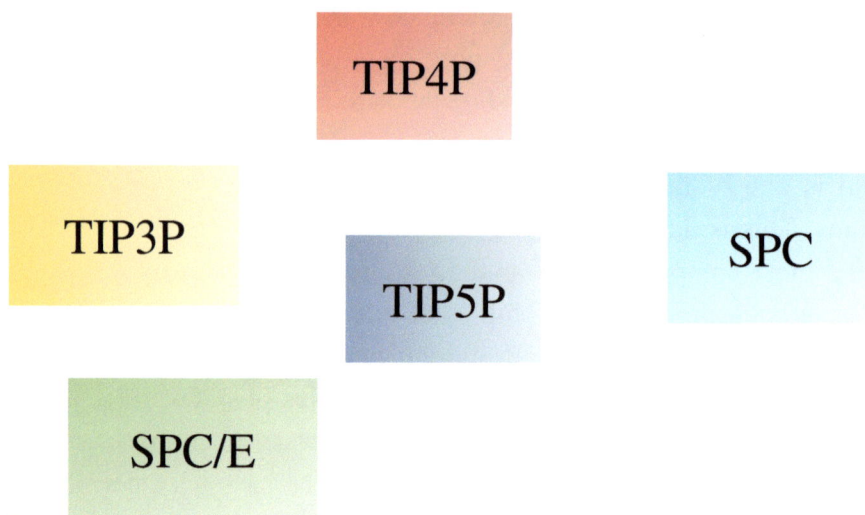

Fig. (4.7). Some of the most popular models widely used currently in molecular simulations. These models are popular for saving computing time significantly.

The second category of water models are known as Flexible models in which the internal changes in the conformation (not rigid geometry like simple point charge models) is allowed. An example for this model is polarisable electropole water model developed jointly by Barnes, Finney, Nicholas and Quinn [46]. Flexible models are very useful for investigating systems containing ions. The third type of water models are *ab—initio* models (for example NCC water model developed jointly by Nieser, Corongiu, and Clementi) [46], which are based on quantum chemical calculations taking account of electronic effects (a brief discussion on Quantum Chemistry can be found in next of this chapter), and are capable of handling polarisation and many body effects such that they can be used for comprehensive investigation of liquids, although they are computationally expensive due to the incorporation of numerous mathematical terms. An interesting fact is that despite all these developments, an accurate water model with which one can sketch all of its properties with a considerable level of accuracy has not been developed yet.

Derivation of the properties from computer simulations depends upon the model, methods & simulation conditions. Care must be taken to ensure sufficient sampling after adequate equilibration, which varies depending upon the thermodynamic conditions, in particular temperature and pressure. For instance, it takes approximately 4 nano seconds (ns) to equilibrate a system of 8000 water molecules at a pressure of 600 MPa at temperature of 100K, whereas for equilibration of the same system at 260 MPa and 280K requires only just over half a nanosecond [117].

FUNDAMENTALS OF QUANTUM CHEMISTRY

Quantum Mechanics, in its current form, originated after pioneering efforts of many brilliant scientists who blended their power of imagination with mathematical ingenuity, most notably Einstein, Planck, Dirac, Heisenberg and Schrödinger, in the twentieth century. With this powerful theory, the structure of atoms (electronic structure) can be explained to a greater resolution. Electronic structure methods of various genres are proved to be very efficient tools that are available to researchers in order to unlock mysterious atomic world where a human eye cannot reach. Electronic structure methods are of two classes – wave functional based and electronic density based. Both of these theoretical methods can either be *ab−initio* or semi empirical, depending upon the extent of parameterisation involved in the computation. Various levels of *ab initio* electronic structure calculations include Hartree−Fock (HF) method, MØller Pesset (MP) Perturbation theory, and Configuration Interaction (CI) methods.

The basis of any quantum mechanical calculations is solving a master equation, popularly known as Schrödinger equation, $H\psi = E\psi$, eponymous of one of the greatest Austrian physicists, Erwin Schrödinger. This mathematical construct tells us that Hamiltonian operator (H) describing the property of a system under investigation acts on a wave function (ψ) by which the location of the system in the space is specified, and extracts the energy levels of the system in return. So simple it may seem, however the calculation is not so straightforward due to multitude of interactions among the electrons, alternatively known as electronic correlation, in the system, and hence we have to resort to various levels of approximations. The ground level approximation one will employ in a quantum mechanical calculation is Born−Oppenheimer approximation; this separates nuclear and electronic motions within an atom, assuming that nuclei are motionless with respect to electrons.

In the remaining part of this section, the foundations of Hartree−Fock calculations are going to be discussed. While Schrödinger equation is considered to be the heart of Quantum Mechanics, solving Hartree−Fock equations serve as the foundation of Quantum Chemistry. Interactions in simple systems such as hydrogen atom (H), with a nucleus and a surrounding electron, can be computed within the framework of Schrödinger equation without resorting to any drastic approximations. When coming to higher electronic systems (from helium (He, atomic number 2) onwards) this situation, however, changes, and we have to employ several more approximations primarily due to the erratic movement of electrons as stated before. Chemists were in search for methods by which multiple electronic interactions can be accounted for. Finally, a breakthrough came in when Douglas Hartree introduced a procedure to calculate the properties of multi−electronic systems. The fundamental idea of Hartree method is to construct a polyelectronic wave function, a product of one−electron wave functions (known as atomic/molecular orbitals); each contains the coordinates of one electron in the system. In Hartree method, the total wave function is represented by the product of one electron wave functions, with each wave function representing one electron each. This means that solving for n−electron system generates n one−electron equations. In every cycle of Hartree calculation, each one−electron wave function is updated by calculating the interaction of each electron in field of all other electrons, which in turn yields a refined poly−electronic wave function. The process is continued until the difference in energy derived from the wave function is negligibly small (consistent with the previous cycle of calculation). Thus, an initial wave function is improved over successive cycle of calculations, and hence the Hartree procedure is known as Self Consistent Procedure (SCF). However, there are two principal limitations inherent in the Hartree method, due to its inability for accounting electron spin and anti–symmetry property of electrons.

In the subsequent development of Hartree's method known as Hartree–Fock method, a mathematical device called Slater determinant, by which both the electronic spin, and anti–symmetric property (corresponding to the exchange of electrons, can be incorporated) were introduced. With these changes, the electronic Hamiltonian takes much more simplified form with three terms (alternatively known as Fock operator): electron core integral comprising two terms (kinetic energy of electron in the field of other electrons and potential energy of attraction of electron to each nuclei), Coloumb integral, representing repulsions between electrons in two different orbitals, and Exchange integral due to the exchange of electrons in orbitals [118]. After the energy derived from the total Hamiltonian is minimised with respect to orthonormal orbitals (after the condition of orthonormality is imposed on orbitals), true Hartree–Fock equations (n equations for a system containing 2n electrons) are obtained. The advantage of Hartree–Fock equations compared to the previous Hartree method is that they can be casted in a matrix form, yielding a characteristic Eigen value (energy) for corresponding Eigen functions (orbitals). This, like Hartree SCF approach, is an iterative procedure until self consistent energies and orbitals are obtained. Although Hartree – Fock procedure is found to be useful for atomic calculations, a more refined approach (Roothaan Hall procedure) must be introduced for performing electronic level calculations on molecules. If the molecular orbitals are represented over the basis sets and with the introduction of Fock matrix, the Hartree–Fock equations can be casted into mathematically more convenient matrix Eigen value problem, with which molecular calculations can be carried out. Next, I explain what basis sets are, which play a pivotal role in this renewed approach, before returning to Rootham Hall approach.

Basis Sets

In this subsection, I venture into discussing an important mathematical construct known as basis sets, judicious choice of which determines the speed and accuracy of quantum chemical calculations. A molecular orbital can be expanded by basis functions, and the set of these functions employed in a given calculation is called basis sets. They are a set of functions that represent the shape of orbitals (physical representation of atoms or molecules), and is cumulatively used to represent the wave function ψ in Schrödinger equation [119]. In order to define a molecule with maximum accuracy infinite functions must be used in the basis set, which is often impossible especially when the molecule is very large. Therefore, one has to compromise in choosing number of functions in a basis set. Two rudimentary types of basis functions are Slater Type Orbitals (abbreviated as STOs) and Gaussian Type Orbitals (GTOs) [112]. The mathematical expressions for STOs and GTOs are similar, which contain a spherical harmonic term, accounting for the angular dependence of the orbital, and an exponential term, accounting for the interaction between the nucleus and the electron. The principal difference between these two orbital models lies in the order of exponential term: in GTO, the order appears to be two compared to one in STOs. Though GTOs have mathematical superior properties than STOs, one of the principal drawbacks of GTOs lies in the fact that there is more number of GTOs must be employed in order to render the calculations within the limits of desired accuracy. Counting limitations and advantages, GTOs are still preferred over STOs in quantum chemical computations [112].

If we contend with a minimal basis set, only one function is used to define each orbital. For the calculation of small molecules such as hydrogen and lithium (first two rows of periodic table), employing minimal basis set gives considerably good results, but additional functions are required for higher members in the periodic table [112]. The application of minimal basis sets is limited due to the factors such as complex bonding environments and electron correlations. Over the years, computational chemists have been developing basis sets of various precision that can handle these problems well. One important step towards dealing with complex bonding environment is employing Double Zeta (DZ) functions. In this mathematical construct, the number of functions to represent an orbital is as twice as that of a standard basis set (STOs), and employing DZ type functions is very effective in treating double and triple bonds [112]. This suggests that DZ type functions could well be suited to organic chemistry in which compounds with double and triple bonds are rife. Variants of DZ, Triple and Quadruple Zeta functions, have also been developed aiming at improving accuracy.

Diffusion properties of electrons effects can now be effectively represented by employing polarisation functions in basis sets. It is evident that in order to represent systems with more electrons, more number of basis functions is required. This makes their evaluation more time consuming, which demands further approximations. One way to limit the number of basis functions in calculations is splitting electrons into valence shell and inner shell regions. By employing contracted basis sets, number of basis functions in chemically unimportant inner shell regions can be reduced (contracted) into half, thereby saving computational time without sacrificing accuracy. Pople style basis sets, Dunning–Huzinaga basis sets, Ahlriches basis sets, MINI, MIDI & MAXI belong to this category.

We can approximate molecular orbitals as a linear combination of basis functions (Linear Combination of Atomic Orbitals (LCAO)). A good review of various basis sets regarding their use and associated problems is given in Jenson's computational chemistry introductory book [112]. The growing list of variety of basis sets indicates that their development is very active research area, aiming at to bring down computational cost of quantum chemical calculations.

Roothan– Hall–Hartree–Fock Approach

Roothaan & Hall improvised the Hartree–Fock procedure, namely Roothaan–Hall approach. In this approach, the total wave function of an atom is expressed as Slater determinants of molecular orbitals (spatial and spin orbitals combined). Inserting appropriate Hamiltonian into the Schrödinger equation and using Slater determinant for the total molecular wave function yield energy in terms of wave functions. Minimisation of energy with respect to the wave function yields Hartree–Fock equations. The Molecular Orbitals (MOs) represented by Linear Combination of Atomic Orbitals (LCAOs) are fed into the Hartree–Fock equations, which is known as Roothaan–Hall (RH) equations. As in the case of Hartree–Fock procedure, Roothaan–Hall procedure requires series of steps.

Step 1: Molecular structure is defined with appropriate choice of basis sets.

Step 2: During this step, several integrals including kinetic and potential energies and overlap integrals are calculated.

Step 3: From overlap integrals (overlap matrix), indicating the strength of the overlap between neighbouring orbitals, an orthogonalizing matrix is constructed.

Step 4: From kinetic energy and potential energy integrals (calculated in step 2), initial Fock matrix is calculated. In addition, a tentative basis set coefficient matrix is constructed at this stage.

Step 5: The Fock matrix is transformed using an orthogonalising matrix, obtained from step 3. The new Fock matrix obtained this manner yields energy levels. Along with the energy matrix, we obtain a coefficient matrix in this step.

Step 6: The coefficient matrix is transformed to the coefficient matrix of the original basis functions. With this step, we are in a position to interpret molecular orbitals in terms of atomic orbitals.

Step 7: Second cycle of this SCF begins by calculating the density matrix from the coefficient matrix obtained in step 5. The density matrices, which can obtained from the coefficient matrices, from two consecutive calculations are compared, if the difference is very negligible, we can say that the SCF calculations converged, and can be terminated. Otherwise, we go to step 5 for further cycles of calculations.

Below shown (Fig. **4.8**) is the diagram indicating the important milestones, starting from Hartree method, in the development of Hartree–Fock LCAO Self Consistent approach.

What we have achieved by the steps mentioned above is the estimation of the total electronic energy. The total Hartree–Fock energy is obtained only after summing total electronic energy with the inter–nuclear repulsive energy. Although considered obsolete now, the HF method was a major breakthrough in exploration of electronic structure, which laid the foundation to other quantum mechanical methods. In a multi–electronic environment, however, electronic correlation is a dominant phenomenon, and this effect is not treated explicitly in HF method; instead, each electron experiences the presence of other electrons in an average way, known as Independent Particle formalism. This is mainly due to the fact that electronic probability in the space is considered with respect to the nucleus, neglecting the presence of other electrons (electron correlation) [119]. Due to this drastic approximation, HF methods yield poor results in most of the cases, such as d block elements in the periodic table, where in electron correlation play a non−negligible contribution to the total electronic energy.

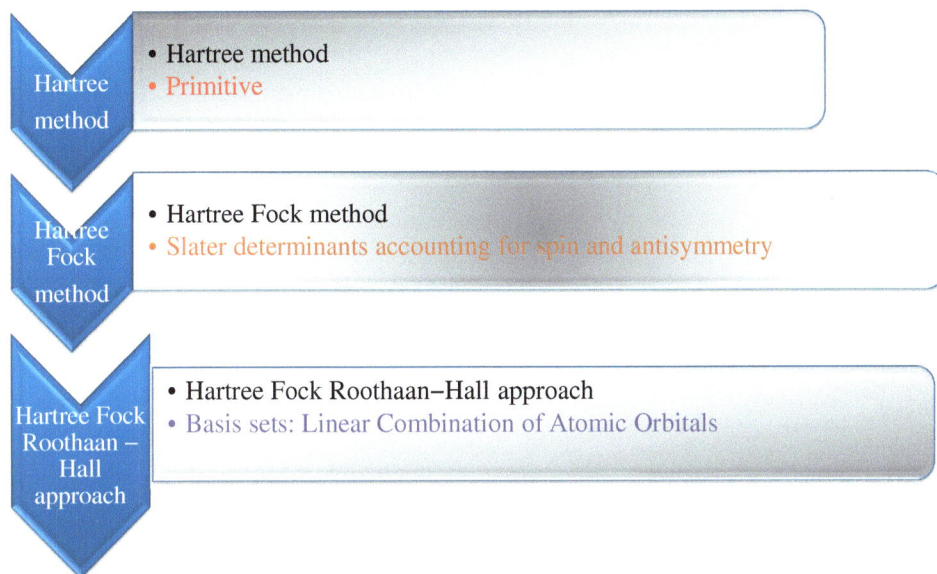

Fig. (4.8). Development of Hartree Fock (HF) method. The Hartree Fock method as we know today is a result of systematic development from the basic Hartree method proposed by Douglas Hartree.

Inefficient treatment of electronic correlation in the standard Hartree−Fock (HF) procedure results in poor estimate of electronic energy. However, electronic correlation can be incorporated as small increments (perturbation) to the standard formalism (HF equation), aiming at to obtain better results: this approach is known as Møller Pesset Perturbation Theory (MPPT). This procedure makes use of the fundamentals of Perturbation Theory, one of the earliest developments of quantum theory. The principle of the Perturbation theory can be summarised in the following way: imagine two physically similar systems A and B. The Schrödinger wave equation of the system A can be solved accurately, but accurate solution for system B is much more complex such that it cannot be found directly. Since both these systems are physically similar, the system B is considered to be evolved from system A. By applying small perturbation corrections (as increments) on system A, the Schrödinger equation for system B can be solved.

Another way of handling electronic correlation is by the idea of Configuration Interaction (CI). In this approach, electrons are promoted to unoccupied level from occupied level that is lower in energy (remember the fact that one orbital can accommodate maximum two electrons). This naturally generates different electronic configurations; each represents a Slater determinant. In a full Configuration Interaction (CI) calculation, application of the linear combination of Slater determinants, representing every possible electronic state, gives an accurate estimation of molecular energy. The main drawback of this method is that the approach works only with very small molecules.

Semi Empirical Methods

Despite the development of these accurate methods and ever increasing availability of computing power year by year, *ab−initio* electronic structure calculations are still expensive due to cumbersome integral evaluations and large basis set manipulations. It is clear that as the number of electrons for a given system increases, the number of integrals to be computed too grow proportionately, the major stumbling block in a standard *ab−initio* procedure. Theorists therefore started to think about remedial methods in order to speed up the calculations. This resulted in the discovery of semi empirical methods, retaining the theoretical frame work with which *ab initio* methods are built of.

Contrary to *ab−initio* methods, which are computationally demanding due to the inclusion of larger basis sets, in semi empirical methods only smaller basis sets are used. In addition, the time consuming part of calculation (the numeration of two electron integrals) are omitted and substituted by parameters obtained either from experiments or full *ab−initio* calculations in semi empirical methods. Due to these methical deviations, semi empirical methods are as not deemed to be consistent as the *ab−initio* methods in several cases. However, these methods provide results of considerable accuracy for certain class of molecules, for example small organic compounds. Most important semi empirical methods include Hückel, Parissor−Parr−Pople (PPP), Austin Model (AM3), Parameterisation Method (PM1) & Fenske−Hall. Another class of semi empirical methods is grounded on a principle known as Neglect of Differential Orbitals (NDO). The principal differences between *ab−initio* and semi empirical methods lie in four major aspects.

1. In *ab−initio* methods, nuclei are considered to be atomic cores, whereas in semi empirical methods, atomic cores constitute nuclei and non−valence electrons, which reduces the computational cost to a greater extent.
2. The Basis Sets used in *ab−initio* methods are complex (with more functions) than semi empirical methods.
3. Many one and two−electron integrals are approximated in semi empirical calculations often set to zero.
4. Overlap matrix is set to unity in semi empirical methods, in doing so the necessity of orthogonalising matrix can be avoided, which makes the calculations much faster.

Density Functional Theory

Density Functional Theory (DFT) is one of the widely used quantum chemical methods. Its popularity increases ever since its inception. DFT is based on Hohenberg−Kohn theorems which posit that ground state properties of an atom or a molecule can be accurately estimated from its electronic density function (first theorem), and there exists one to one correspondence between electronic density and energy (second theorem) [118]. The latter proposition means that with a guess of trial electron density function, the true ground state electronic density can be found. The variation in the estimated electronic density with respect to the true electronic density can be annulled if exact functional, which is unknown, is employed. Kohn−Sham theorems offer an "efficient" way to find energy and electronic density. The ground state electronic energy, as in the case of Hartree−Fock wave function theory, is the sum of two potential energy terms (the attractive interaction between electron and nucleus, and the repulsive interaction between electron and electron) and a kinetic energy term (of electrons). Kohn and Sham framed the expression for total electronic energy within the framework of Density Functional Theory, four terms contribute to the ground state electronic energy: potential energy between nuclei and electrons represented in the form of charge cloud, electronic kinetic energy of a reference system of non−interacting particles, classical electrostatic repulsion energy (please refer chapter two), and exchange–correlation term which is the sum of the kinetic energy deviation from the reference system of non–interacting electrons and electron–electron repulsion energy. The following step is to write this energy expression in terms of orbitals, which is then differentiated with respect to the orbitals under the condition that they be orthonormal to each other (similar to what we saw in deriving Hartree–Fock equations). The important steps in Density Functional calculations are given below [118].

Step 1: The geometry of the system of interest is given.

Step 2: A suitable basis set is specified.

Step 3: Initial guess of electronic density. Exchange correlation function is calculated from this electronic density.

Step 4: Kohn – Sham operator is calculated from electronic density and exchange correlation.

Step 5: A Kohn–Sham matrix is constructed from Kohn–Sham operator and basis functions.

Step 6: In this step, Kohn–Sham matrix is orthogonolised, and is subsequently diagonalised. The resulting matrices are known as coefficient matrix, and energy matrix.

Step 7: The coefficient matrix is transformed in order to obtain the Kohn Sham orbitals as the weighted sum of the original basis functions, prior to the condition of orthogonality is applied.

Step 8: From the Kohn Sham Molecular orbitals (KS MOs), one can calculate the electron density, and with this, first iteration is completed.

The second cycle begins with the step 3 (here the guessed electronic density is replaced by calculated electronic density), and the subsequent steps are continued until step 8. If there is no significant changes from the energy levels obtained from the previous step, and the calculation can be stopped. Otherwise, the cycle of calculations (from strep 3 to step 6) has to be continued until the specified convergence criteria is met. The following schematic diagram (Fig. **4.9**) depicts various stages in the standard DFT algorithm.

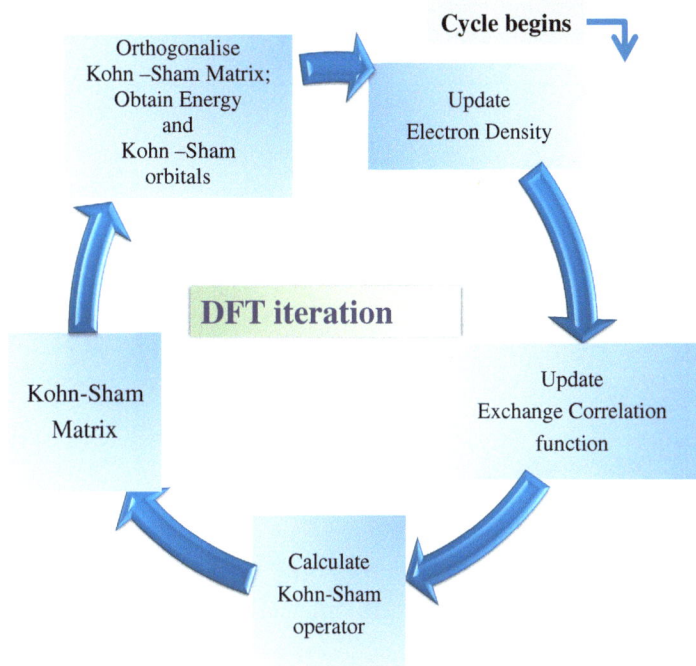

Cycle begins

Orthogonalise Kohn –Sham Matrix; Obtain Energy and Kohn –Sham orbitals

Update Electron Density

DFT iteration

Kohn-Sham Matrix

Update Exchange Correlation function

Calculate Kohn-Sham operator

Fig. (4.9). Standard density functional approach. Five important steps in the standard DFT protocol begin with updating the electron density, followed by updating Exchange Correlation function. This enables us to calculate Kohn−Sham operator and Kohn-Sham matrix. After orthogonalising Kohn−Sham matrix, one can obtain energy and Kohn−Sham orbitals.

Various levels of approximations are widely employed within the framework of DFT. One of such approximations is Local Density Approximations, abbreviated as LDA. Other approximations include Local Spin Density Approximation (LSDA), and Generalised Gradient Approximation (GGA). A brief review of these approximations can be found in [118] and other DFT books. Although electron correlation effects are embedded in Density Functional Theory, the exact form of which is still unknown, which poses a difficulty in systematic improvement of DFT results, and scientists often rely on experience and intuition to arrive at accurate results [118].

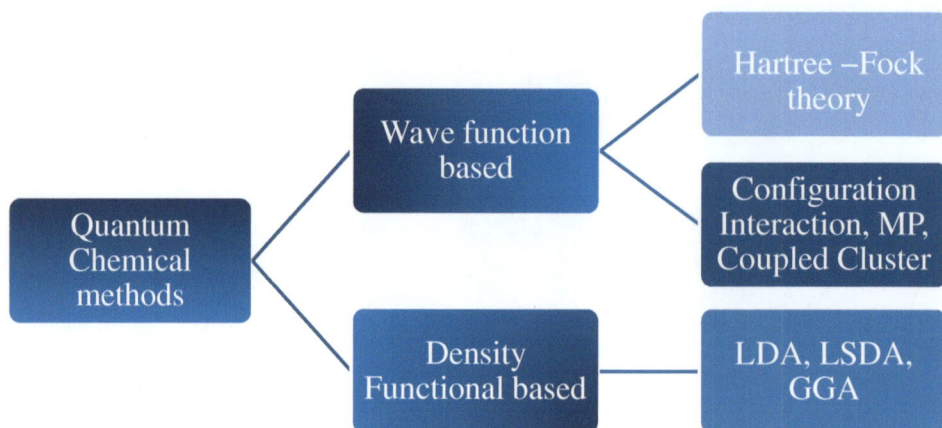

Fig. (4.10). Classification of Quantum Chemical methods. Wave function and Density Functional methods are the two principal classes of Quantum Chemical methods.

A diligent reader finds similarities between Hartree−Fock methods and Kohn−Sham approach to Density Functional methods. However, there are several advantages of DFT based methods over wave function based methods. Electron correlation is in principle an integral component in Density Functional methods, contrary to Wave Function methods in which additional treatments on Hartree−Fock theory have to be made in order to incorporate this phenomenon [118]. Examples for such add−ons include Møller–Plesset perturbation, and Configuration Interaction (CI) methods. Because of its explicit treatment of electron correlation, DFT based methods give better estimate of energy for transition metal compounds [118]. Furthermore, electronic density is a measurable quantity by experiments, for example by X−ray diffraction, whereas wave functions are not [118]. Hence, results from DFT theoretical calculations can be compared with experimental findings. As the system size increases, the complexity of wave functions increases, compared to electronic density [112]. For example, the wave function of a n−electron molecule depends upon 4n variables, whereas electron density, of which DFT is dependent on, is a function of only three variables regardless of the size of the molecule.

Fig. (**4.10**) above demonstrates the major quantum chemical computer methods.

CONCLUDING REMARKS

I take this opportunity to render the reader only a brief account of major molecular and atomistic level simulation techniques, due to the pace at which the subject continues to advance, and also due to the space limit of this book. Molecular Dynamics and Monte Carlo methods still remain to be the principal computational modes of investigation at molecular level. In order to explore deep into the electronic structure of matter, wave function methods and density functional based methods are considered to be the major workforces. Hartree method has earned historical importance in quantum chemistry being the first multi−electronic approach that was developed in order to obtain total electron energy of a system containing many electrons, and its improvised version Hartree-Fock-Rooth-m-Hall procedure serves a basis for post Hartree-Fock methods. On the other hand, DFT based methods provide a greater opportunity for a comparison of theoretical and experimental modes of investigation *via* electron density. Can quantum mechanical methods be applied in simulations of liquids such as water? Since the inception of Car−Parinello Molecular Dynamics (CPMD) method into computational chemistry, many more serious attempts have been made in order to combine quantum mechanical calculations with the classical Molecular Dynamics or Monte Carlo methods, and these methods yielded success at varying degrees [120]. It is expected that with ever increasing computational power such methods will be more frequently be employed in molecular simulations in near future.

CONFLICT OF INTEREST

The author confirms that he has no conflict of interest to declare for this publication.

ACKNOWLEDGEMENTS

Declared none.

CHAPTER 5

Water Between Its Freezing and Boiling Points

Abstract: Structural elucidation of water is so fundamental in understanding its roles as a solvent as well as a reagent in facilitating multifarious chemical reactions. The internal structure of water molecule is very "simple" to explain yet the physical and chemical properties of this liquid remains to be elusive in spite of tremendous theoretical and experimental efforts till date. Several propositions have been made in order to account for water structure, in particular its enigmatic hydrogen bonding environment that accounts for its exceptional properties. The concept of uniform distribution of tetrahedral network in water has been emerged from various experimental investigations. Water structure as equilibria of large number of clusters formed by varying number of water molecules has also been proposed based on computer simulations and Raman spectroscopy. Percolation model provides a quantitative picture of hydrogen bonding in liquids. The nature of hydrogen bond is dynamic in nature, spurring sporadic changes in its local structure, which can effectively be probed by various spectroscopic and scattering techniques. Local structure of water molecules is influenced by thermodynamic changes, most notably in temperature and density. Both computational experimental findings reveal that density plays a vital role in determining average number of hydrogen bonds a water monomer can have across wide temperature domain. More importantly, water undergoes a cascade of morphological changes upon alteration in temperature, which is still a fascinating subject for many researchers.

Keywords: Bifurcated hydrogen, Clusters, Coordination number, Density, Electrostatic interaction, Exchange repulsion, NMR, Percolation theory, Raman spectroscopy, VSEPR, Walrafen pentamers.

INTRODUCTION

Liquid water, the substance what we call 'water' in our daily life, exists usually in temperature ranging from 273 to 373 Kelvin. Water, one of the simplest hetero-

Jestin Baby Mandumpal

atomic substances, consists of one 'heavy' oxygen atom (atomic mass = 16g/mol) and two lighter hydrogen atoms (atomic mass =1.008 g/mol). The static model, as shown in Fig. (**5.1**), for water is very "simple".

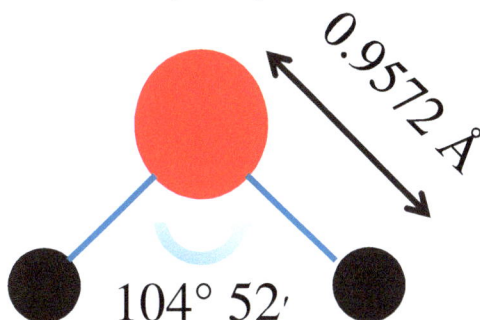

Fig. (5.1). The static geometry water. Water consists of one oxygen atom and two hydrogen atoms connected by two single bonds. The bond length between oxygen and hydrogen atoms in water is 0.9572 Å and the angle between the two oxygen − hydrogen bonds is 104.52°.

A water molecule has two oxygen−hydrogen bonds, and there are two types of oxygen−hydrogen bonds in water too (one of these is an example for strong covalent bond). The distance between hydrogen and oxygen atoms is approximately 0.96 Å with the angle between them 104.52° [121]. This angular deviation in water from a pure covalent bond angle (90°), as a result of repulsion arising from partial ionic character of oxygen−hydrogen bond, may offer a glimpse of its enigmatic properties [122]. It must also be noted that the deviation from the ideal bond angle comes about due to the repulsion between two lone pairs on oxygen atoms, commonly referred as Valence Shell Electron Pair Repulsion (VSEPR) theory.

The static model is however a very crude approximation due to three different modes of vibration of oxygen − hydrogen bonds [121]. They correspond to symmetric and asymmetric stretching (3657 cm^{-1} and 3756 cm^{-1}) and bending vibrational motions (1595 cm^{-1}). The schematic diagrams of these three modes of vibration are shown in Fig. (**5.2**).

The second type of oxygen−hydrogen bonds is weaker intermolecular hydrogen bonds. This is much more significant than the covalent bond in relation to the properties of water, as we will see throughout this book. What is a hydrogen bond? In order to have hydrogen bond established between two neighbouring molecules, there must be a hydrogen bond donor (D) and hydrogen bond acceptor (A), and it is expected that the angle between DH (donor (D) –hydrogen bond (H)) and AH (acceptor – hydrogen bond) be approximately 180° [87, 92] (An alternative definition to hydrogen bond based on electrostatic interaction has already been given in the introductory chapter).

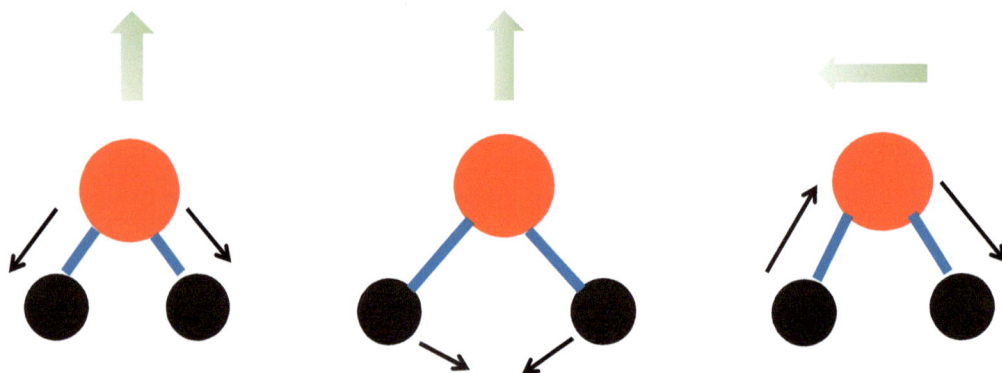

Fig. (5.2). Three different modes of vibration in water molecules. In the left is shown symmetric stretching vibrations occurring at a frequency of 3657cm^{-1}. In the middle and on the right are shown bending vibration and asymmetric stretching with a frequency of 1595 cm^{-1} and 3756 cm^{-1} respectively.

Hydrogen bond is a distant dependant, and several quantum mechanical forces contribute to it. These forces include Exchange Repulsion (ER) due to electron exchange from two neighbouring molecules, and Electron Delocalisation (ED) due to charge transfer from an occupied orbital of one molecule to unoccupied orbital of its neighbouring molecule [123]. It has been found that hydrogen bonds are replicated due to electronic redistribution, a signature of cooperative coupling in water [124].

Hydrogen bonds are formed in variety of chemical species, organic compounds including carbohydrates, proteins, nucleic acids and other numerous biological molecules, as well as a wide array of inorganic compounds. Based on strength of bonding, hydrogen bonds have been classified as weak (with energies between −2.4 and −12 kcal/mol), strong (with energies between −12 and −24 kcal/mol), and very strong (with energies more than −24kcal/mol) [125]. The range of hydrogen bond energy, the energy required to break one mole of hydrogen bonds in water, is so broad (1 kcal/mol – 50kcal/mol), reflecting structural flexibility of the hydrogen bonded systems [126]. In fact, the energy of hydrogen bonds greatly depends upon the chemical environment: it has been estimated that the hydrogen bond energy of two uncharged groups is in between 1.0 kcal/mol to 1.4 kcal/moles. This amount doubles when one of the participant of hydrogen bonding is a charged group [127]. If both the donor and the acceptor are the charged species, the energy further increases up to 4 kcal/mol.

The complex character of hydrogen bonds is reflected in its varying strength. In the case of water, in one end, its strength can be compared with that of ice, and in other end, its strength is comparable with that of non – hydrogen bonded oxygen–hydrogen bonds [128]. This can be evidenced from different peaks obtained in the Infra–Red (IR) spectrum with varying frequencies (3300, 2440, 3277, 2421, 3705, 2719, 3670, 2695 cm^{-1}), in addition to the differences in the estimated energies of hydrogen bonds explained before.

Water is a structured liquid, in which molecules are held together either by strong polar bonds or hydrogen bonds as opposed to regular liquids where in much weaker disperse interactions dominate. Notably water has less number of nearest neighbours (in the range of four to five) than regular liquids do [129]. Hydrogen bond in water is an example of a "classical hydrogen bond" since both hydrogen bond donor and acceptor belong to group 16 in the periodic table [125]. Other sub–categories within this group are hydrogen bonds between with unconventional donors (for example C–H) and unconventional acceptors such as halogens, and dihydrogen bonds [125]. For a better overview of classical hydrogen bonds, readers are advised to go through [125].

According to the classical picture, water molecules are attracted to each other by the combination of electrostatic and covalent forces [122]. Dipole–dipole (a type of electrostatic interaction) and covalent forces account for 95% and 5% respectively for the formation of a hydrogen bond. Although contribution of covalent character to the hydrogen bond is negligible in comparison with the electrostatic forces, it does however play a vital role in intermolecular forces. This is because of the fact that the generation of dipole moment is determined by the extent of covalent bonding [122]. In this chapter, I would like to discuss two important aspects pertaining to water, in particular its molecular and electronic structure, and various models suggested for accounting for hydrogen bonding that underpins the interactions between water molecules.

WATER STRUCTURE

The structure of water can be elucidated by obtaining pair distribution functions from either simulations or experiments as explained in chapter 3. Fig. (**5.3**) shows three possible pair distribution functions of water, obtained from computer simulations, at ambient conditions (at 298 K, 1 atmosphere & experimental density). At this temperature as the peaks in reducing amplitude (at 3 Å, 4.5 Å & ~7 Å respectively) in oxygen–oxygen pair distribution functions indicates, a water molecule surrounded by at least three layers of other water molecules. The first peak of oxygen–oxygen pair distribution function gives the extent of hydrogen bonds existing in water, whereas the second peak indicates the interaction between the reference molecule and water molecule beyond its first hydration shell. The taller and sharper peak indicates stronger bonds at short distances [128].

The oxygen – hydrogen pair distribution functions provide a vital information regarding the intermolecular hydrogen bonding existing in water molecules. The first peak appeared at ~ 1.8Å in the pair distribution function is an indication of hydrogen bonds, and from the second peak we infer the presence of neighbouring water molecules in the first hydration shell [131]. Coordination number, calculated by integrating the area under the first peak in the pair distribution function until first minimum, is estimated to be two which indicates that the oxygen atom of a water molecule forms two hydrogen bonds with hydrogen atoms of the neighbouring water molecules [132]. Subsequent integration until second minimum yield a coordination number four [133]. The invariability of coordination number (four) across a wide range of temperature implies the strength of tetrahedral structure in water [134]. It can be see that the local order in water vanishes after 7–8 Å, implying that a water molecule can interact with approximately 70 other water molecules, according to the estimate of Narten and Levy [135]. The fact that a change in peak by 0.01Å means a difference of 0.2 kJ necessitates a better resolution for radial distribution curves that can be obtained from computer simulations and experiments alike [135].

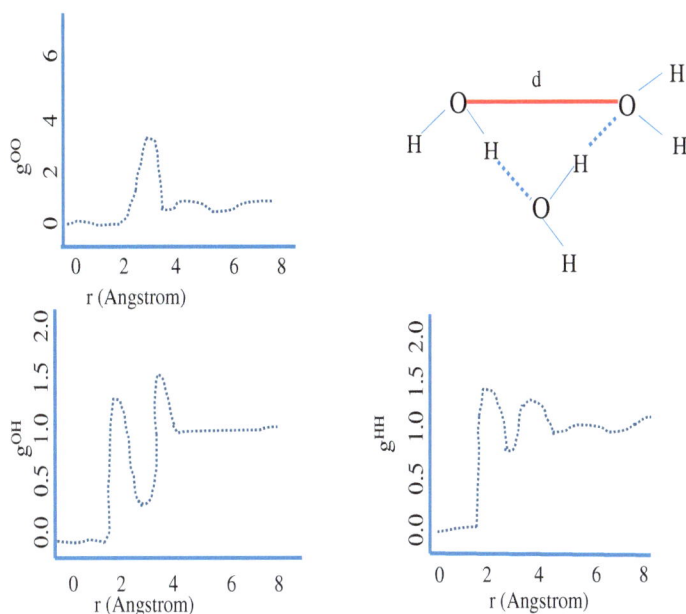

Fig. (5.3). Pair Distribution of water at ambient conditions (298 K and 1 atmospheric pressure). Oxygen-oxygen pair distribution function has been shown in top left. The first peak appeared at around 3 Å corresponds to oxygen atoms of neighbouring water molecules, while the second peak seen at around 4.5 Å indicates the presence of second nearest neighbour as the sketch on the top right indicates. The oxygen–hydrogen pair distribution functions provide an estimate of hydrogen bonding. The coordination number up to second minimum suggests that a water monomer can indulge in two hydrogen bonds and two oxygen–hydrogen covalent bonds. Two significant peaks are observed in the case of hydrogen–hydrogen pair distribution functions (bottom right): first peak at around 2.3 Å, while the second peak indicates hydrogen atom in the nearest water monomer. These ideas have been adapted from [130].

A three dimensional picture of coordination, as shown in the Fig. (**5.4**), in water provides a better understanding of how water molecules orient with respect to each other. In addition to the four neighbours at the four corners of a tetrahedron, there can be several other water molecules present in the space between the regular tetrahedral sites [134]. This can be lead to the coexistence of tetrahedral and smaller non–tetrahedral water structures at ambient conditions. Computer simulations performed by Vallauri *et al.* also underpin this finding [136]. They have found out a significant proportion of water molecules with coordination numbers three and five, with a smaller population of water molecules with coordination number six at room temperatures. As temperature increases, preponderance of water molecules with lower coordination numbers is observed in water. These findings have important connotations with one of the two important classes of theories on the structure of water, Mixture model, which will be discussed later in this chapter.

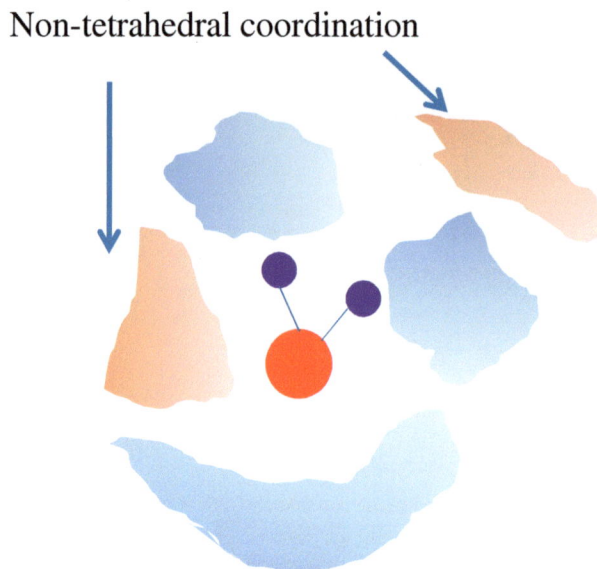

Fig. (5.4). Three dimensional picture of oxygen density in water. Water molecules are four fold (tetrahedral) coordinated. Additional water molecules, as shown by the arrows, imply a strong layer of water molecules around each water monomers at ambient conditions. The idea has been adapted from [134].

Electronic Structure

X−ray emission spectroscopy provides vital information regarding the influence of hydrogen bonding on the electronic structure of water molecules [137]. A water molecule has 10 electrons altogether (8 from oxygen atom, whose electronic configuration is $1s^2\ 2s^2\ 2p^4$, and one each from two hydrogen atoms, with electronic configuration $1s^1$. The order of energies of the molecular orbitals (MO) in water molecule is $1a_1 < 2a_1 < 1b_2 < 3a_1 < 1b_1$, where a_1, b_1 and b_2 are irreducible representations pertaining to C_{2v} group, the symmetry of water according to group theoretical norms. The ground state electronic configuration for water molecule is $(1a_1)^2, (2a_1)^2, (1b_2)^2, (3a_1)^2, (1b_1)^2$. The orbitals ($1b_2$ and $3a_1$), oriented in the plane of the nuclei, are responsible for the oxygen – hydrogen bonding. A strong mixing of $3a_1$ states of water (the two peaks belonging to $3a_1$ states submerged into one), attributed to the covalent character of hydrogen bonding, has been observed from the spectroscopic measurements (The reader unfamiliar with Group Theory is requested to go through chapters 23 through 25 in [44] or any other books on the subject).

Hydrogen Bonds in Water: A Closer Look

To unlock the mysterious properties of water it is necessary to obtain a clear understanding of the fundamental unit of water structure. It is important to note that hydrogen bond in water is longer than non−hydrogen−bonded oxygen − hydrogen bonds by about 1Å. There exist several theories regarding the form of hydrogen bond network in liquid water. Earlier developments of these theories are solely based on experimental techniques, some of which, for example X−ray diffraction, may be inadequate due to the short times scales (in the range of 10^{-13} to 10^{-14} seconds) of the hydrogen bond dynamics [45]. Most of these experiments are not simply "good enough" to capture the short time scale dynamics.

With the advent of Molecular Dynamics computer simulation techniques, the dynamics at the femto seconds scale (at the scale of 10^{-15} seconds) can effectively be monitored employing a water model, and the "accepted" theories of water have been challenged more frequently ever since. This is because of the fact that introduction of computers into scientific research in the latter half of twentieth century has provided researchers to test various mathematical constructs embodied the water structure. Thus viewing water from a molecular perspective was possible, and the most rudimentary way of doing so is by calculation of configuration integrals, which accounts for pairwise interaction of all particles in a specified volume. At this juncture, a diligent reader will recollect the discussion on theories of water discussed in chapter 2.

Variety of models that claim to represent interactions among water molecules have been put forward by chemists and physicists, which can be categorised into two classes: Uniformist and Mixture models. Their supporting arguments were primarily based on experiments. Water structure is homogenous according to Uniformist models, whereas in the mixture model advocates heterogeneous structure, *i.e.* the coexistence of several species with different bonding patterns [45].

The foundation of Uniformist model, also known as Continuum model, is based on Bernal and Fowler's original idea of tetrahedral coordinated water molecules, as shown in Fig. (**5.5**). According to their proposition, each water molecule on average participates in four hydrogen bonds, with its neighbours towards forming a strong extensive homogenous three dimensional network wherein a water monomer acts as a double donor and a double acceptor [138]. This model could explain the longer life time of local structure of liquids such as water than simple liquids (for example ammonia) where in hydrogen bonds do not play any role in their structure [139]. What we must understand is that in order to have hydrogen bonding established between two molecules, an atom from one molecule must possess both lone pairs and hydrogen atoms. A comparison of water with ammonia or hydrogen fluoride (HF) will be helpful to understand this fact. Ammonia (NH_3) has three hydrogen atoms, but has only one lone pair on average (as opposed to two for water), and therefore it can form only one hydrogen bond. Hence, a three dimensional network cannot be formed in ammonia. Similarly hydrogen fluoride can form only one hydrogen bond since the molecule possess only one hydrogen atom despite the fact that fluorine atom owns three lone pair of electrons.

Although a large chunk of the scientific community in principle did agree with this proposition, several authors have challenged the idea of four hydrogen bonds per water molecule, the coordination number of water monomer. The irony is that it was Bernal himself first suggested that coordination number of a water monomer can be three (not four!) in bulk water [138e]. Fleming and Gibbs estimated this number to be 3.9, while King and Barletta suggested 3.1 [140]. The Lattice model suggested by Angel gives the coordination number just below 3, and Rahman *et al.* suggest that this value can be as low as 2.1. In Scheraga's model, average hydrogen bonds per water molecule was estimated to be 1.84. In fact these observations are very much related to the energy of hydrogen bonds as demonstrated by Rahman *et al.* [140], and hence we need to have a closer look at the energetics of the hydrogen bonds.

The variation of hydrogen bond energy as the function of average number of hydrogen bonds is shown in Fig. (**5.6**). From the figure it is clear that average number of hydrogen bonds influences hydrogen bond energy directly: as this number becomes as large as eight, corresponding energy becomes one third to that of water dimers.

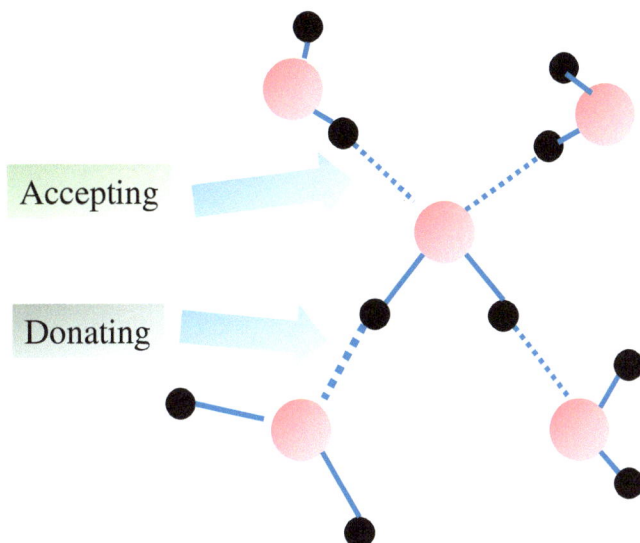

Fig. (5.5). Tetragonal coordination of a water molecule. Each water molecule is coordinated to four other water molecules (tetragonal structure). The oxygen atom of the central water molecule acts as a double donor (by donating its two hydrogen atoms to oxygen atoms of neighbouring water molecules) and an acceptor (by accepting hydrogen atoms of two other water molecules).

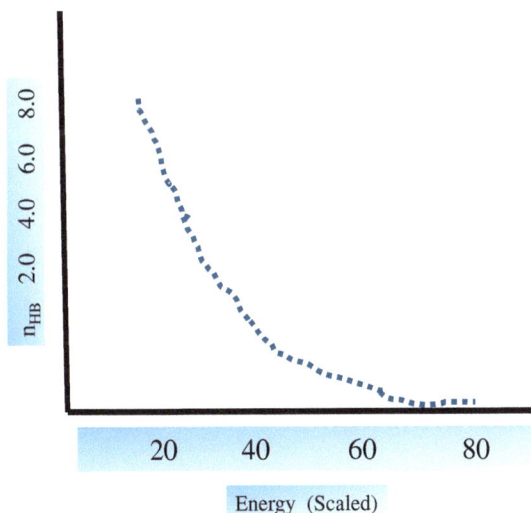

Fig. (5.6). Variation of average number of hydrogen bonds relative to changes in energy of the hydrogen bond. A water molecule can be hydrogen−bonded to as many as eight other water molecules at the lowest hydrogen bond energy shown in the graph. The idea has been adapted from [140].

Different numbers estimated for the water's coordination led to scepticism on universal applicability of tetragonal network model, and several researchers became ardent proponents of Mixture model [128]. The model promotes the idea of coexistence of two or more structures having different number of hydrogen bonds per molecule [45,141]. A brief survey through literature yields variety of Mixture models including Pople's Distorted Bond model, and Scheraga's Flickering Cluster model [135]. Flickering cluster model is an interesting proposition: according to this model, water can be viewed as two states differing in density, namely bulky phase in which water molecules are well ordered, and dense phase wherein disordered molecules predominates. Several authors have come up with varying estimates to the percentage of broken hydrogen bonds that can be present in the liquid water [128]. Leighton *et al.* have calculated that as many as 50% of hydrogen bonds present in the liquid water are broken based on the ultrasonic absorption experiments [128]. But Levy *et al.*, based on X−ray diffraction experiments, suggested that there can be only 25% broken hydrogen bonds present in liquid water at same temperature (see the references in [128]). The proponents of Cluster theory point into the fact that the energy of hydrogen bond in water is an arbitrary value, as we saw before, to support their rejection of the concept of existence of an extensive three dimensional tetragonal network. Davis and Litovitz have proposed a Two State model in order to account of hydrogen bonding in water [122]. According to this model (a typical example for Mixture models mentioned above), water exists as puckered hexagonal rings as in the case of chair form of cyclohexane, and some of these rings are joined in open packed structure (analogues to ice crystals), while others are arranged in more closely packed as shown in the Fig. (**5.7**). The Two State model advocates equilibrium between open packed structures and closely packed structures, and these two structures are inter−convertible, accounting for the dynamic nature of constant breaking and forming of hydrogen bonds in water. When the hydrogen bonds indicated by the dotted line in structure b is broken, the structure reverts back to structure a after a rotation of top layer by 30 degree.

The following figure (Fig. **5.8**) provides a better snap shot at Two State model, suggested to explain the constantly changing local structure of water, in which a molecule ideally has two hydrogen bonded neighbours. As you can clearly see in the figure, the hexagonal structures fit with each other in close pack configuration, creating only limited space between them. The hexagonal layers exist as the units of three layers.

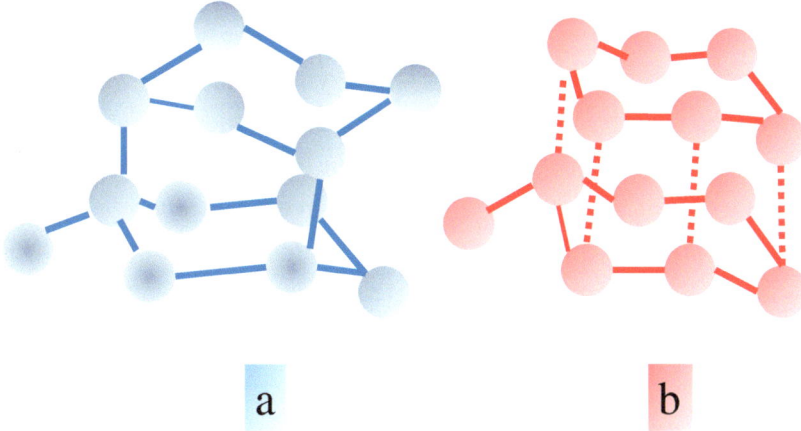

Fig. (5.7). The open and closely packed structures in water. The hydrogen bonds are indicated by thick lines. Two forms, (**a**) and (**b**), are inter–convertible. It has to be noted that in structure a there are several hydrogen bonds that connect two layers, whereas the structure b is closely packed without hydrogen bonds between two layers. The idea has been adapted from [122].

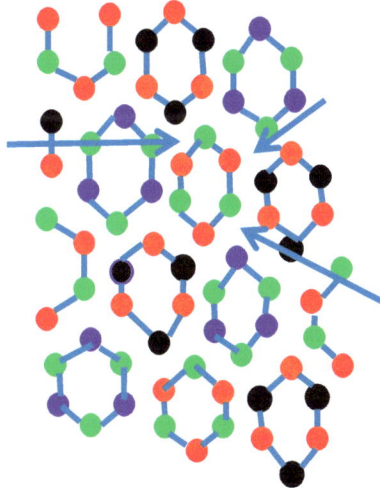

Fig. (5.8). Three dimensional (3-D) close packing of hexagonal rings. Different coloured circles (black, red, green and purple denote third, second, first and zero layers respectively) denote the corresponding number of layers. The arrows indicate the free space between the hexagons. The idea has been adapted from [122].

Modelling of hydrogen bonding in water is not limited with tetragons and hexagons alone. Chaplin came up with a complex pattern to explain hydrogen bonding in water [127]. According to his theory, water can form large clusters of diameter 30 Å containing as many as 280 water molecules. This structure is in fact a complicated network comprised of pentagons and hexagons. Interestingly this proposed structure, icosahedral, is highly symmetrical, in which each water molecule can form four hydrogen bonds as double donor and double acceptor. Notably, the proposed icosahedral structure can maintain the structural stability without breaking any hydrogen bonds by interconverting to two forms: expanded structure (ES) and a high density collapsed structure (CS). In the following figure (Fig. **5.9**), a sketch of the enigmatic icosahedral model of Chaplin is provided. A close inspection of Chaplin's model along the centre of two overlapping pentagons reveals that the models constitute several concentric circles, with varying number of water molecules on each of them.

One can see that Chaplin's model has been a revolutionary approach, providing a different outlook from the traditional view of intermolecular bonding pattern in water.

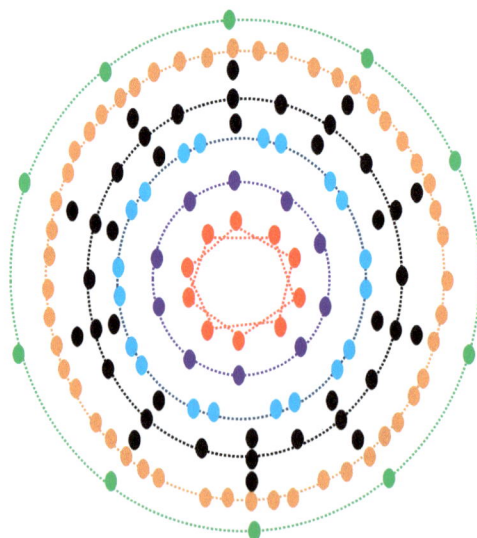

Fig. (5.9). Icosahedral model proposed by Chaplin. This model contains 280 water molecules, composed of fourteen flattened tetrahedral units. Each of these units contains 14 water molecules. The model can be represented by several consecutive circles, in each of them different number of water molecules are arranged. For better clarity, the connectivity between water molecules has been omitted.

Another setback for the Tetragonal Network theory is the presence of a fifth water molecule (a defect) in the four coordinated hydrogen bonded network, confirmed by experimentally and theoretically alike [142]. The idea of Bifurcated Hydrogen Bond (BHB) has been introduced to explain this "defect", which can be formed either due to the sharing of one of the hydrogen atoms of the central water molecules by two neighbouring oxygen atoms or acceptance of two hydrogen atoms from two different water molecules by a lone pair of the central oxygen atom. The fifth molecule in the first coordination shell of water molecules facilitates reorganisation of hydrogen bond network by reducing the energy barrier for translational and rotational motions [142]. Fig. (**5.10**) depicts a Bifurcated Hydrogen Bond (BHB) shared by three oxygen atoms. Earlier observation of the appearance of tiny patches of four–bonded molecules, with local density lower than overall density might be attributed to this pseudo tetragonal network [143]. The fluctuation in density has profound significance in explaining several thermodynamic anomalies of water such as isothermal compressibility, specific heat at constant pressure, and thermal expansivity [144].

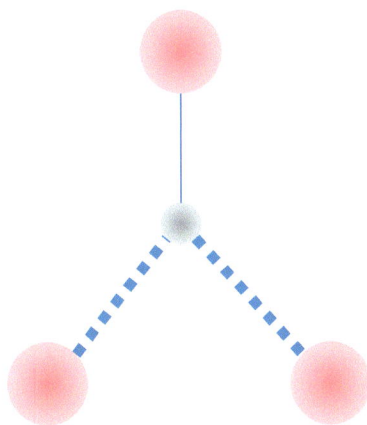

Fig. (5.10). A Bifurcated Hydrogen Bond (BHB). As opposed to linear bonds, a hydrogen atom is shared by two more neighbouring oxygen atoms (shown in red) in a Bifurcated Hydrogen Bond.

In addition to the aforementioned theories, Stanley and Teixeira have proposed an analytical construct known as Percolation Model (PM) to account for the quantitative estimation of hydrogen bonds in water [138d]. The model has been demonstrated in Fig. (**5.11**). According to this model, the probability of water molecules being connected is based on several factors such as oxygen – oxygen inter atomic distance, energy of the "probable" bond and orientation of molecule. A water molecule can have either of five states (0 through 4), depending on the number of intact hydrogen bonds it makes. Compared to the other theoretical propositions, Percolation theory provides a more quantitative picture such as cluster size, number of water networks and number of water molecules of a water cluster. The maximum number of water molecules that can be a part of a network (gel network as Stanley *et al.* termed it) in a given sample is observed when the average number hydrogen bonds per water molecule is one, whereas maximum cluster size (*i.e.* number of water molecules in a cluster) is calculated when average number of hydrogen bonds per water molecule crosses a threshold value, between three and four [140]. In addition, they have noted that the number of networks that can coexist in water at a particular time decreases as cluster size increases. Other consequences of their theory are that the number of networks takes extreme values when average number of hydrogen bonds per water is one and four respectively, and that the hydrogen bond should possess a minimum amount of energy (threshold energy). Hydrogen bond characteristics of aqueous solutions such as alcohols can also be explained by Percolation theory [37].

The most notable challenge to the Tetrahedral Network theory was posed a decade ago by a group of multinational researchers [145]. According to their combined X− ray absorption spectroscopy and X− ray Raman scattering studies, they could obtain at least four different spectra with varying intensity and phonon energy from the bulk sample [145b]. This clearly suggests that there can be observed at least four different local water environments, identified by different coordination numbers. They concluded that these are broken hydrogen bonded (distorted) structures, which can form rings and chains [146, 147]. If the sample had been homogeneous, its spectra would have been unique. They also suggested that the equilibrium between the tetrahedral structures and other non−tetrahedral clusters tilts in the favour of the latter (80%), and this equilibrium is expected to shift further towards the formation of weakly bound clusters upon the rise in temperature. Furthermore, the maximum number of hydrogen bonds that each water molecule can have is two, with a water monomer being single donor and single acceptor, as opposed to four in the traditional 3−D network model. There can also have numerous terminated loosely connected hydrogen bond networks of varying sizes. Cluster population analysis suggests the presence of clusters containing five, six and eight water molecules in normal water [148]. A detail micro level analysis of such water clusters can be found in the chapter on water at higher temperatures.

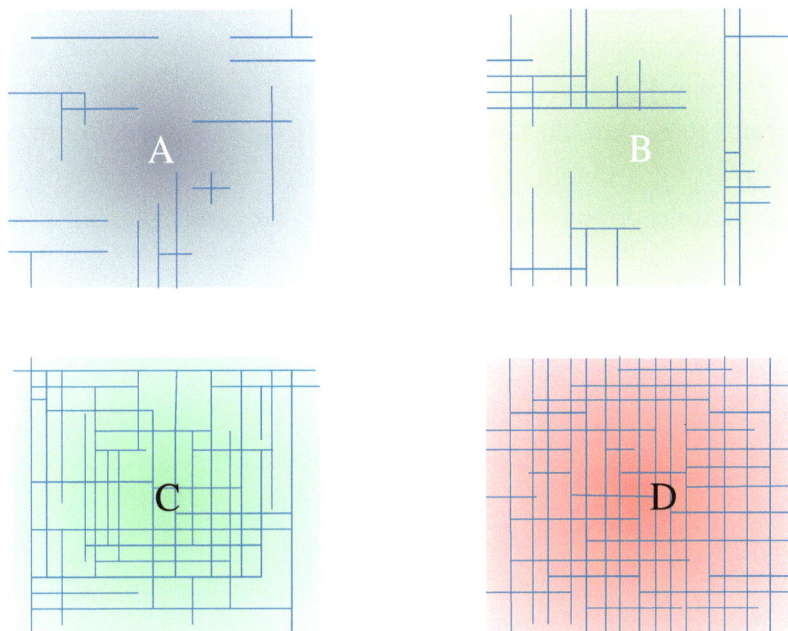

Fig. (5.11). Percolation Theory illustrated. One can clearly see that number of networks, and number of hydrogen bonds increase from A through D. The idea has been adapted from [138d].

However, taking a new twist to the everlasting debate over the nature of hydrogen bonds in water, Suchomel *et al.* have suggested, based on higher resolution X–ray diffraction experiments, that the ring type configurations reported in water can be considered as the continuation of the extensive tetrahedral network. They believe that this continuous network provides flexibility to the water structure [149].

The Dynamics of Hydrogen Bonds

Hydrogen bonds in liquids are incessantly formed and broken, and therefore their lifetimes are so short such that by ordinary means it cannot be properly tracked. This underpins the fact that longer hydrogen bonds (which are therefore weak) can become shorter and stronger [150]. The process of bond breaking and forming occur in short time scales ranging from femtoseconds (10^{-15} seconds) to picoseconds (10^{-12} seconds), and hence is experimentally very hard to detect [126]. However, with Molecular Dynamics simulations, the process of the breaking and forming of a bond can be tracked without losing accuracy as mentioned previously. In an interesting computational investigation using more recent TIP5P water model, Parinello *et al.* have obtained the lifetimes of hydrogen bonds, as shown in Fig. (**5.12**) [139]. The figure shows a sharp increase in life times between 270 K and 280 K, which is close to the Temperature of Maximum Density (TMD) for the model, indicating the connection between hydrogen bonds and some anomalous properties of water near this region. A detail discussion on this matter you can find in chapter 9.

Water molecules orient itself about an arbitrary axis swapping their hydrogen bond partners during the process. This plays a fundamental role in several physical processes such as proton transport [151]. The partner switching occurring in a group of three water molecules in three facile steps: first, the central water monomer connected to two other water monomers by two hydrogen bonds start to move away from one of them. In the meantime, a hydrogen bond is established between the two water monomers that were not connected previously, resulting in a hydrogen bond complex. The mechanism is complete upon the disintegration of the complex. The following figure (Fig. **5.13**) summarises the swapping mechanism.

Hydrogen bond dynamics also play a significant role on collective motions in liquid water. As a result, water molecules have different coordination numbers [145b]. These motions (translational, and hindered rotations alternatively known as librational motions), can occur simultaneously in different regions in bulk sample. Naturally, the rate of movements of water clusters varies from region to region. Among different modes of motions in clusters, vibration motions are prominent and can accurately be characterised by Normal Mode Analysis (NMA) [83]. In the case of water, three distinct modes can be traced from NMA: two accounting for the vibrational motions at 60 cm^{-1} (attributed to O...O...O bending) and 200 cm^{-1} (corresponding to O...O stretching motions), while the remaining accounts for librational motion at a higher frequency. The schematic diagram shown below (Fig. **5.14**) provides a general idea to the reader as to how water molecules move in bulk sample.

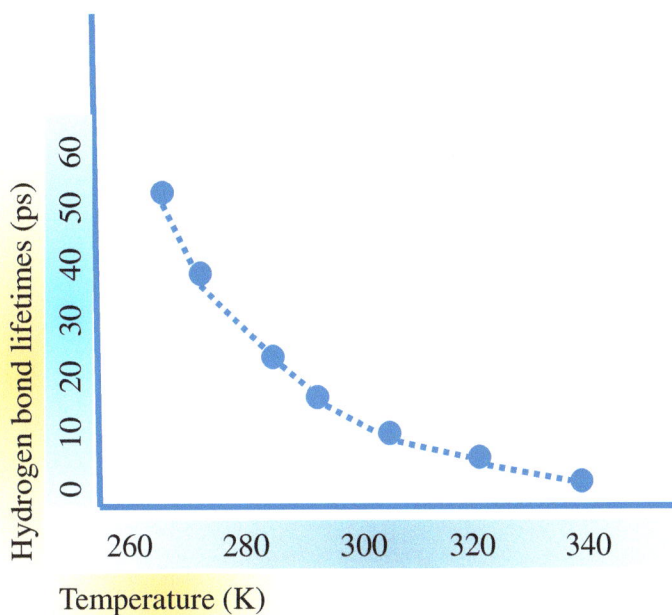

Fig. (5.12). Hydrogen bond life times *versus* temperature. Hydrogen bond lifetimes and temperature (in Kelvin) have been shown in y and x axis respectively. At higher temperatures beyond 340 K, the lifetimes are reduced to less than one sixth of that of at 260 K. The idea has been adapted from [139].

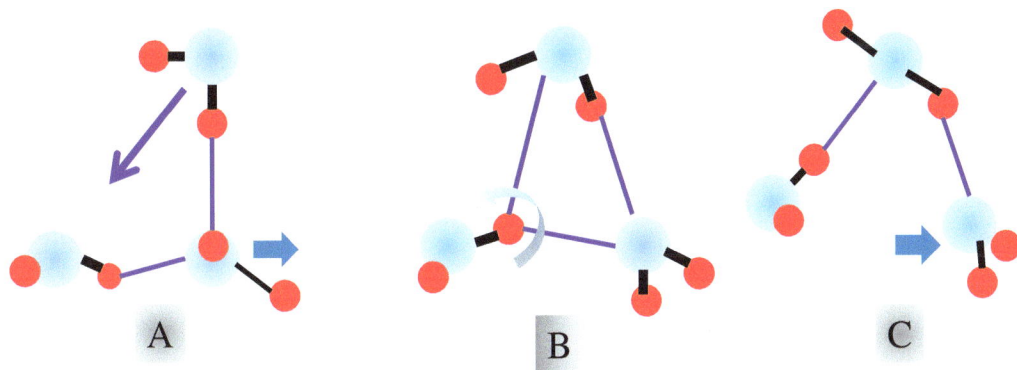

Fig. (5.13). Hydrogen bond switching mechanism. In the left (figure **A**) a water monomer (shown bottom right guided by an arrow) moves away from the water monomer placed its left, which is approached by another water molecule from above. The figure **B** shows the resulting hydrogen bond complex. As shown in the figure **C**, the mechanism is completed upon the swapping of the central water monomers. The idea has been adapted from [151].

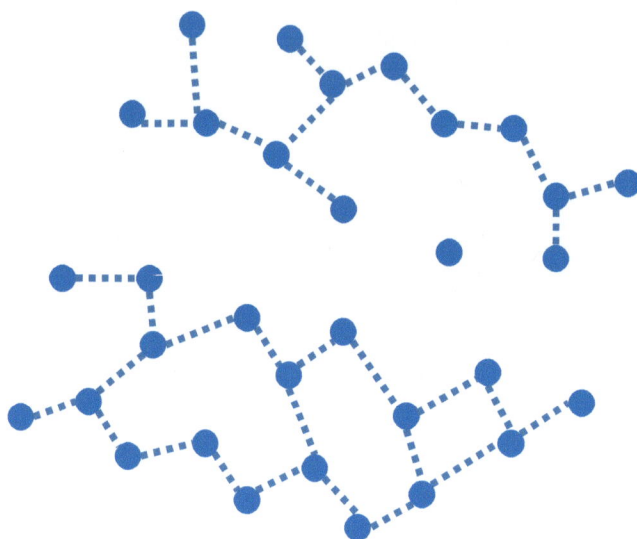

Fig. (5.14). Collective motion observed in liquid water. A snapshot of the movements of water molecules in bulk sample is shown in the figure. Water molecules form clusters and move together. The displacement of individual water monomers depends upon the size of water clusters. The blue circles represent oxygen atoms, and the hydrogen atoms are omitted for clarity. The idea has been adapted from [83].

The discussions on the dynamics of the hydrogen bonds and its impact on the collective motions in water are incomplete without mentioning the symmetry in energy profile of hydrogen bonds, which is based on the location of hydrogen atoms between the donor and acceptor oxygen atoms [87]. Hydrogen atoms in fact shuttle between the donor and acceptor atoms along hydrogen bonds. Based on the location of hydrogen bonds, we obtain different potential energy profiles as shown below, Fig. (**5.15**). When hydrogen atom is closer to one of the oxygen atoms the profile appears to be highly asymmetrical as you can see in the figure (left). Another possibility is that when hydrogen atom moves across (delocalised) the bond. In this case, the zero point energy (the lowest possible energy a system can have) is equal to the energy barrier height. The practically–little energy cost makes the movements of hydrogen atoms very frequent. The third case occurs when hydrogen atom is located exactly at the centre of oxygen–oxygen distance. In this case, the energy barrier is disappeared, as you can see in Fig. (**5.15**).

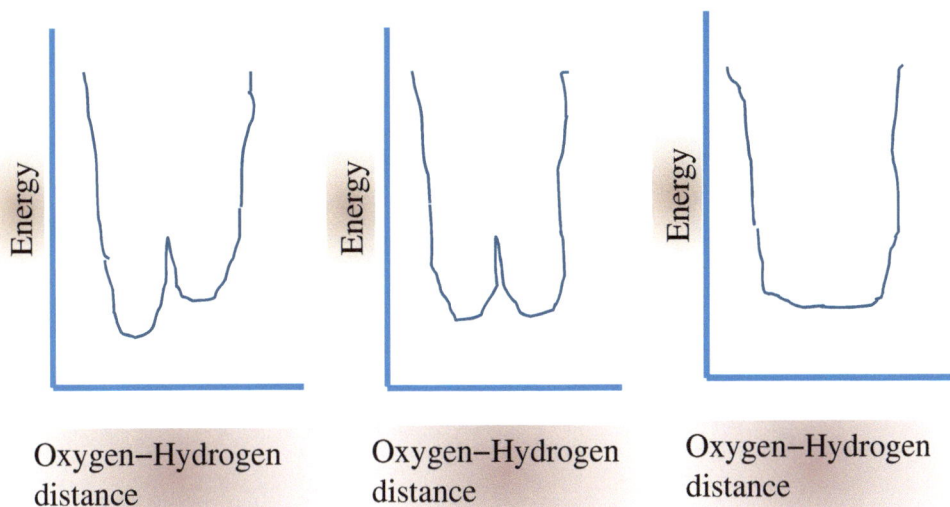

Fig. (5.15). Potential energy curves for hydrogen bonds demonstrated. The potential energy profile corresponding to the situation where hydrogen is located (denoted by oxygen–hydrogen distance in x axis) is shown in the diagram. The hydrogen atom is very close to one of the oxygen atoms (shown in the left). The potential energy profile for the state corresponding to the free movement of hydrogen atoms across the distance between the two oxygen atoms is shown in the central figure. The situation in which hydrogen atom is located in the middle of donor–acceptor distance is represented by the figure shown in the right. The idea has been adapted from [87].

Thermodynamics of Hydrogen Bonds in Water

It is interesting and important to discern the thermodynamic effects on the hydrogen bonds in water. Both pressure and temperature have profound impact upon the quantity and quality of hydrogen bonds in water. The variation of both of these thermodynamic variables offers structural heterogeneities in liquid water. In a fascinating computational study, Stanley *et al.* investigated the hydrogen bonding pattern in Walrafen pentamers, a water monomer surrounded by four other water molecules [152]. In Walrafen pentamers, the number of inter–pentamer hydrogen bonds with which two of such pentamers can be connected to each other is different, as shown in Fig. (**5.16**).

It was found that energetically favourable pentamer structures are those with zero and one intermolecular hydrogen bonds. This indicates that despite being closely packed, water molecules prefer not to have a homogenous structure; rather they like to be "free" without being part of a large extensive network.

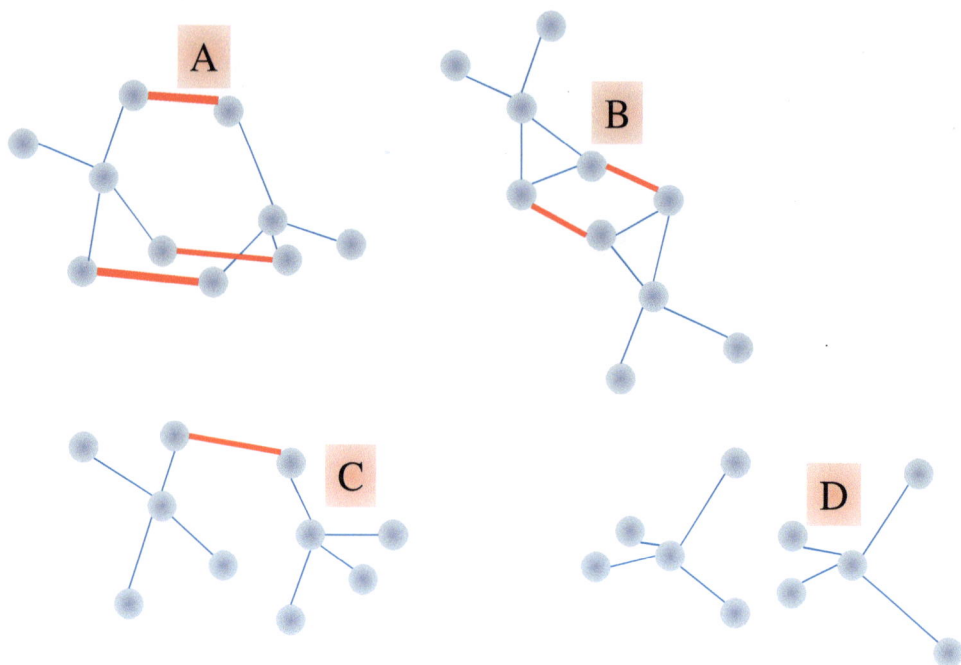

Fig. (5.16). Connectivity between two Walrafen pentamers. In figure **A,** two pentamers are connected by three hydrogen bonds (denoted by three red lines) to form a cage structure; in figure **B,** the pentamers are connected by two inter–molecular hydrogen bonds (shown as two red lines) to form a structure akin to chair form of cyclohexane. In structure **C**, there is only one inter–molecular hydrogen bond between the pentamers. Hydrogen bonding between pentamer structures is absent in structure **D**. These ideas have been adapted from [152].

Lagana *et al.* have investigated the effect of temperature, specifically in the range of 283 K to 348 K, on the hydrogen bond distribution in water using Raman spectroscopy and computer simulations [153]. Two peaks has been observed in the Raman spectra of water, pertaining to the collective vibrations of the hydrogen bonds in the three dimensional tetrahedral network at 3250 cm^{-1}, and non–collective reduced vibration of hydrogen bonds that are not part of a large network at 3400 cm^{-1}. The vibrations at the higher frequency can be attributed to the contributions from disordered non–tetrahedral clusters. The spectra at various temperatures reveal that the peak at lower frequency is diminished as temperature increases, whereas the peak at 3400 cm^{-1} is appeared to be more pronounced.

Order parameter analysis is another way of testing structure compactness. According to the modified version of original formulation put forward by Chau and Hardwick, order parameter for tetrahedron is 1 and for ideal gas, it is zero [153]. It has been shown by Lagana *et al.* employing SPC/E water model that the tetrahedral ordering decreases upon an increase in temperature over the interval from 273 K to 373 K, and more and more tetrahedral structures are converted to more disordered states. These results in principle support the findings of Wernet *et al.* suggesting the coexistence of different types of structures at a time in water [145a].

Parinello *et al.* has investigated computationally the impact of temperature upon the stability of the local structure of water, as summarised in the following figure [139]. As temperature increases from 274 K to 294 K, three major differences can be noted. Firstly the lifetimes of hydrogen bonds increase gradually as temperature decreases up to 280 K, then increases sharply at around 260 K. This has been correlated with anomalous properties of water (for a lengthy discussion on the anomalous properties of water, please see chapter 9). Of the two models Raiteri *et al.* investigated, this effect is more profound with simulations using TIP5P model than in SPC model; secondly, the lifetimes of the more stable tetrahedral coordinated water cluster increase slightly at lower temperature and thirdly, as temperature decreases, fraction of water molecules with higher lifetimes increases. At this juncture, however, we have to note that several authors have demonstrated experimentally the role of density than temperature in determining the number of hydrogen bonds in water.

We have just discussed about the "invisible" microstructural dynamics of water, which spur on "visible" macroscopic transformation of this liquid over a wide range of temperatures. Beyond its boiling temperature, water becomes superheated (chapter 8 exclusively devoted to this subject). On the other hand, below its melting temperatures, water first becomes supercooled state (discussed in the following chapter), a metastable state which leads to more stable form of water at low temperatures, ice (detail account of this is given in Chapter 7). The following figure (Fig. **5.17**) summarises these changes.

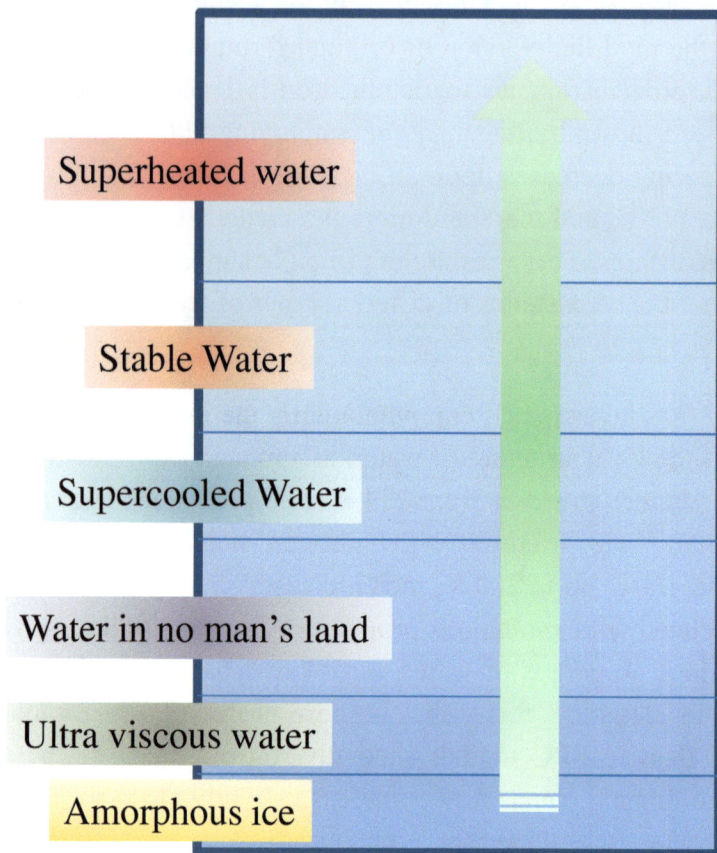

Fig. (5.17). Water across wide range of temperatures. The morphological changes in water observed across a wide range of temperature are shown in the figure. Water exists as liquid under normal conditions, from 273 K to 373 K. Beyond its boiling temperature, water can be transformed into superheated state. Water can attain supercooled state until 232 Kelvin, below which water it normally forms crystalline ice. Ice in amorphous forms can exist below 150 K with varying pressures.

The variation of other thermodynamic variables such as pressure and density and its implications on transformations of water will be discussed elsewhere, in the chapters covering other phases of water.

CONCLUDING REMARKS

The properties of water depend heavily on the network among water molecules *via* hydrogen bonding. Several theories on this mysterious nature of the network have been proposed over the years in order to comprehend water's enigmatic properties, most notably two classes of theories: Uniformist models and Mixture models. Uniformist models advocate the existence of an extensive three-dimensional water network, and Mixture models propose the coexistence of chainsand rings of water of varying magnitude along with the three dimensional tetrahedral network. It can be seen that the latter has emerged to mount a serious challenge to the Uniformist model. Among the models proposed for the understanding of water structure stands out Chaplin's icosahedral water model. Monumental experimental and theoretical (through computer simulations) efforts are required to test the validity of this model, due to its unique shape and large number of water molecules participating in the network. Although it might be cumbersome to detect such a gigantic structure, computational investigations on its stability can be verified using both classical and quantum models. Hydrogen bond complexes such as Bifurcated Hydrogen Bonds (BHB) have been suggested to explain the structural reorientation of water networks that leads to important physical processes such as proton transfer. Normal water undergoes remarkable transformation upon changes in thermodynamic variables such as temperature. Most notably lower temperature physics and chemistry of water is interesting, which will be discussed in the following two chapters.

CONFLICT OF INTEREST

The author confirms that he has no conflict of interest to declare for this publication.

ACKNOWLEDGEMENTS

Declared none.

CHAPTER 6

Supercooled & Glassy Water

Abstract: The roles of two low temperature and non−crystalline forms of water, (supercooled and glassy water) are very pivotal in supporting the existence of several microorganisms below 0°C, although they are very metastable with respect to the stable crystalline form of water, ice. In the supercooled regime, the hydrogen bond lifetime of a single hydrogen bond and water clusters are found to be significantly higher than in higher temperatures. Diffusion coefficient and configurational entropy show a distinct maximum at density 1.15g/cm^3. Two inter−convertible forms of supercooled water, known as Low Density Liquid (LDL) and High Density Liquid (HDL), are found to coexist at temperatures below the freezing point of water. If water is cooled at very fast rate, it becomes glassy, the most profound form of water in the universe, bypassing the formation of ice. Polyamorphism is one of the characteristics observed in glassy water. Glass transition temperature in water has sparked debate in the scientific community. Different experimental procedures as well as water models produced varying values for the glass transition temperatures in water. It has been experimentally monitored and computationally simulated the transition between the two glassy phases of water, HDA and LDA. The transition is terminated at a critical point, according to Liquid−Liquid Critical Point (LLCP) theory. The concept of strong and fragile glasses is very powerful tool in furthering our understanding of the dynamics of glassy materials. It is interesting to note that a transition from strong to fragile occurs in water.

Keywords: Aerodynamics, Aviation, Crystallisation, Desterilisation, Diffusion, Glass transition, HDA, HDL, Inherent structure, LDA, LDL, Nucleation temperature, Polyamorphism, SANS.

INTRODUCTION

In addition to its three well known, standard and stable states (solid, liquid and gas) water also exists as supercooled and glassy (two of its known metastable forms) which are vital ingredients in sustaining life to numerous microorganisms at low temperatures. Investigation of supercooled water is gaining attention among the scientific community mainly due to the following reasons: firstly, water's anomalies are more pronounced in supercooled regime, and therefore a detail understanding of supercooled water can shed more light on its elusive behaviour in this temperature range, and secondly, supercooled water impacts on certain technologies including pharmaceutical and food industries, and cryopreservation, the technique of preserving organs and other biomaterials in low temperature (a discussion of various aspects encompassing cryopreservation can be found in [37]). Further, supercooled water instils a growing interest among aeronautical engineers due to its pivotal role in aviation industry which transfers millions of people annually from place to place worldwide. The growth of supercooled water droplets outside the aeroplane results in the formation of larger ice crystals (as shown in (Fig. **6.1**)), known as ice accretion, which causes numerous malfunctions in aircraft engine and the aircraft itself [154]. These include increase in the weight of aircraft resulting in altering the aerodynamics of side and rear, malfunctioning of landing gear and communication systems, decreasing flight lift, increase in propeller vibrations, making errors in instruments which provide vital information about aircraft such as air speed, altitude, and vertical speed, and reduction in visibility. The problems related to the formation of ice on aircrafts are severe in winter than in summer. It is interesting to note the fact that it is supercooled water in the subzero conditions, not the tiny ice crystals present in clouds, that causes the aforementioned disastrous effects.

The first part of this chapter examines the structure of supercooled water, which is followed by the important methods of its preparation. The discussion ends with a reference to diffusive motions of H_2O molecules. In the second part of this chapter, a detail discussion on glassy water is provided. This includes various glassy forms of water, experiments on the transition between the two known phases of glassy water, and glass transition temperature. The chapter is concluded with an important property that occurs in water at the glass transition temperature, Fragile Strong Crossover (FSC).

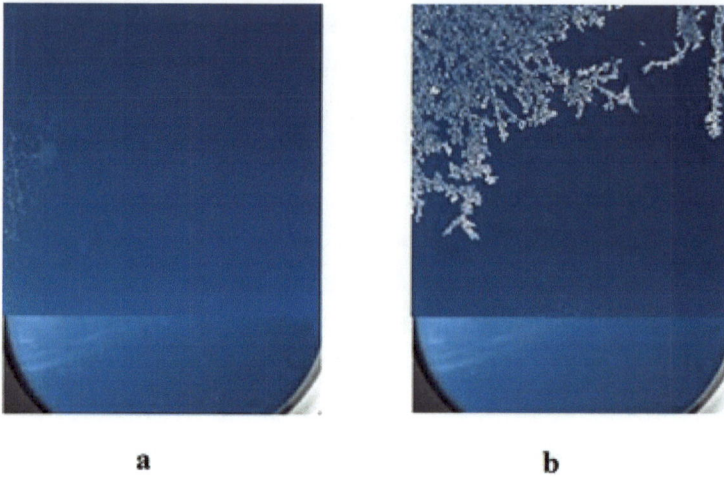

a   **b**

Fig. (6.1). Ice formation on aircraft body. Ice formation on the aircraft can result in serious malfunctions during the flight. Aircraft window without (left) and (right) with the formation of ice (picture trimmed for clarity).

STRUCTURE OF SUPERCOOLED WATER

Water forms a strong tetrahedral network between the nucleation temperature, T_H (the temperature at which nucleation of water molecules occurs) and the melting temperature, T_m [155]. Computer simulations carried out by Stanley *et al.* reveal the fundamental differences in the structure of water at lower temperatures and higher temperatures [156]. Sciortino *et al.* have investigated the relationship between the fraction of molecules and coordination numbers employing traditional Molecular Dynamics and Inherent Structure (IS) simulations. Inherent Simulation techniques are based on Energy Landscape Theory (ELT), discussed in chapter 2. In IS simulations, the structure corresponds to the local potential energy minimum can be accurately mapped. They noted that as the temperature approached 210 K, a large number of water molecules attain a stable four –coordinated state (Fig. **6.2**). On the contrary, within the temperature range of 300K and 700 K, there is little difference in the fraction of tetragonal coordinated water molecules.

It is very difficult to discern the impact of the density upon the number of neighbours around a water molecule from these simulations. However, it appears that as the density increases, more water molecules attain four–coordinated state in all temperatures investigated. A diligent reader may remember the discussion about the temperature dependence on tetrahedral ordering and the life times of hydrogen bonding, discussed in chapter 5.

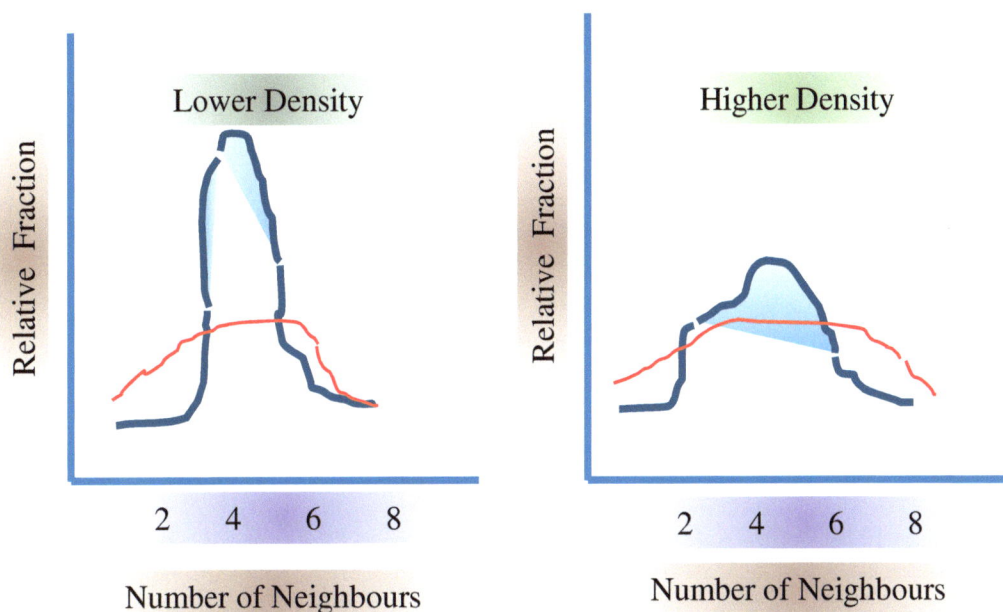

Fig. (6.2). Relative fraction of water molecules (at lower density (left) and higher density (right)) with different number of neighbours is shown in the schematic diagram. The statistics shown here is based on traditional Molecular Dynamics (MD) simulations using SPC/E model. As temperature decreases to 210 K (shown in black), number of water molecules with four neighbours is found to be predominant in water regardless of the density. Redlines correspond to 300 K.

The structuring of water is indicated by the sharpening of the first and second peaks in the pair distribution functions derived from Inherent Structure (IS) simulations (see Fig. **6.3** for a schematic diagram). In the supercooled water (at 210 K) it can be seen that the intensity of the peak around 3.2 Å increases from the lowest density to the highest density (1.40 g/cm^3) simulated [156]. This has been seen as the signature of distorted hydrogen bonds in the system. The peaks around 2.8 and 4.5 Å signify tetrahedral network as explained elsewhere in the book. It is not difficult to imagine the situation of distortion of perfect tetrahedral structure upon hike in density, as the perfectly ordered hydrogen bonds are crushed in order to make up for reduced volume.

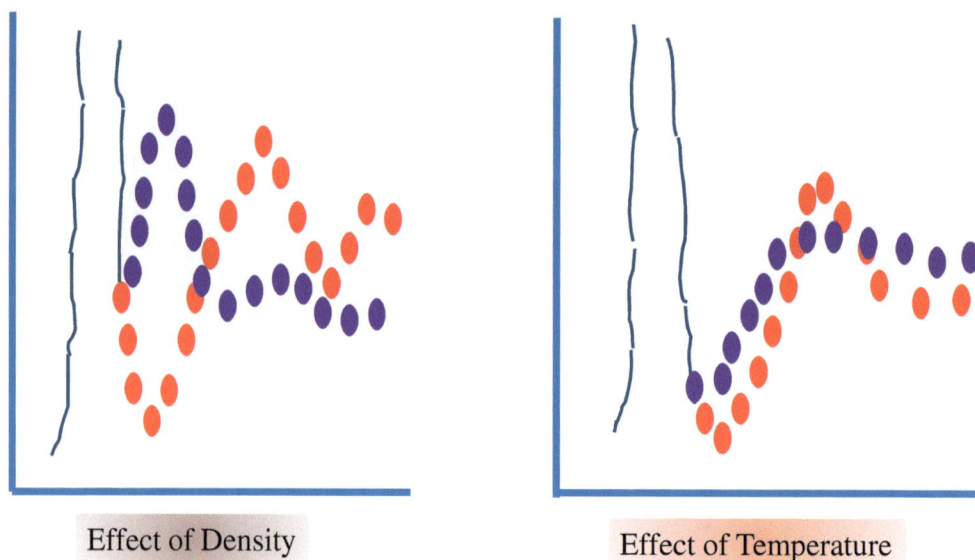

Effect of Density   Effect of Temperature

Fig. (6.3). Oxygen–oxygen radial distribution functions obtained for water (from Inherent Structure simulations) for various densities (left at 210 K) and temperatures (right at 1.00 g/cm³). Curves with violet dots represent rdfs at the highest density (1.4 g/cm³) and temperatures (700 K) simulated. The curves with red dots represent the lowest densities (0.95 g/cm³) and temperatures (210 K) simulated. As the temperature decreases, the peak gets shaper (right), while the decrease in density (left panel) corresponds to the strengthening of second and third peaks. The idea has been adapted from [156].

At low pressures and temperatures, in the region between the nucleation temperature, T_H, and freezing temperature, 0°C, water takes open structure. This structure is characterised by ideal tetrahedral angle (the angle between three neighbouring oxygen atoms), 109°.47′ [155]. What is the effect of an increase in pressure upon the structure of cold water under 273 K? One can quickly note that the open tetrahedral structures in water begin to collapse. First, the second and third shells around the central water molecule are broken, prompting an increase in the number of water molecules in the first shell. The consequence of this is an increase of density around the central water molecule [155]. From the computer simulations performed by Stanley *et al.* using ST2 water model, that the number of water molecules in the first hydration shell exhibit a sharp increase from 3.9 at 0.80 g/cm³ to 4.6 at 1.00 g/cm³ [142]. In addition, the presence of extra water molecules (generating an interesting bonding scenario called Bifurcated Hydrogen Bonds (BHB)) around the central water molecule (in first coordination shell) lowers the barriers for translational and rotational motions [142]. The reader may note that a discussion on bifurcated bonds was presented in the previous chapter. The increase in the number of water molecules in the first hydration shell has a major role in altering dynamics of water in low temperatures, since hydrogen bond energy is higher than the thermal energy in low temperatures, which promotes the restructuring of hydrogen bond network [142].

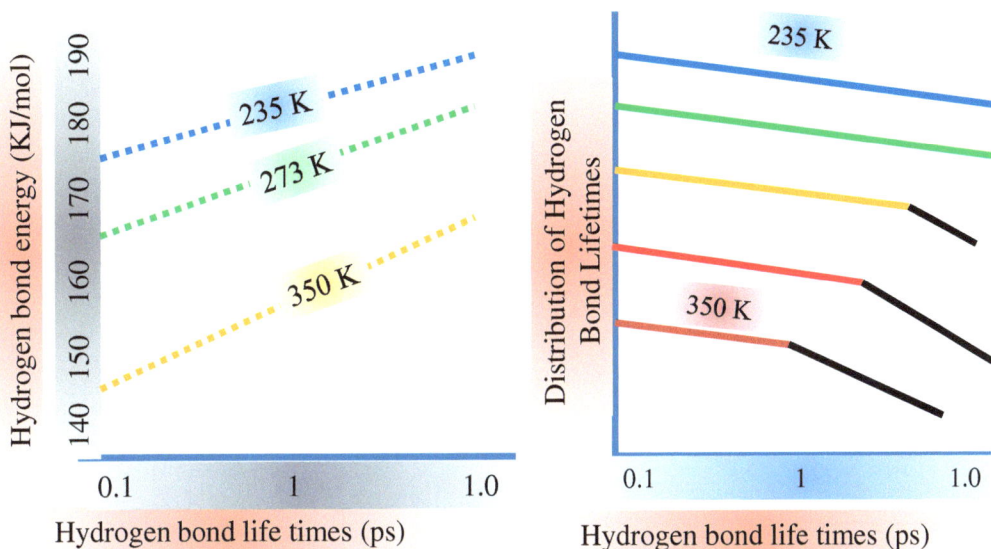

Fig. (6.4). Hydrogen bond lifetimes *versus* average energy (left), and Hydrogen bond lifetime *versus* hydrogen bond lifetime distribution (right). Please note the black lines which get longer as temperature increases in the right panel. The idea has been adapted from [157].

Computer simulation by Havlin *et al.* also indicate higher hydrogen bond life times in supercooled water as compared to normal and superheated water [157]. This is in good agreement with the computational findings of Raiteri *et al.*, discussed in the previous chapter [139]. The average energy of hydrogen bonds between two water molecules shows a higher value for supercooled water in comparison with water at higher temperatures. In the figure shown below (Fig. **6.4**), we note that hydrogen bond lifetime and strength of the hydrogen bonds increases as temperature decreases. It can also be observed that as temperature increases, the slope of average energy increases. On the other hand, the distribution of hydrogen bond lifetimes suggests two different characteristics of lifetimes. It can clearly be seen that as temperature decreases only a straight line, suggestive of slower decay, is observed (see Fig. (**6.4**), right). On the contrary, at higher temperatures, a different pattern, indicative of faster decay (as shown by black lines), emerges.

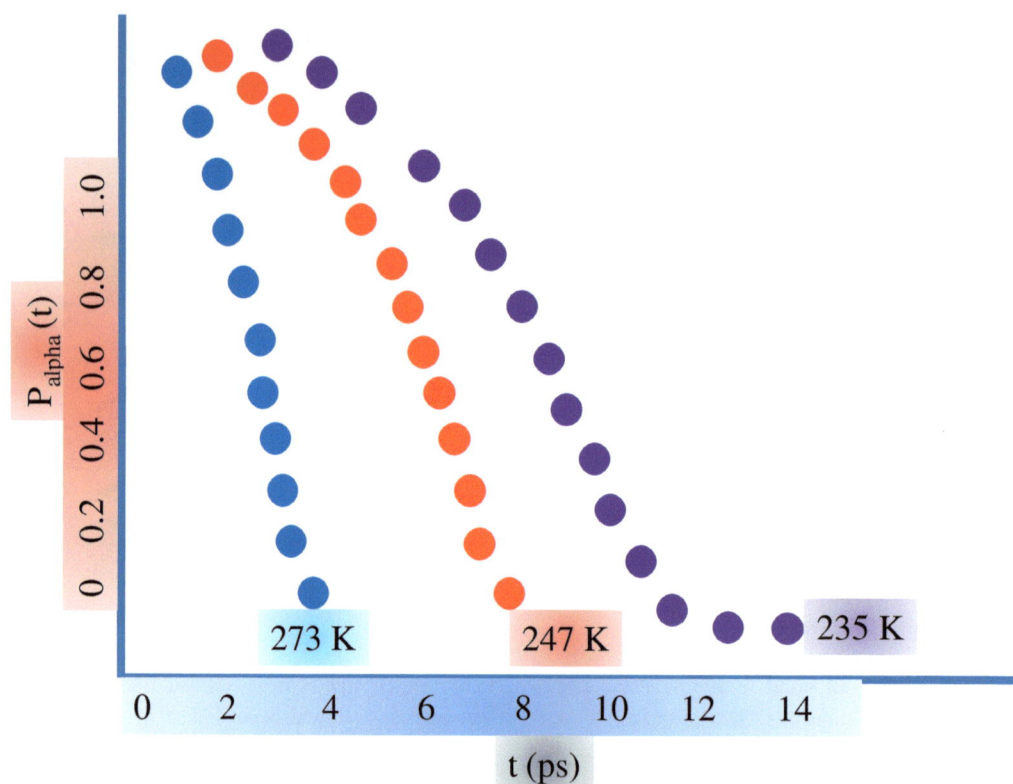

Fig. (6.5). Hydrogen bond lifetime distribution over wide range of temperatures. As temperature decreases, more number of water clusters can have higher hydrogen bond life times, and distribution of water clusters with different lifetimes increases. The idea has been adapted from [157].

Why does this two–step decay occur? This is primarily attributed to the size of water clusters which vary across the temperature domain investigated. Water molecules move preferentially in clusters in the supercooled regime, and these clusters tend to grow in size and are found to be very compact as temperature decreases [158]. As stated before, the number of water networks with higher life times increases as temperature decreases. At the freezing temperature of water (273 K), one can hardly find water cluster with lifetimes higher than 3 ps. On the contrary, at 235 K water clusters with lifetimes less than 3 ps are found to be rare, and at least 10% of the water clusters formed at this temperature have the lifetimes as high as 14 ps. It is interesting to note that at as opposed to higher temperatures (beyond 273 K), at which the hydrogen bond lifetimes converge to a single number, in the supercooled regime one can observe water clusters with assorted lifetimes, as shown in Fig. (**6.5**).

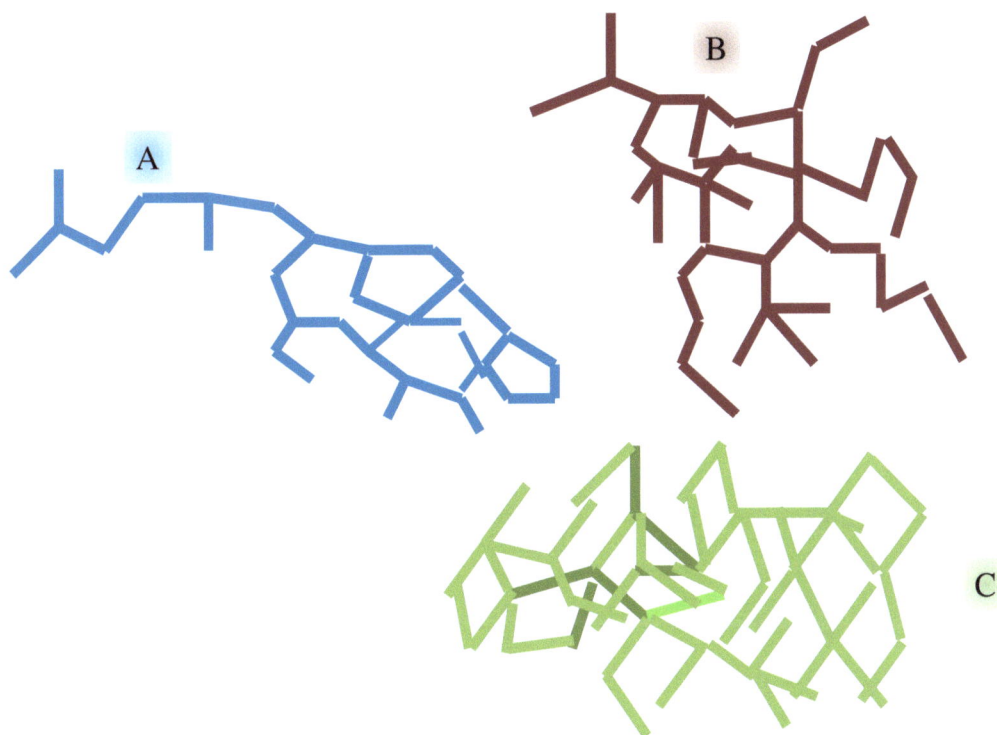

Fig. (6.6). Clustering in supercooled water. Schematic diagram of clusters with varying sizes that are formed in supercooled water at 200K: As the number of participating molecules increases (**A** through **C**), the cluster takes a three dimensional form. The idea has been adapted from [159].

By comparison with normal liquid water, a larger proportion of the water molecules in the supercooled liquid are appeared to be more ordered, and there is a marked difference in size and shape of clusters found in supercooled water. Clusters in the supercooled water are temperature sensitive both in terms of their structure and local ordering [159]. They are found to be of varying sizes from 2 to 300 molecules per cluster, with smaller ones taking a chain–like and larger ones adopting a three dimensional spherical shapes respectively. It has been suggested that several anomalous thermodynamic properties such as singularities in isobaric specific heat are observed in the supercooled water due to the clustering of H_2O molecules [160]. A schematic representation of the supercooled water clusters with varying number of water molecules is shown in figure, Fig. (**6.6**).

WATER BELOW SUPERCOOLING TEMPERATURE

Water below its supercooling temperature is suggested to exist as two inter−convertible forms: Low Density Liquid (LDL), where in tetrahedral order of water is preserved, and High Density Liquid (HDL) with the disappearance of such order [155, 161]. This transition was first observed in a computer simulation performed by Stanley *et al.* using two different water models, ST2 and TIP4P [155]. This is the basis of Liquid − Liquid Critical Point Theory (LLCT), developed for explaining several anomalous properties observed in supercooled water [162] (a detail discussion on this theory, and a comparison with competing theories can be found in chapter 9). These two phases can be distinguished by a threshold coordination number: LDL has coordination number less than 4.4, while HDL has coordination number above 4.4 [163]. In the following figure, Fig. (**6.7**), the difference in molecular arrangement between high density and low density water is shown.

Based on computer simulations, it has been found that there occur two types of Liquid−Liquid transitions in supercooled liquid: a transition from Low Density Liquid (LDL) to High Density Liquid (HDL) at around 210 K and 1.025 g/cm^3, and a High Density Liquid (HDL) − Very Low Density Liquid (VHDL) transition at 165 K and 1.3 g/cm^3. The higher density transition (HDL to VHDL) is accompanied by large changes in coordination number (seven to eight water molecules around the central water). Using robust water models such as TIP4P−Ew, these transformations have been extensively investigated, as shown in the following diagram, Fig. (**6.8**), and these transformations have been identified with two critical regions (denoted by C* and C** in Fig. (**6.8**)), one at lower density (~1.05 g/cm^3)and other is at higher density(~1.3 g/cm^3).

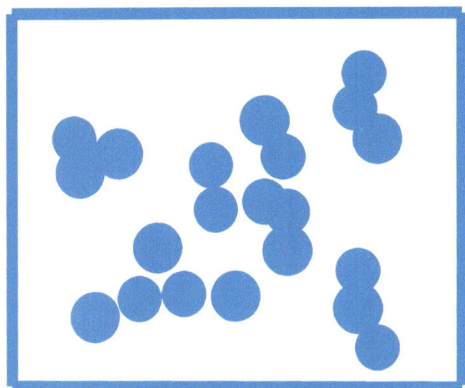 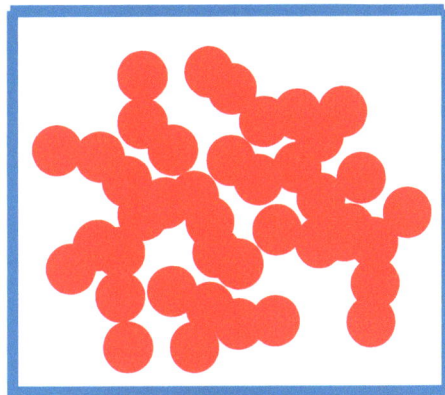

Low Density   High Density

Fig. (6.7). A Schematic diagram of the molecular arrangement in low–density (left) and high–density (right) phases of water. Particles clustered together in the high–density form resulting in coordination number higher than four, while in the low density form particles are positioned far than its higher density analogue, resulting in fewer coordination numbers. The idea has been adopted from [164].

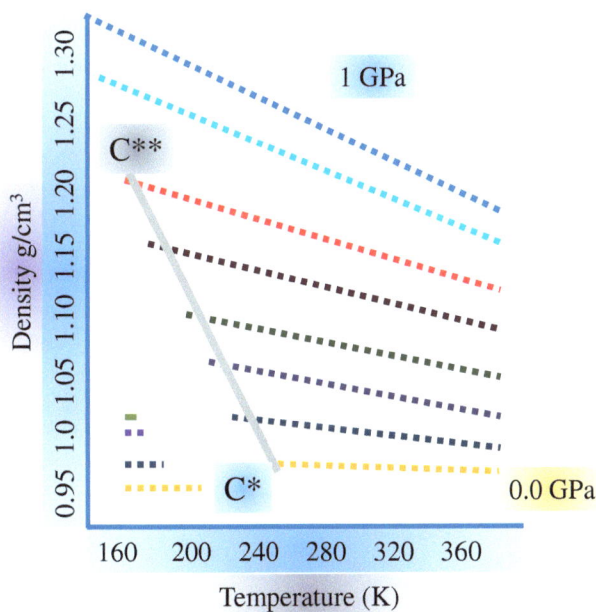

Fig. (6.8). High Density to Low Density liquid transformation in supercooled water. Isobars obtained using TIP4P-Ew water model. C* and C** are critical temperatures corresponding to LDL – HDL and HDL–VHDL transitions respectively. The idea has been adapted from [165].

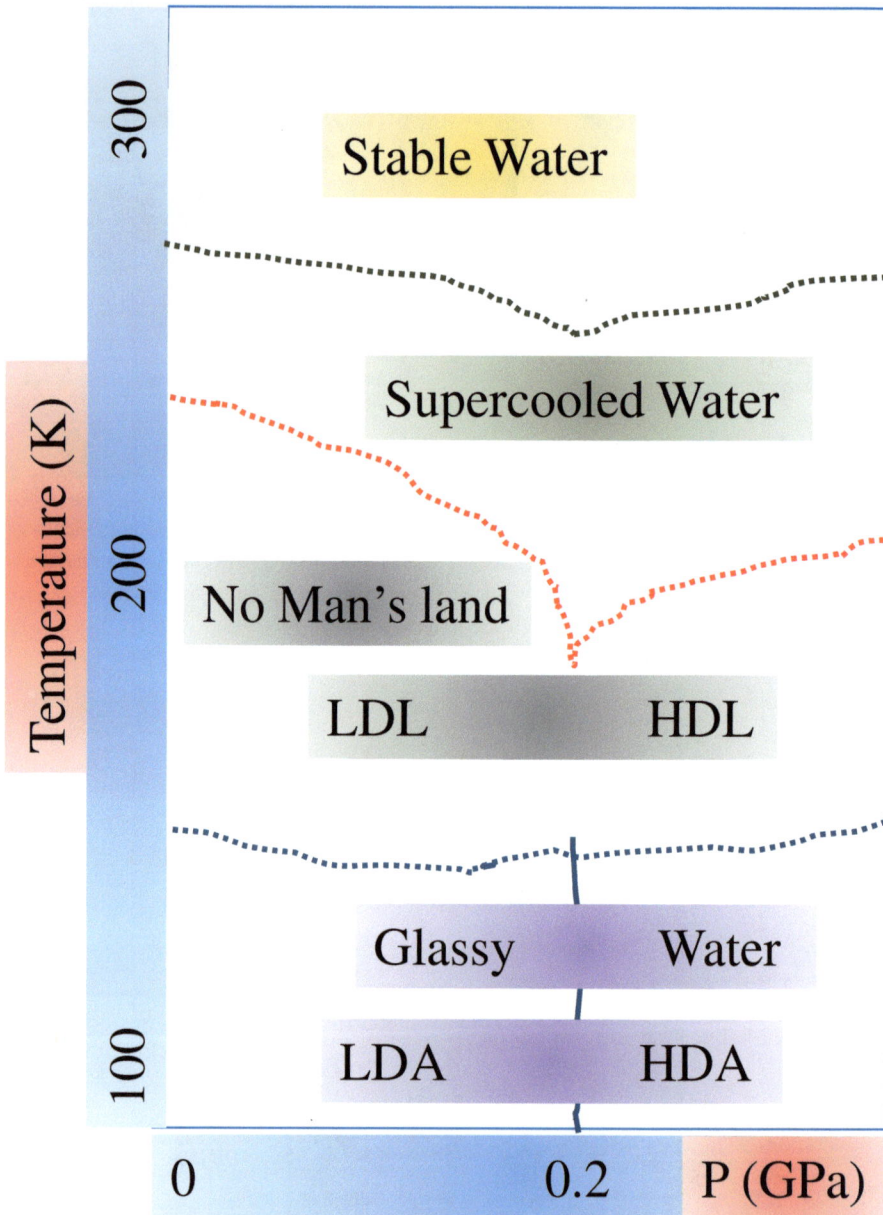

Fig. (6.9). Hypothetical phase diagram of water in low temperature. Water can supercooled to even under 200K, well below the stable region of normal water. Between supercooled and glassy regions, water does not exist as liquid. Low Density Liquid (LDL) and High Density Liquid (LDL) are amorphous phases of solid water. The temperature (y axis) and pressure (plotted on x axis) have not been scaled. The idea has been adapted from [166].

In Fig. (**6.9**), an 'approximate' phase diagram of water depicting two phases of supercooled water (LDL and HDL) have been shown. It is important to note from the figure that water can be supercooled even below nucleation temperature, T_H, down to beyond 200 K with appropriate pressure. It is largely controversial to assign the nature of water in the region between supercooled and glassy state (about which more will be said), and hence the region is known as "No Man's land", wherein performing experiments is impossible by available techniques today.

How Diffusive Supercooled Water is?

Stanley *et al.* have extensively studied the density dependence of diffusion coefficient in supercooled water, using SPC/E water model. Diffusion coefficients of water molecules in supercooled state are highly density dependent, recording the maximum when density equals to 1.15 g/cm³ [167]. Two competitive effects that directly affect the movements of particles operate in supercooled water: disruption of hydrogen bond network and increased packing. The former occurs at higher density that results in freely moving water molecules. Increased packing is also the net effect of hike in density. At lower density, particles in general, can adopt many different configurations. At extremely high density, on the other hand, a constraint in the extent of configuration space that water molecules can cover is imposed. This means that when density increases beyond a threshold value, the number of different conformations that a molecule can have (configurational contribution to entropy, $S_{conf,}$ which is the difference between total entropy and vibrational entropy) is further reduced significantly. This is reflected in the reduction in mobility, evident by a decrease in diffusion coefficient [168]. The following figure, Fig. (**6.10**) demonstrates the relationship between diffusion coefficients with density in supercooled water. Although temperature does influence the maximum is diffusion coefficient only to a small extent, it does affect the shape of the profile very clearly: at lower temperature (supercooling range) the curve becomes more sharp with a clear diffusion coefficient maximum emerges at 1.15 g/cm³.

Diffusion in supercooled water has been suggested to follow two distinct mechanisms: homogenous and heterogeneous [169]. In homogenous diffusion process, the relaxation of molecules in the sample is identical, whereas in heterogeneous diffusion process, mobile molecules have much greater tendency towards the formation of clusters and aggregates [169]. The following figure (Fig. **6.11**) demonstrates the difference between these two mechanisms.

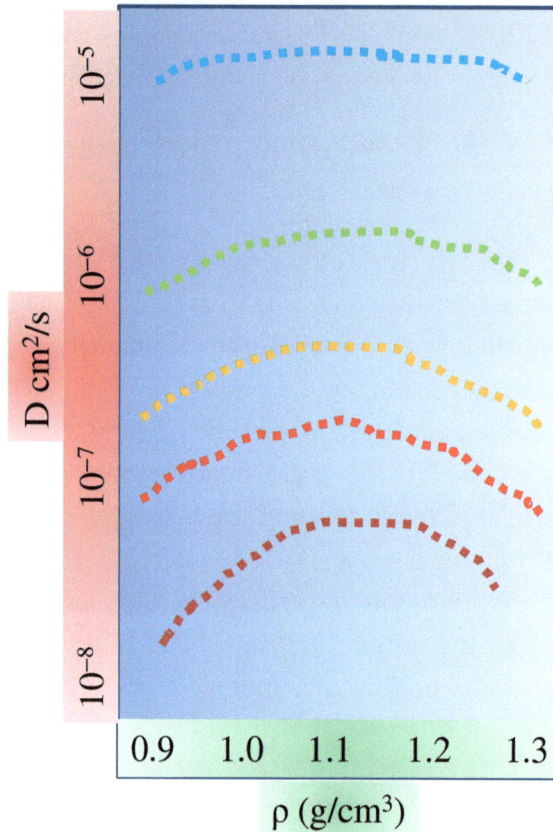

Fig. (6.10). Diffusion coefficient *versus* density in supercooled water. At 1.15 g/cm³ the maximum value is recorded for diffusion coefficient. This effect is more prominent in lower temperatures. The five temperatures recorded are 250, 240, 230, 220 and 210 K from top to bottom. The figure has been modified from [167].

It is clear from the Fig. (**6.11**) that heterogeneous diffusion process is multi−step process wherein the movements of individual water molecules is accompanied by the expansion of clusters they belong to. In the light of Continuum−Mixture water model controversy, one can clearly see that homogeneous diffusion aptly represents the Continuum model, whereas the Mixture model is supported by the heterogeneous diffusion.

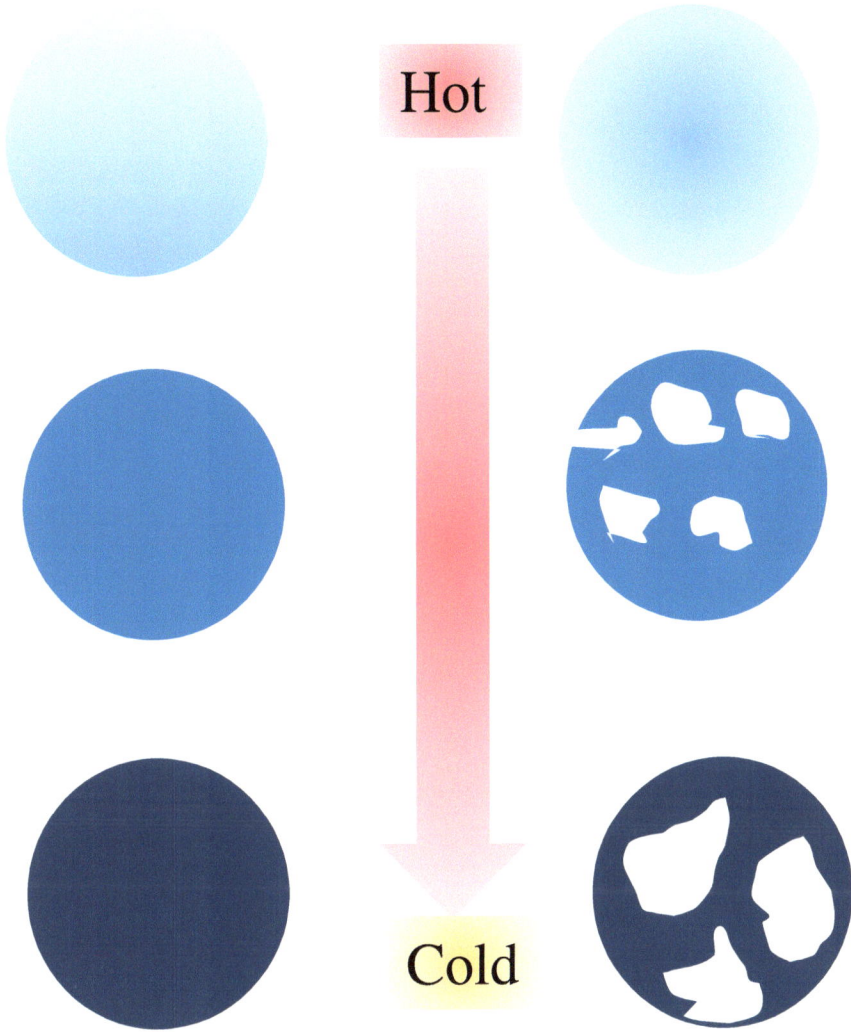

Fig. (6.11). Homogenous and heterogeneous diffusion in supercooled liquids. On the left shown is homogenous diffusion process in which molecules relax uniformly throughout the whole sample. On the right shown is heterogeneous diffusion process, in which molecules undergo diffusion at different rates. Some of the molecules are more mobile than the other, which results in the clustering. As temperatures decrease, the size of these clusters increase.

Relaxation Times

The notion of strong and fragile in condensed systems such as water can aptly be described by Arrhenius equation, as explained in one of the earlier chapters. As you can see in Fig. (**6.12**), water does deviate from the strong characteristics

(Arrhenius behaviour as shown by the straight line in red) [170]. In the supercooled regime, water follows non−Arrhenius behaviour (fragile).

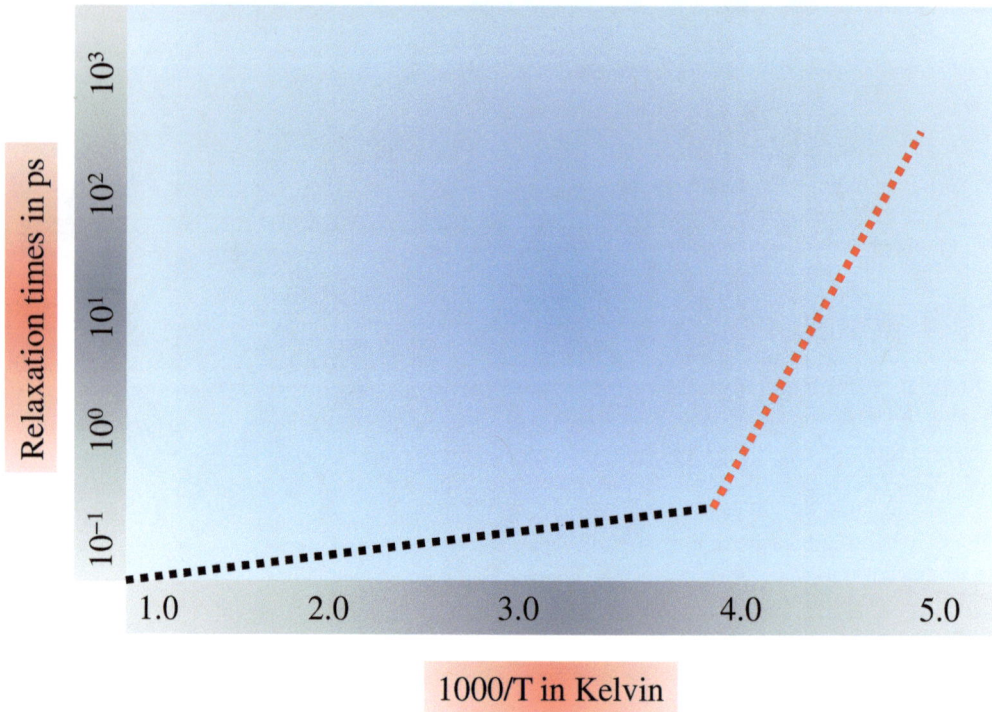

Fig. (6.12). Relaxation times *versus* Temperature in liquid water. Changes in relaxation times upon variation in temperatures (scaled in x axis) is shown in the plot. The black dotted line denotes Arrhenius behaviour, which at lower temperatures is violated as shown by red dotted lines. The idea has been adapted from [170].

HOW SUPERCOOLING CAN BE ACHIEVED?

Bulk water can be supercooled to 242 Kelvin bypassing the formation of ice, but the sample must be 'CLEAN', otherwise it will result in immediate crystallisation. This temperature is the natural limit to which what can be supercooled, and is known as homogeneous nucleation temperature of water, T_H. Good glass formers (for example B_2O_3) resist crystallisation to a greater extent and undergo supercooling. On the contrary, in water, regardless of the state of being pure or impure, its crystalline form, ice, is formed spontaneously below the nucleation temperature. This is exemplified by the fact that with mere presence of a steel rod in the container wherein supercooled water is kept, it can quickly be transformed to ice, without the need of the rod being inside the sample [171]. This situation makes water a "bad" glass former [110, 172]. Nevertheless the imminent crystallisation can be delayed or altogether avoided if several measures are taken. Interestingly the sample of water forming to ice can be delayed by keeping the sample still or its opposite process, constant agitation. It has been reported that water could be supercooled to 254 K upon vigorous stirring [171]. More effectively, crystallisation of water can be deferred to a lower temperature by reducing the concentration of impurities that cause the formation of ice, by dividing the sample into very tiny droplets [173]. Since the concentration of impurities so small in these droplets, rate of crystallisation will be negligible such that critical mass for the ice formation can never be achieved. Successful laboratory preparation of supercooled water up to 201 Kelvin has been reported by Rau using the same principle, but this work has largely been forgotten by contemporary researchers [174]. This method is a variant of Desterilisation technique, in which several cycles of freezing and melting of a particular water droplet is performed, preventing ice nucleation. Continuing the process for other droplets achieves a greater level of supercooling even up to temperature as low as 200 Kelvin.

In research laboratories, however, two alternative means are employed in order to avoid the formation of ice: by applying huge pressure of the order of five & employing high rate of cooling. Invoking these conditions, supercooled water are efficiently converted to glass [175]. Notable experimental methods either "avoiding" or delaying crystallisation includes condensation of water vapour in supersonic flow through small nozzles, and expansion of D_2O with nitrogen in a Laval nozzle. Spectroscopic investigations on supercooled water have recently been appeared in the literature [101].

GLASSY WATER

There are several methods by which glass can be formed, including Chemical Vapour Decomposition (CVD), cold decompression of high pressure stable crystal, cold electrochemical decomposition, polymerisation reactions in liquid phase, heavy particle bombardment of crystals, chemical reactions such as hydrolysis followed by drying and reactions in gaseous phase and ensuing condensation [172].

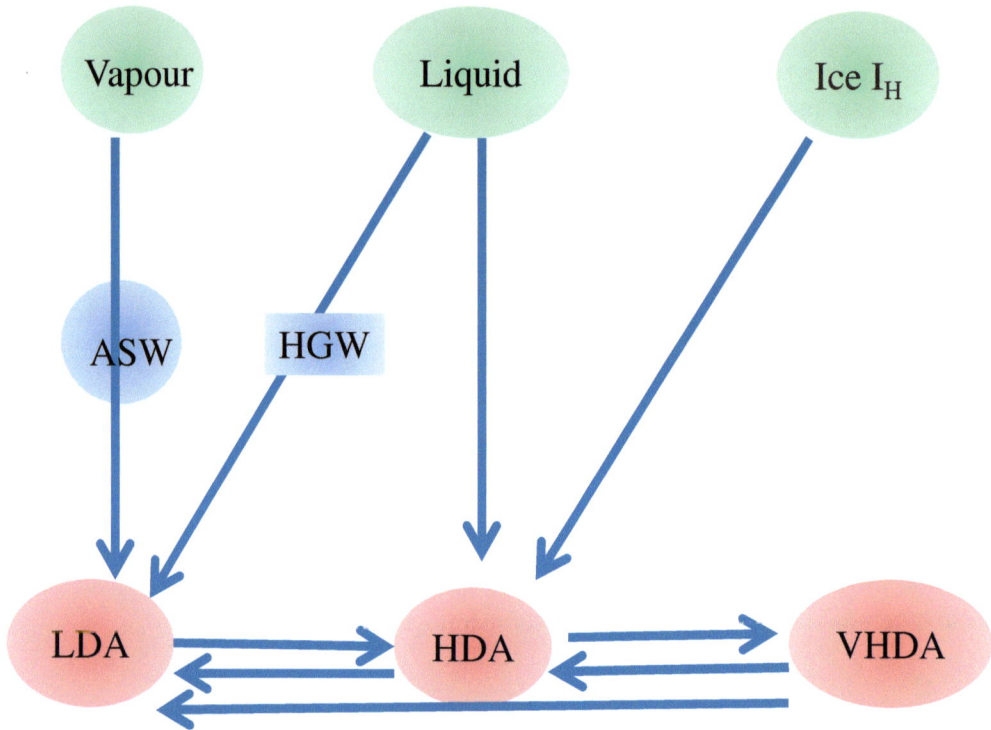

Fig. (6.13). Various routes to the preparation of glassy water. Low Density Amorphous (LDA), which can be directly obtained from vapour through Amorphous Solid Water (ASW), liquid through Hyperquenched Glassy Water (HGW), and Very High Density Amorphous (VHDA) *via* polyamorphic transition through High Density Amorphous (HDA) ice. The idea has been adapted from [173].

It has been theoretically predicted that glassy water can exist in four different phases: namely LDA, Normal Density Water (NDW) analogue, HDA, and VHDA [176]. When water is subjected to high pressure at low temperature, it transforms into a form of glassy water, known as Low Density Amorphous (LDA)(considered to be plentiful in comets), which when subjected to further pressure while lowering the temperature is transformed to High Density Amorphous (HDA) [110]. But this higher density form, HDA, was prepared first from I$_h$ ice by compression and bombarding with electron beam [177]. LDA can also be produced by several other means, for example, low−vapour deposition onto a cold target, rapid quenching of liquid at the rate of 10^6 Kelvin per second (known as hyper quenching) or heating of HDA [82]. The terms glassy water and amorphous ice are found to be used in tandem in the scientific literature [178]. Various ways by which three forms of glassy water (LDA, HDA & VHDA) can be produced as in the scheme shown above, Fig. (**6.13**). Hyperquenched Glassy Water (HGW) and Amorphous Solid Water (ASW) are the two intermediate forms while liquid and vapour are converted to Low Density Amorphous (LDA) water. However, the difference between LDA, HGW and ASW is so subtle such that they can be considered as same "substance" [173]. On the other hand, the experimental confirmation for the analogue of Normal Density Water (NDW) is in waiting.

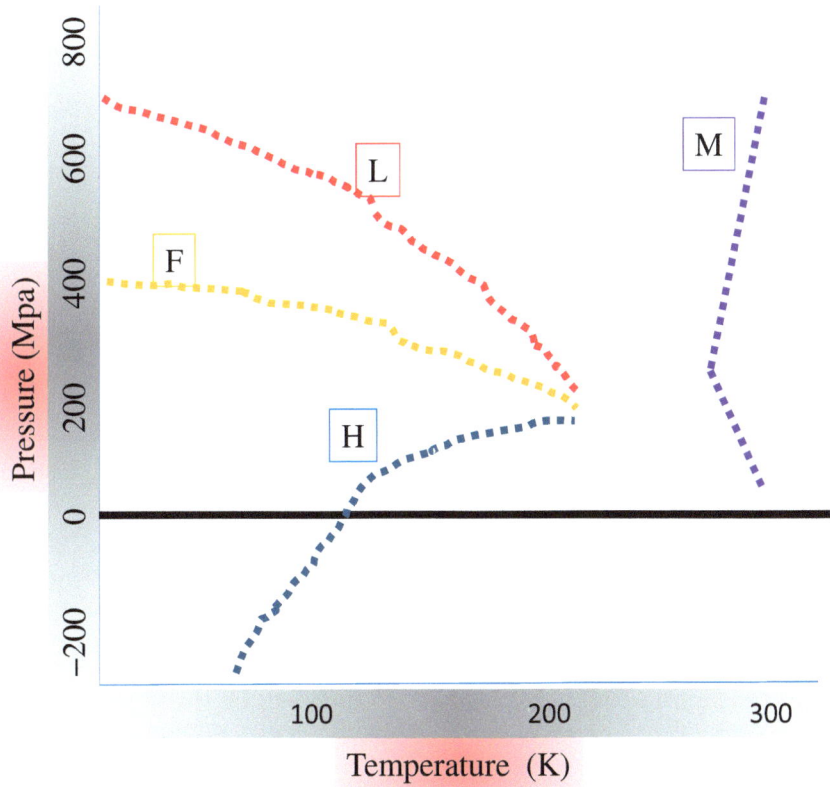

Fig. (6.14). The stability of HDA and LDA, depicted across wide temperature and pressure. Above the line F High Density Amorphous (HDA) is stable, and below Low Density Amorphous (LDA). H and L are spinodal lines that are originated from the critical point, shown as a black dot at around 200 K. M denotes the equilibrium melting point of ice. The idea has been adapted from [179].

Which amorphous form of water is stable: Low Density Amorphous (LDA) ice or High Density Amorphous (HDA) ice? Based on computer simulations, Stanley *et al.* have obtained a phase diagram by which the stability of these amorphous phases can be mapped in terms of pressure and temperature (Fig. **6.14**) [179].

First, as shown above (Fig. **6.14**), there exists a coexistence line (shown by F in the figure), above which HDA is more stable. Below this line, which corresponds to the phase transitions between LDA and HDA, LDA is more stable. The thermal and mechanical stability limits (metastability) of HDA and LDA can be identified by two spinodal lines, H and L (as indicated by these letters in the figure) respectively.

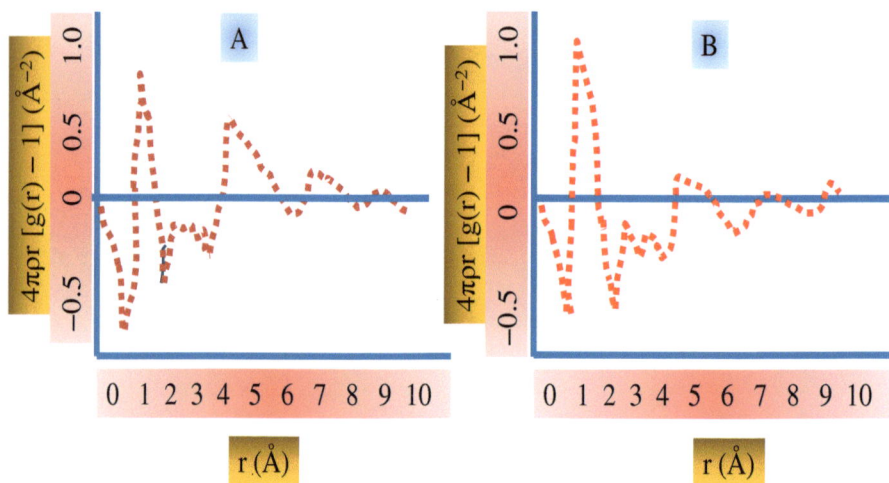

Fig. (6.15). Pair correlation functions of HDA (**a**) and LDA (**b**). The structural differences between the two forms of glassy water can be noticed in longer distance from the molecule of reference. The idea has been adapted from [75].

The structural difference between HDA and LDA can be illustrated from the pair correlation functions as in the case of other phases of water. It can readily be seen from Fig. (**6.15**) that the first noticeable effect is the reduction of intensity of peaks immediately after 2Å in HDA, indicating weaker correlations than LDA. Also at higher distance (beyond the first nearest-neighbour distance) the peaks in HDA are found to be broader, which accentuates this observation. The X−ray diffraction findings indicate that in these both forms of glassy water the nearest−neighbour environment are similar [71]. Furthermore, the coordination numbers in both forms are found to be four, suggesting near tetrahedral coordination (alternatively known as open structure) with their nearest neighbours. Very High Density Amorphous (VHDA) ice can be obtained from HDA ice upon annealing above 130 K and 0.8 GPa pressure [180].

It is interesting to note the striking similarities between the structure of supercooled water at very low pressures and Low Density Amorphous (LDA) ice. Similarly the structure of supercooled water at elevated pressures resembles High Density Amorphous Ice (HDA), illustrated in Fig. (**6.16**) [117].

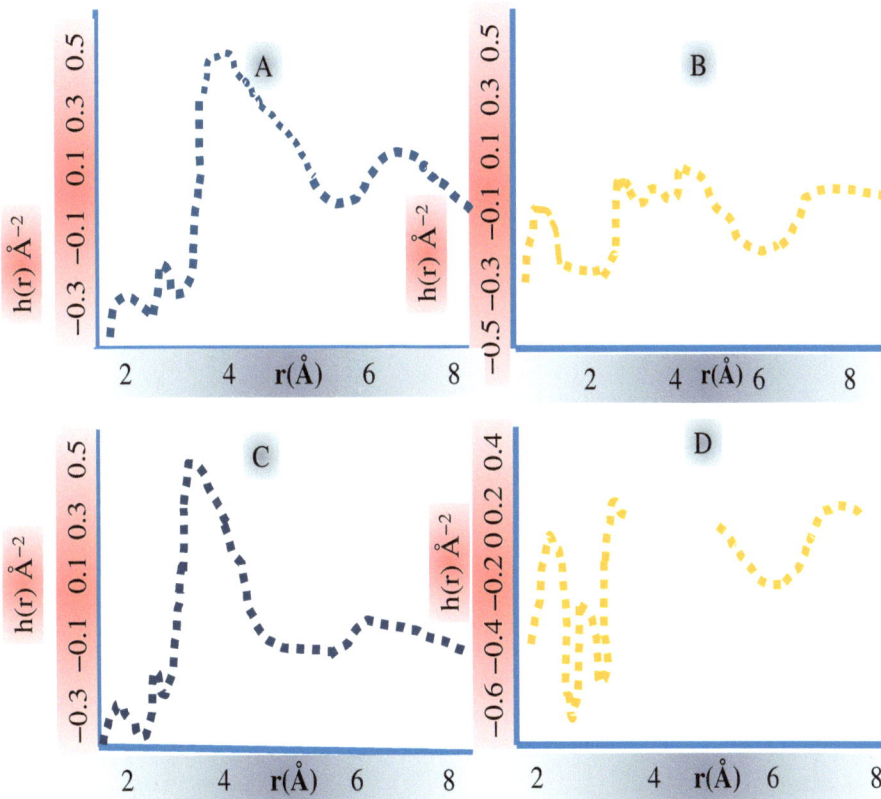

Fig. (6.16). Weighted pair correlation functions (h(r)) of supercooled and amorphous phases of water, obtained from experiments. Weighted pair correlation functions have been shown in each of the semi plots. In (**A**) shown is h(r) of supercooled water at 600 MPa and 243 K; in (**B**) shown is supercooled water at -200 MPa and 243 K; in (**C**) shown is the h(r) of HDA at 128 K and in (**D**) h(r) of LDA at 128 K and atmospheric pressure. The idea has been adapted from [117].

Suzuki *et al.* have investigated the HDA – LDA transformation using Raman spectroscopy, despite the fact that it is very difficult to record experimental measurements of these subtle thermodynamic transformations [181]. At 115K, 50% of the High Density Amorphous (HDA) ice is transformed to Low Density Amorphous (LDA) form in 125 minutes. In comparison, complete conversion to low density form is quickly made in just 90 seconds, as Fig. (**6.17**) indicates.

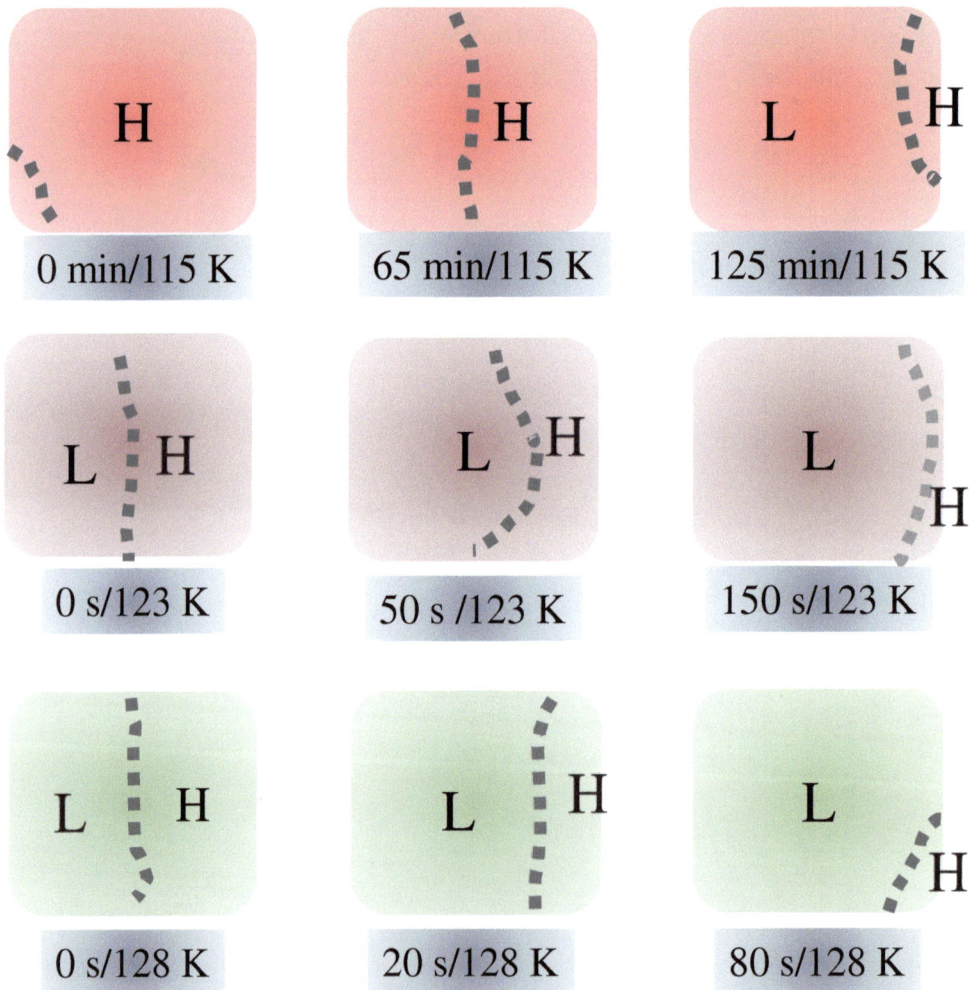

Fig. (6.17). A sketch showing HDA–LDA transformations occurring at three different temperatures. As clearly seen from the sketch, an increase in temperature enhances HDA–LDA transformation. The boundary region is shown with grey line. The idea has been adapted from [181].

However, recent experiments and computer simulations have challenged the existence of High Density Amorphous (HDA) water. Potential Energy Landscape (PEL) theory, as we have seen in a previous chapter, has been emerged as an important theoretical tool for material characterisation. A cursory glance at Fig. (**6.18**) reveals that the PEL profile corresponding to LDA is highly symmetrical. On the contrary, it can be clearly concluded from the unsymmetrical profile shown on the right indicates that High Density Amorphous (HDA) gives way to more stable Very High Density Amorphous (VHDA) at higher pressure, while LDA remains the stable phase at lower pressures.

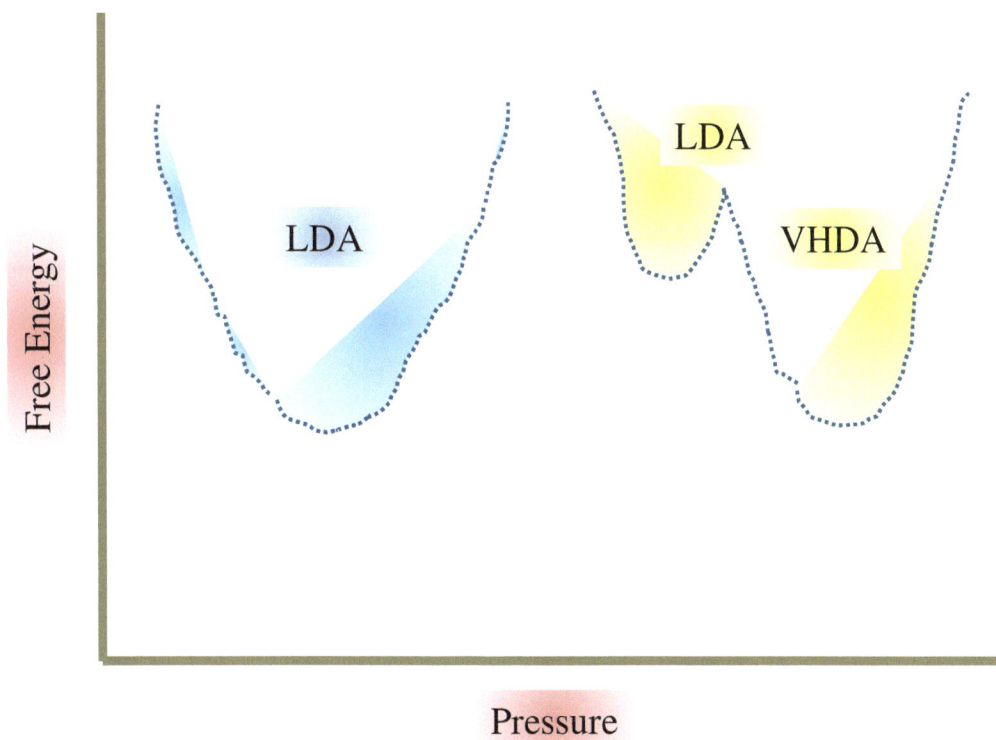

Fig. (6.18). Potential Energy Landscape (PEL) of the three forms of amorphous ice. At low pressure LDA is clearly found to be the stable amorphous phase of water, whereas at higher pressures High Density Amorphous (HDA) phase is appeared only to be metastable, before eventually being converted to Very High Density Amorphous (VHDA).The idea has been adapted from [182].

GLASS TRANSITION TEMPERATURE OF WATER

Unlike boiling point and melting point of water, its glass transition temperature

(T_g) has not been simply defined, and accurate identification of glass transition temperature of water is one of the most debated themes in material science [56]. Glass transition refers to the transition from liquid phase into glassy water upon cooling, or glassy phase into liquid phase upon heating [56,183]. Within the context of water, glass transition is very central to the understanding of polyamorphism exhibited by amorphous solid water, and to the structural elucidation of water between glassy and supercooled phases [183]. More importantly, accurate measurement of glass transition temperature is required for better exploitation of the properties of supercooled water.

The identification of "true" glass transition temperature in water is difficult to achieve due to the existence of several amorphous forms of water (such as Hyper–quenched Glassy Water (HGW), Low Density Amorphous (LDA) water and High Density Amorphous (HDA) water) between its supercooled liquid and glassy phases. Furthermore, the experimentally determined glass transition temperatures to a greater extent depend upon the way by which the samples are prepared. There have been many attempts in the past to identify the glass transition temperature in water. Nearly 50 years back McMillan and Los estimated its glass transition temperature 139 K with a very small rate of cooling 20 K/minute [183]. Sugisaki *et al.* observed a transition from glass to liquid near 135 K. Johari *et al.* has obtained T_g 136 K upon very high rate of cooling (greater 10^5 K s^{-1} known as hyper–quenched) of liquid water, while Angel *et al.* proposes that glass transition temperature in water be at around 165 K [56]. Handa and Klug have observed glass transition temperature of Low Density Amorphous (LDA) water upon its heating at 124 K at a cooling rate of 0.167 K per minute [106].

Using several computer water models including the most widely used, Oleinikova *et al.* have estimated glass transition temperature of water using the traditional molecular dynamics simulations. As predicted, due to the differences in parameters such as partial charge and model geometry, these values show a great variation: TIP4P (180 K), ST2 (235 K), ST2RF (255), SPC/E (220) and TIP5P (215) [184]. Using annealing simulation technique, Stanley *et al.* have obtained the glass transition temperature for TIP4P water model, 188 K [185].

This does not suggest that all these efforts just mentioned above are worthless;

many efforts have been put in to understand the dependency of other physical variables on glass transition temperatures, and from these, vital information on the dynamics of water at low temperatures have been deduced. In one of such interesting studies, Starr *et al.* have unearthed the pressure dependence of glass transition temperatures as shown in the following sketch [186]. Fig. (**6.19**) shows a glass transition minimum at pressure 300 MPa, from which the diffusion coefficient maximum emanates (200 Kelvin onwards).

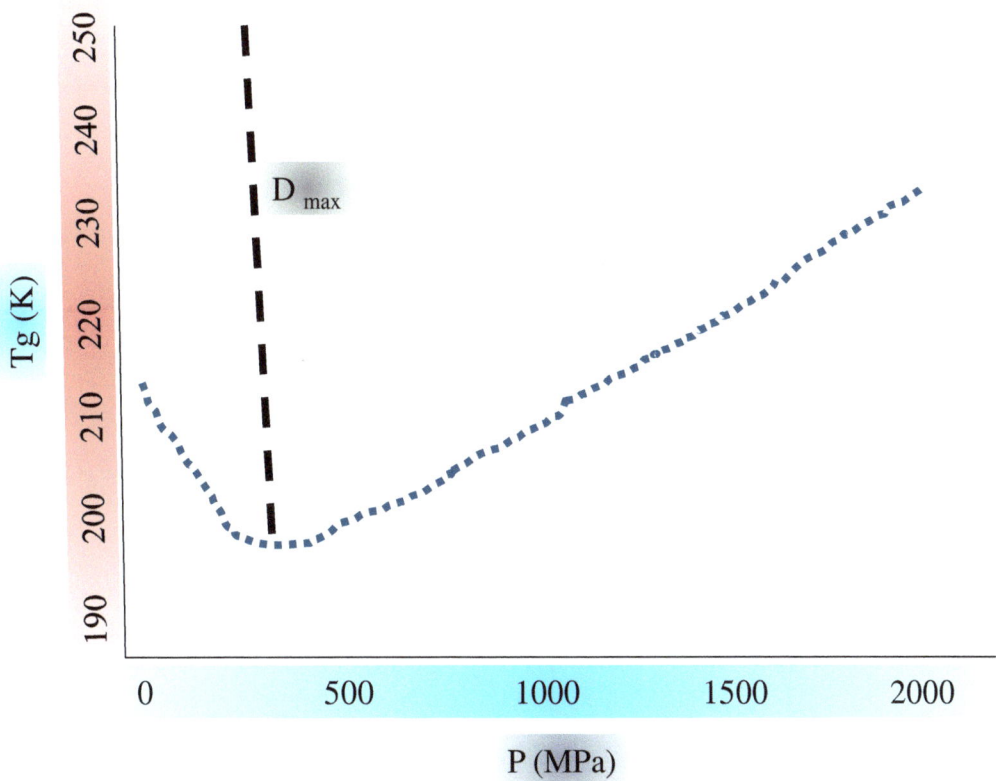

Fig. (6.19). Schematic plot of glass transition temperatures (T$_g$) with respect to the changes in pressures. The data has been obtained using SPC/E water model by Starr *et al.* The blue line shows the diffusion coefficient maximum. The idea has been adapted from [186].

A mathematical explanation in terms of diffusive relaxation time (τ) has been given to account for the correlation between T$_{g\,minimum}$ and D$_{maximum}$ [186]. Since the diffusive relaxation time is an explicit function of temperature and pressure, its derivative (dτ) can be written as total differential form (as the sum of temperature (dT) and pressure (dP)). We also note that the cooling rates employed in all of their simulations are same, and this renders dτ equal to zero along the glass transition curve. The variation of τ with respect to temperature is negative due to its monotonous decrease upon decreasing temperature, and from basic calculus, it can be inferred that the minimum in the glass transition temperature (T_g) curve corresponds to its derivative with respect to pressure $(\frac{dT_g}{dP})$ being equal to zero. Mathematically this means that pressure and temperature derivatives of the relaxation times $(\frac{d\tau}{dP}$ *and* $\frac{d\tau}{dT})$ inversely correlated, recording their maximum and minimum at glass transition temperatures respectively.

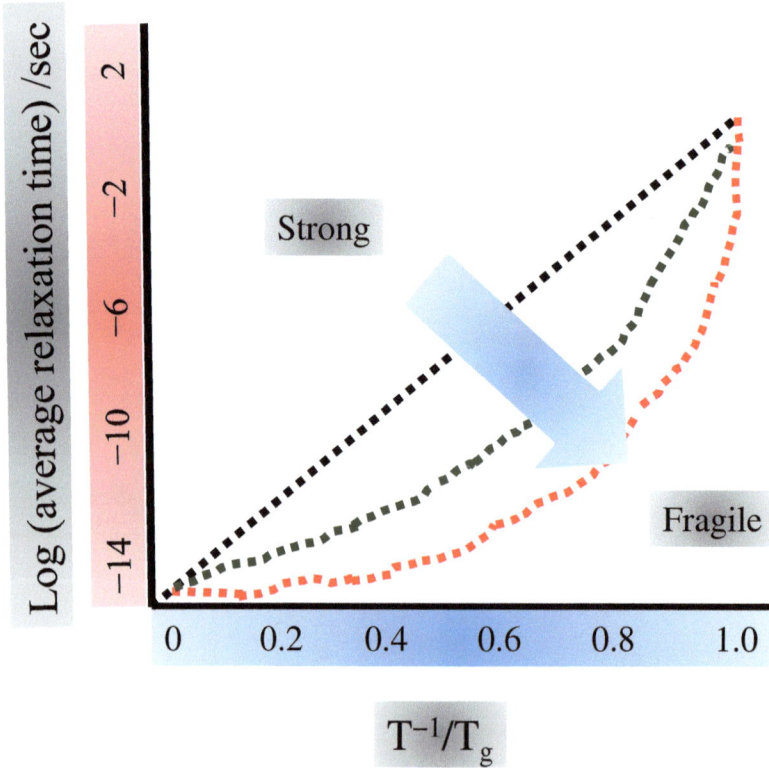

Fig. (6.20). Relaxation time *versus* Temperature. The characteristic difference between strong and fragile glass is sketched in the diagram. Strong glass undergoes smooth changes over the complete temperature range, whereas the system undergoes abrupt changes near the glass transition temperature as fragility increases. The broad blue arrow indicates the deviation from strong to fragile behaviour. The idea has been adapted from [187].

Now let me go back to discuss the qualitative aspects of the glass formation. As shown in Fig. (**6.20**), we plot T_g against average relaxation time for gaining a deeper understanding of strong–fragile classification of glass. The materials exhibiting strong character show very linear behaviour (Arrhenius) irrespective of temperatures. On the contrary, as materials become more fragile, the more they deviate from Arrhenius behaviour. Furthermore, as temperature approaches T_g, a sharp increase in relaxation time is observed.

We came across the fragile character of water near the nucleation temperature in one of the previous sections. One fact we must note that this characterisation is not thermodynamic, rather a kinetic nature. This means that the rate of cooling by which a normal liquid brought under supercooling condition dictates whether water is strong or fragile. The fundamental differences between strong and fragile glassy materials can also be understood by noting the variations in the excess heat capacity at glass transition temperatures. In strong glass formers such BeF_2 and SiO_2 the change in heat capacity at the glass transition is very little (strong), as indicated by a little increase in heat capacity at the transition, after which these materials exhibit fragile behaviour. On the contrary, in systems like water, silicon and germanium, the variation in heat capacity is negligible prior to the heat capacity jump, after which these systems exhibit fragile characteristics, and hence in these type of materials, an interesting phenomenon known as Fragile–to–Strong Crossover (FSC) occurs [110]. The third class of systems is the organic substances such as toluene and butene that form a fragile glassy system, indicated by a sharp increase in the response function (relaxation times). Similar behaviour can be observed in the case of entropy as well: only a smooth change in entropy can be observed in the aforementioned strong glasses (mostly inorganic compounds) such as SiO_2. On the contrary a drastic entropic change at the onset of glass transition occurs in very fragile systems (mostly organic compounds). Systems such as water exhibit an intermediate variation in entropy. The thermodynamic "region" in which the changes in these quantities, in particular heat capacity, influences the stability of glasses formed as shown below: a close inspection of the following figure (Fig. **6.21**) reveals that systems such as water is crystallised very easily, with the sudden change in heat capacity lying between T_m (melting point temperature) and T_g (glass transition temperature), whereas in the stronger glassy systems the corresponding change is observed beyond T_m, rendering crystallisation almost impossible.

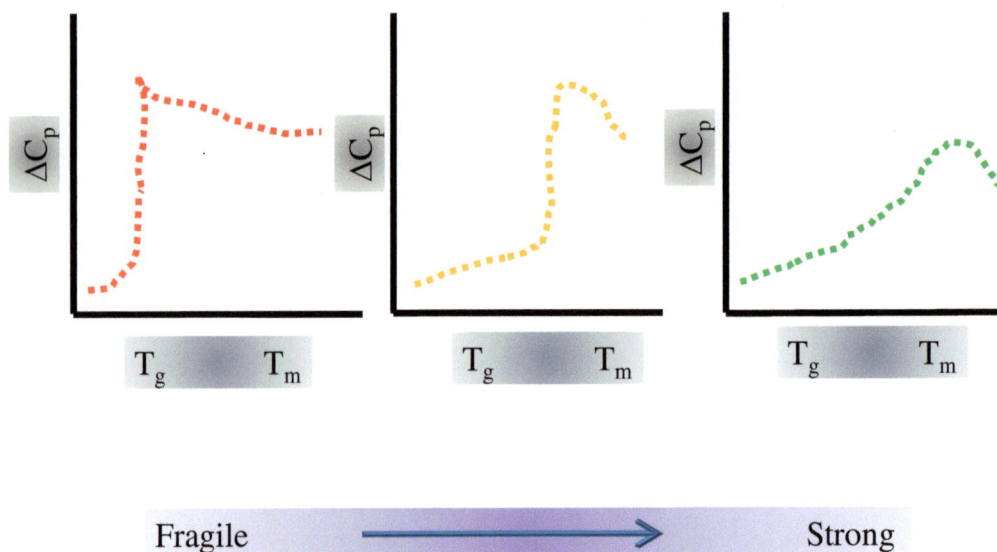

Fig. (6.21). Thermodynamic characterisation of strong and fragile glassy systems. The diagrams show the variation of heat capacities with respect to the changes in temperature. On the left shown is the fragile glassy systems (for example organic compounds such as butene) for which sudden changes in heat capacity occurs below the glass transition temperatures, while on the right shown is the strong glass formers such as SiO_2. The heat capacity hike occurs above the melting temperature, T_m. Both of these systems are glass formers. In the sketch drawn in the middle of the diagram, the profile corresponding to systems such as water, Si and Ge is shown, where in the heat capacity hike occurs between T_g and T_m, suggesting that that these systems are very much prone to crystallisation. The idea has been adapted from [188].

Angel notes that the heat capacity change marks the transition from one liquid phase to the other upon drop in temperature [188]. One of the liquid phases is less diffusive than the other, and is very prone to crystallisation. An elaborative discussion on liquid–liquid transition can be found in a later chapter.

CONCLUDING REMARKS

A brief discussion on supercooled and glassy water has been given in this chapter. There is a profusion of assumptions still existing in the scientific community over the nature of phase separation in the supercooled and amorphous water. Different water models have yielded varying glass transition temperatures, rendering our understanding of water in a vague manner. We can say that only an accurate water model, which is believed to account for all its peculiarities, will unlock the mysteries shrouding supercooled and glassy water. An important theoretical development in last decades about glassy behaviour of water is that it is neither a strong nor a fragile liquid; rather it behaves like an intermediate substance, and an inter–conversion from Fragile to Strong Crossover (FSC) occurs in supercooled water. In the next chapter I am going to discuss another important phase of low temperature water, ice.

CONFLICT OF INTEREST

The author confirms that he has no conflict of interest to declare for this publication.

ACKNOWLEDGEMENTS

I express my gratitude to the following people and organisations for granting me permission to use figures under their copyright in the chapters: Dr. Dietmar Paschek, American Institute of Physics, Nature Publishing Group, and American Physical Society.

CHAPTER 7

Ice, The Crystalline Phase of Water

Abstract: Transition from water to ice is very crucial in many natural and artificial processes on which our lives depend. No other substance exhibits more crystalline forms than ice, the solid phase of water. Several ice polymorphs are found to exist in pairs, corresponding to high temperature proton disordered state and low temperature proton ordered state. Hexagonal ice is the dominant form of ice at ambient conditions. Ice X is highly symmetrical ice polymorph with hydrogen atoms exactly positioned equidistant to the two adjacent oxygen atoms. Five and seven membered rings of water molecules are observed in ice XII. The orientation of hydrogen bonds plays important roles in assigning the geometries of various forms of ice as in the case of normal and supercooled waters. The largest hydrogen bond bending is observed in ice VI. Orientations of hydrogen atoms result in Bjerrum and ionisation defects in ice crystals, which are responsible for dielectric effects. Auto ionisation, leading to the generation of hydronium and hydroxyl ions in water, promotes ionisation defects. Catalytic properties of ice are found to be remarkable in large number of reactions. TIP4P water model and its variants seem to be the popular models for simulating ice phase of water. Rotational motion of oxygen–hydrogen bonds is responsible for the destruction of ice lattice as temperature increases, leading to the melting of ice. Ice exhibits excellent electrical, optical, mechanical, thermal & surface properties. Thanks to its exceptional thermal properties, ice has been successfully employed as a better alternative to the traditional air cooling systems.

Keywords: Auto ionisation, Bjerrum, Clusters, D defect, Ferroelectric, Heterogeneous, Homogeneous, imidazole, L defect, polymorph, Polytype, Rectifier, Semiconductor, Thermoluminescence, Tyndall.

INTRODUCTION

When liquid water freezes, its crystalline polymorph ice forms, and this transition has profound impact on many processes that have direct relevance to our lives. Like water, ice is also subjected to immense scholarly activities due to its critical

Jestin Baby Mandumpal

role in many fields including our planetary system, food preservation, *etc.* [189].

A wide variety of structures of ice, have been identified, as water molecules orient themselves in specific patterns in order to minimise energy upon changing physical conditions (pressure and temperature) [190]. The presence of ice in atmosphere and its role in the creation of cloud formation have made ice an interesting topic for geologists and other environmental specialists [72,189]. It has also been found that rheology of ice has profound impact upon the existence and appearance of several planetary objects [189b]. In food technology, ice has already occupied a prime position in preserving food materials domestically and commercially since ancient times [191]. Ice morphology dictates texture and physical properties of frozen materials and the popular cold sweet food ice cream [192]. Ice also has excellent adhesion properties on solid surface, which is relevant to the aviation industry and ship navigation as ice crystals are developed on aeroplanes and ships during their journey through supercooled clouds and storms respectively, which was discussed earlier [72].

Thus, it is very crystal clear that further multi–dimensional investigations on ice are very much needed for the betterment of the standard of our daily lives. Following a discussion on supercooling and supercooled water in the previous chapter, we are going to discuss the structure, properties and applications of ice in depth in this chapter. Although I mentioned about the connection between supercooling and crystallisation in the previous chapter, a repetition is unavoidable sometimes as these two states (metastable states) are inextricably linked to each other.

The supercooled state is a metastable state, highly unstable and can give away to more stable state, ice in the case of water, which is much more ordered solid species [193]. We saw in the previous chapter that water droplets can form clusters, which are so abundant in supercooled water, and the decrease in temperature promotes clustering of water molecules [160]. Clusters are grown in size as temperature drops by, and the movement of water molecules becomes sluggish. Some of the clusters attain a critical size by growing around themselves (a process known as nucleation). This leads to the transformation of supercooled water to large block of ice, and this process is known as crystallisation. Three other alternative routes to crystallisation have also been suggested as shown in Fig. (**7.1**). The diagrams indicate that crystallisation can occur as a result of the contact between ice nucleus and water droplet, which can be inside or outside the water droplet [194]. Here, we must remember that a water droplet consists of thousands of water molecules.

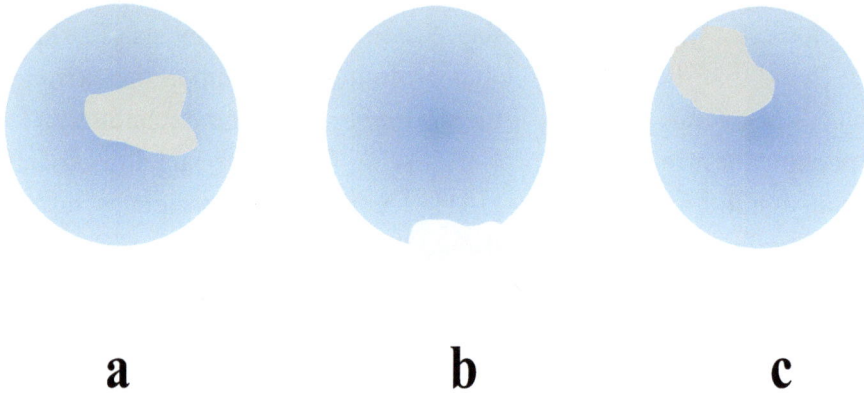

a   **b**   **c**

Fig. (7.1). Crystallisation of water droplet. Three ways by which ice nucleus promotes crystallisation of water droplet (shown in sky-blue circle): (**a**) ice nuclei inside a water droplet, (**b**) nominal contact between ice nuclei and water droplet (**c**) ice nuclei inside the water droplet.

Thus, the formation of ice is preceded by a process known as nucleation [195] which can either due to the electrostatic interaction of polar part of water (known as homogenous nucleation) or due to the presence of substance other than water considered as impurities (known as heterogeneous nucleation). The homogenous temperature of water is 231 K, below which it readily undergoes crystallisation under normal conditions (at very low cooling rate and at low pressures). It is very noteworthy that at higher temperature (close to the freezing temperature of water), the critical radius for homogenous nucleation is five times higher than the critical radius required for nucleation at 231K. Consequently, the number of particles required for attaining critical radius is also significantly less shown in Table **7.1**. At the same time, as mentioned earlier, the number of water molecules participating in the process of crystallisation increases as temperature drops by. Computer simulations show that for a drop in temperature from 240 K to 235 K, the number of water molecules that are part of "ice building" increases by a whopping amount of 40% [196].

Table 7.1. Dependency of critical radius on temperature.

Temperature (K)	Number of Water Molecules	Critical Radius (nm)
263	15943	4.20
253	1944	2.08
243	566	1.38
233	234	1.03
223	122	0.83

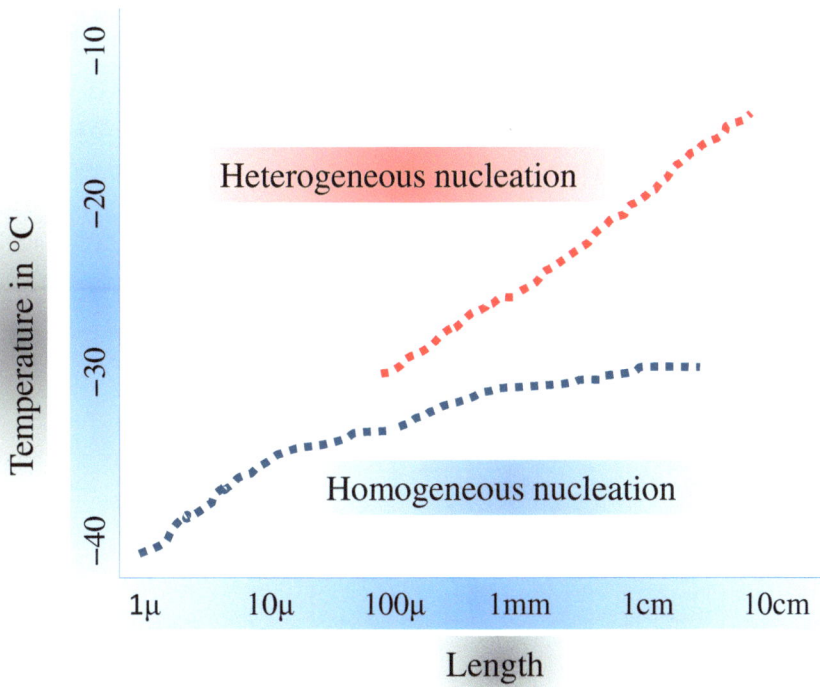

Fig. (7.2). Nucleation of water molecules as function of temperature and the size of the sample. Red line indicates the domain of heterogeneous nucleation, which is very higher than the volumetric condition for homogenous nucleation (shown in blue). The steepness is sharper in the case of heterogeneous nucleation suggesting that this type of nucleation is greatly affected by changes in the size of the sample and temperature.

Even though the nucleation temperature of water is 231 K, it does not mean that it cannot be nucleated before this temperature. If the nucleation of the ice phase occurs by the presence of foreign particles (that is by means of heterogeneous nucleation) which may be affected by size and impurity content in water such as silver iodide (AgI), and rate of cooling. If foreign particles do not play a role, the system is said to be formed by homogenous nucleation. Normally in the bulk sample, water is nucleated by heterogeneous nucleation, which takes place at relatively high temperature [97a, 195]. In bulk samples, the presence of impurities reduces the free energy deficit between microcrystals and macroscopic crystalline ice for the nucleation. Fig. (7.2) demonstrates the relationship between the temperature and the two nucleation processes explained above.

As the title indicates, this chapter is exclusively devoted to ice. Firstly, the discussion is centered on various forms of ice, stressing on their structural features. This is followed by several inter–conversion routes between different forms of ice, and a brief comparison of water/ice models with respect to experimental density measurements. The chapter is concluded with important properties and applications of ice.

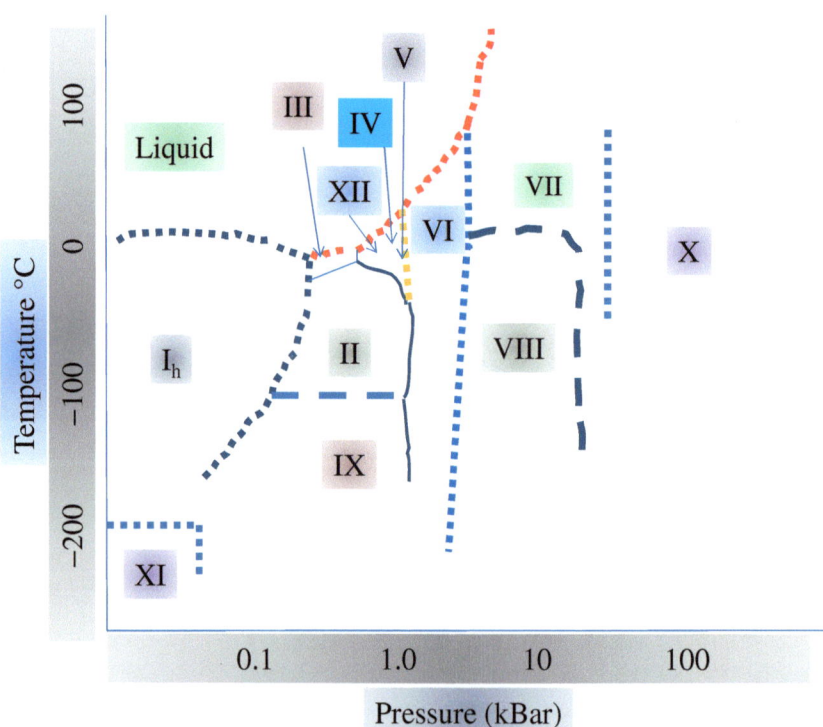

Fig. (7.3). Phase diagram of ice. Twelve different forms of ice have been shown in the diagram. Temperature range in which ice exists is so enormous, extending from as low as 50 Kelvin to 350 Kelvin. The pressure also exhibit large variation from 0.1 to 100 K bar. The idea has been adapted from [121].

ICE IN VARIOUS FORMS

Over 15 crystalline forms of ices have been discovered and experimentally recognised, which are denoted systematically using Roman numerals, and the number of crystalline forms of ice is growing [77,104,197]. Quite recently, Grigorieva *et al.* have discovered a new form of ice, square ice, using high resolution electron microscopy technique [104]. Interestingly, it has been quoted by Loerting *et al.* that in every six years a new polymorphic form of ice is discovered [198]. So far, this "tradition" has not been broken. Fig. (**7.3**) shows the phase diagram of ice, indicating ten ice polymorphs, within the range of 0 Kelvin to 400 Kelvin and 0.1 kbar to 100 kbar.

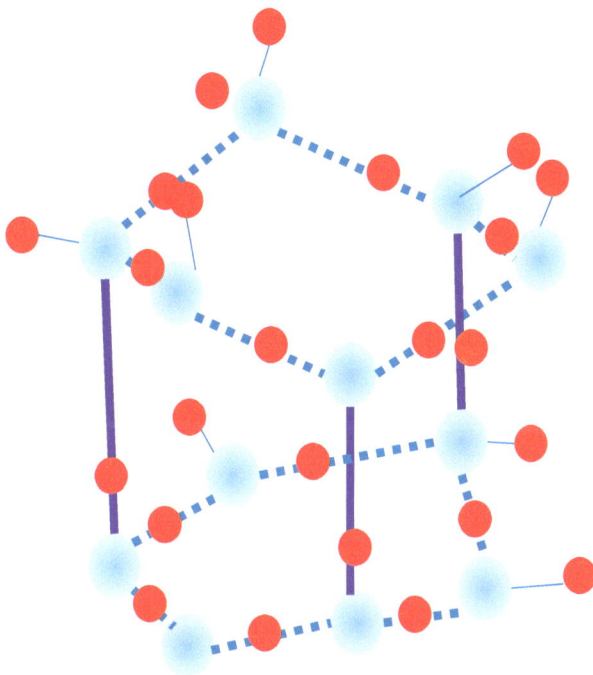

Fig. (7.4). The molecular arrangement in I_h. The alternative water molecules are configured in same fashion. The sky blue and red coloured spheres indicate oxygen and hydrogen atoms respectively.

I_h, the Most Common Type of Ice

The most common type of ice, and the most stable under normal conditions, is hexagonal ice, commonly denoted by I_h. Hexagonal ice is obtained when liquid water freezes at atmospheric pressure, resulting in an ordered arrangement of oxygen atoms [198, 199]. Oxygen atoms of the hexagonal ice are arranged on an interconnected hexagonal bilayer in such a way that each of these atoms is surrounded by four neighbouring oxygen atoms making a "perfect" tetrahedral arrangement, and the distance between two oxygen atoms is estimated to be 2.74 Å [148]. Three of the four hydrogen bonds are between water molecules that belong to the same layer, while the remaining hydrogen bonds are formed between water molecules belonging to different layers.

Computer simulations reveal that hexagonal ice can be produced by pressure induced amorphization. Hexagonal ice (I_h) can be transformed to High Density Amorphous (HDA) ice and High Density Liquid (HDL), at high and low temperatures respectively [200]. The molecular arrangement of hexagonal ice is shown in Fig. (**7.4**).

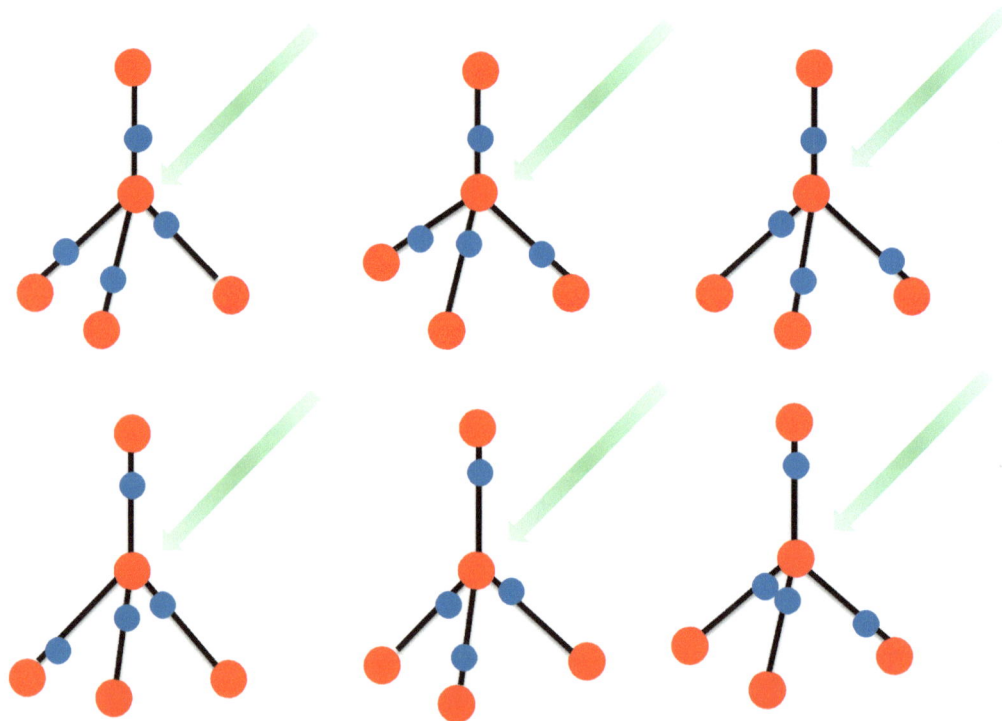

Fig. (7.5). Orientation in Walrafen pentamers. Six possible arrangements of hydrogen atoms around the oxygen atom of the central water molecule (shown by the green arrow) are shown. The oxygen and hydrogen atoms are shown in red and blue respectively. In each of the six configurations, two hydrogen atoms are closer to the central oxygen atoms (shown by an arrow) than the other two.

I_h represents the limiting case of perfectly ordered tetrahedral network of hydrogen bonds, and is the most experimentally investigated phase of ice [116]. As mentioned in Chapter 2, spectroscopic measurements are highly useful in tracking hydrogen bonding in liquids. Its stretching vibrations can be characterised by the red shift, as indicated by a theoretical analysis made by Kryachko [201]. Ice in general is disordered crystal with water molecules orienting themselves in numerous configurations, based on six possible arrangements of the hydrogen atoms that participate in bonding (chemical and hydrogen bonding) with the oxygen atom of each water molecule as shown Fig. (**7.5**) above. In the figure, orientations of water molecules in Walrafen pentamers are shown.

This suggests that there can be variations in the bond angles and bond lengths in water molecules, corresponding to each specific configuration [202]. It has been shown that there can be numerous distinct configurations for ice [199]. For example, over 40 symmetrically distinct arrangements have been observed for ice VI [202].

Bernal−Fowler Rules

At this point, we need to think about the arrangement of H_2O molecules in the ice crystal lattice; this will certainly provide us better insights into numerous configurations by which the ordinary ice can exist under normal conditions. Bernal and Fowler codified the following four principal rules on the position of hydrogen atoms in water molecules in the ice crystal [72].

Rule 1: Two hydrogen atoms are attached to each oxygen atom in water molecules with a distance of 0.95 Å.

Rule 2: Each water molecule is surrounded by four other water molecules in a perfect tetrahedral arrangement with two hydrogen atoms of each water molecule orienting towards oxygen atom of neighbouring H_2O molecules. Each oxygen atom has four possible hydrogen sites. This warrants that each water molecule can act as donor and acceptor of hydrogen bonds at the same time.

Rule 3: Only one hydrogen atom can occupy the space between two adjacent oxygen atoms.

Rule 4: The difference in the distributions of hydrogen atoms around the oxygen atom in each water molecule result in numerous configurations.

Defects in the Ice

In ice, H_2O molecules are stranded by four hydrogen bonds, with the oxygen lattice tetrahedral coordinated. Several experiments suggest that ice is not a perfect crystal, and certain violations are "permitted" in this perfect tetrahedral arrangement, known as defects. Interestingly this results in several physical properties including mechanical relaxation and electrical conductivity [203]. There are two types of defects that can occur in the ice crystal: Bjerrum and ionisation defects.

An important observation on the defects in the crystalline arrangement of ice was first made by a Scandinavian, Bjerrum, who suggested that there can be two major defects that can exist in the crystal, known as D and L defects, due to the rotation of oxygen–hydrogen bonds. The D defect refers to the situation wherein two hydrogen atoms position between oxygen atoms of the two neighbouring water molecules, a clear violation of the third Bernal–Fowler rule which restricts the number of hydrogen atoms that can be on line between two adjacent oxygen atoms to one [72]. On the other hand, the L defect points into the situation wherein there can be no hydrogen atoms on line between the oxygen atoms of two adjacent water molecules. It is interesting to note that the occurrence of these defects, which are concomitant, in a neighbouring pair of oxygen–oxygen bonds can induce D and L defects throughout the crystal. The following Fig. (**7.6**) demonstrates the origin of the D and L defects and their propagation throughout the crystal.

These defects are induced in other water molecules, due to the rotation of hydrogen atoms of these molecules. The movements of hydrogen atom are accompanied by changes in individual dipole moments of water molecules. The D and L effects can also be induced in the ice lattice by doping chemicals such as ammonia (NH_3), hydrofluoric acid (HF), ammonium hydroxide (NH_4OH) and ammonium fluoride (NH_4F) [72]. Out of these, ammonia and ammonium hydroxide induce D defect in the system, and hydrofluoric acid creates L defect in the system.

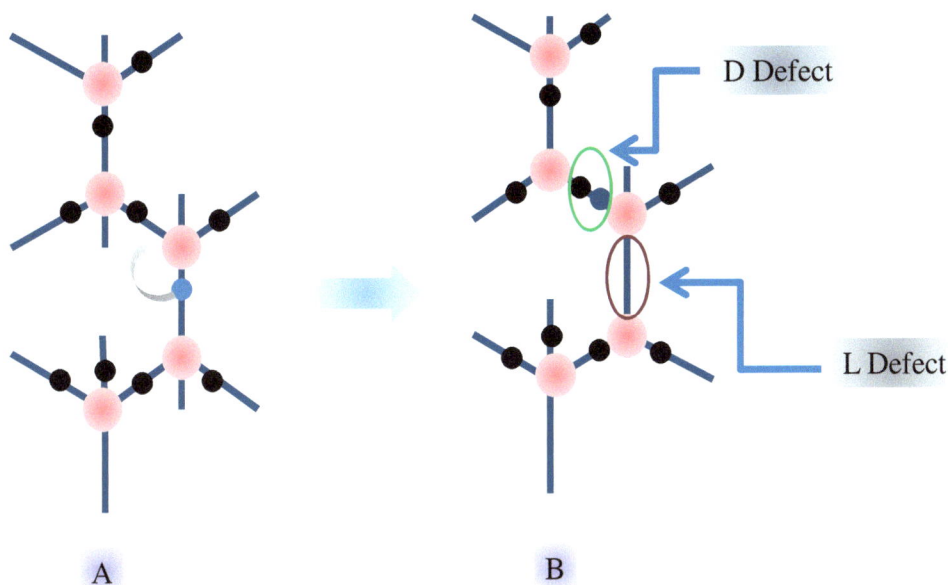

Fig. (7.6). Bjerrum defects in ice. In each water molecules, oxygen atoms (shown in red circles) are bonded to two hydrogen atoms (shown in black circles) as shown in the figure a. Rotation of any one of the hydrogen atoms of a water molecule (as shown by the direction of arrow in the panel A) in the ice crystal generates D (with two hydrogen atoms on line between two adjacent oxygen atoms as shown by green circle as shown in panel B) and L defects (with no hydrogen atom on line between two adjacent oxygen atoms as shown by brown circle).

Water undergoes auto ionisation, generation of pairs of positive and negative ions by itself, which results in another defect, namely ionisation defect. The chemical equation representing auto ionisation is shown below.

$$2H_2O \leftrightarrow H_3O^+ + OH^-$$

Once these ions, hydronium (H_3O^+) and hydroxyl (OH^-) ions, are formed from water, they are separated, and transferred to other sites by proton jump, according to the following two chemical equations.

$$H_3O^+ + H_2O \leftrightarrow H_2O + H_3O^+$$

$$OH^- + H_2O \leftrightarrow H_2O + OH^-$$

In the first equation, a proton is transferred from the hydronium ion to water, and in the second, hydroxyl ion abstracts a proton from water, the processes in which both hydronium and hydroxyl ions are regenerated. The ionisation defect is demonstrated in the following diagram, Fig. (**7.7**).

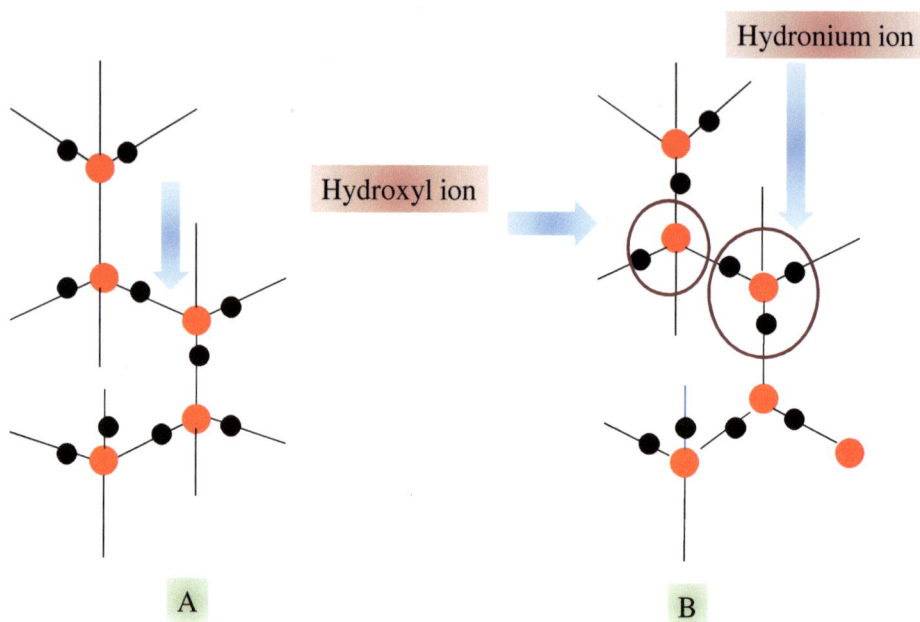

Fig. (7.7). Ionisation defect. In the panel **A**, auto ionisation of two water molecules occurs at the bond indicated by the blue arrow. This creates two ions: an anion (hydroxyl) and a cation (hydronium) as shown by brown circle in the panel **B**.

Bjerrum and ionisation defects in ice enhance its conductive properties. Another change in the ice crystal associated with these defects is the alternation in the interatomic distance between two adjacent oxygen atoms: D effect increases the normal oxygen–oxygen distance from 2.76Å to 4.4 Å. L defect, on the other hand, raises the bond length only about 0.2 Å [72]. Wang *et al.* have notified another type of defect in ice, in which oxygen atoms are found to be deviated, up to about 1Å, from the lattice points of ice [204].

Molecular Structure of Ice Polymorphs

The molecular structure of hexagonal ice, I_h has been investigated extensively by computer simulations [205]. It has been revealed that pressure and temperature alter the oxygen–oxygen interatomic distance considerably. The distance increases as pressure decreases, whereas a rise in temperature increases the bond length between hydrogen and oxygen atoms [205b]. In addition, the difference in mean square displacement, a measure of thermal vibration in a molecule, of hydrogen and oxygen atoms at the bulk and surface of ice decreases as temperature drops below the melting temperature of ice [206].

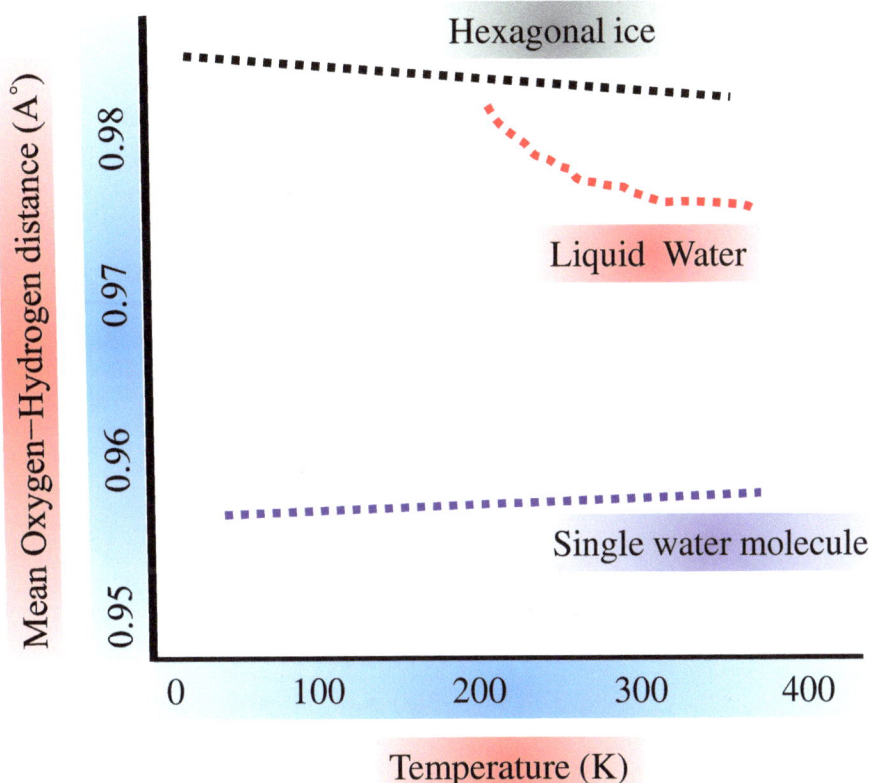

Fig. (7.8). The intramolecular distance between oxygen and hydrogen atoms across a wide range of temperature, calculated by Path Integral Molecular Dynamics simulation. The bond lengths in hexagonal ice, liquid water and an isolated water molecule are indicated by dots in black, red and violet respectively. Variation of the bond length is clearly visible in the condensed phases of water.

The oxygen – oxygen radial distribution function analysis of ice VII shows five distinct peaks within the distance of 1 nanometer as opposed to three for I_h, echoing its sharp crystalline character, which is in stark contrast of the RDF curves amorphous ice forms, HDA and LDA which show only one distinct peak [200]. As in the case of different forms of water (normal and supercooled) discussed previously, hydrogen bonds play a decisive role in ice structures as well. In an isolated water molecule, the bond length between oxygen and hydrogen atoms shows little variation upon changes in temperatures. At the same time, intramolecular distance between oxygen and hydrogen atoms shows a sharper decrease as temperature drops, as Fig. (**7.8**) indicates. On the contrary, 3% of increase in the bond length is recorded in I_h, as indicated by Path Integral Molecular Dynamics simulation (PIMD) [207].

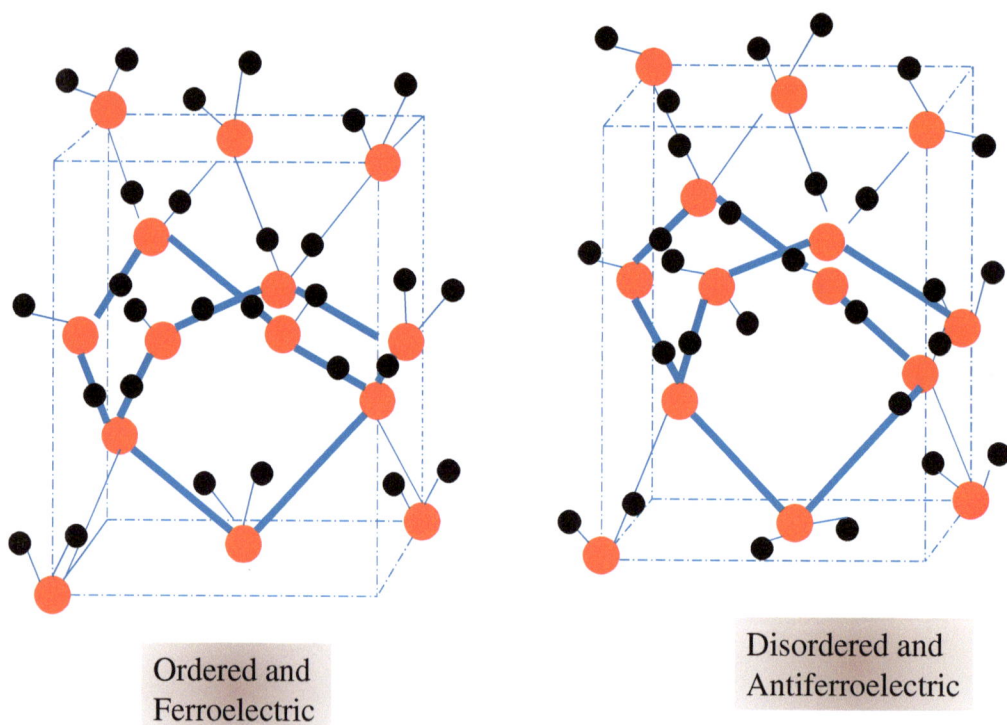

Ordered and Ferroelectric

Disordered and Antiferroelectric

Fig. (7.9). Proton ordered−disordered transition occurring in the ice crystal. On the left, an ordered proton arrangement characterised by the same orientation of hydrogen atoms results in ferroelectric behaviour, whilst a disordered arrangement of protons in the ice crystal yields disordered state.

The decrease in bond length is more prominent in the hexagonal ice than in liquid water, which is attributed due to the enhanced rigidity of the crystalline phase of water and the less mobility of H_2O molecules in ice than in water [207]. Ice can have different proton orientations while satisfying Bernal−Fowler rules, and this can result in either proton ordered or proton disordered phase [208]. Proton disordered phase of hexagonal ice, observed in higher temperature, are interconverted to proton ordered phase, which is also known as ice XI, at temperature around 70 K [208]. Ordered−disordered transition occurring in ice corresponds to ferroelectric−antiferroelectric switching, demonstrated in the schematic diagram of cubic ice, in Fig. (**7.9**) [109]. In the ordered states, the protons are said to be in polar arrangement, with negative poles and positive poles are oriented towards each other, which can be destroyed beyond certain temperature, known as Curie temperature.

I_c (cubic ice), as shown in Fig. (**7.10**) is formed in a wide range of temperatures, ranging from 120 Kelvin to 273 Kelvin by several means including dissociation of gas hydrates, warming of amorphous ices and high pressure ice polymorphs, and freezing of gels, electrolytes and aqueous molecular solutions [103].

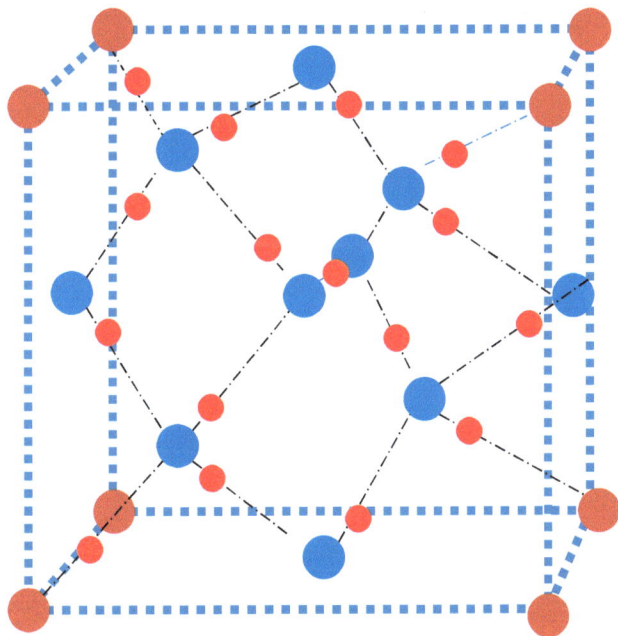

Fig. (7.10). The geometry of cubic ice (I_c) shown in the unit cell containing eight water molecules. Oxygen atoms within the cell and at the corners of the cube are shown blue and brown respectively. Hydrogen atoms are shown red.

The oxygen–oxygen radial distribution curve of cubic ice shows similar features of hexagonal ice. Both I_h and I_c, jointly referred to ice I, share several physical properties such as density (0.92 g/cm^3) at atmospheric pressure and interatomic distance [198]. Their spectroscopic and radial distribution profiles also indicate similar structural features [209]. Hence they are known as polytypes since the only principal difference between these two forms of ices is how different the hexagonal rings by which they form crystals are packed, in Face Centered Cubic (FCC) lattice and hexagonal lattice respectively [198].

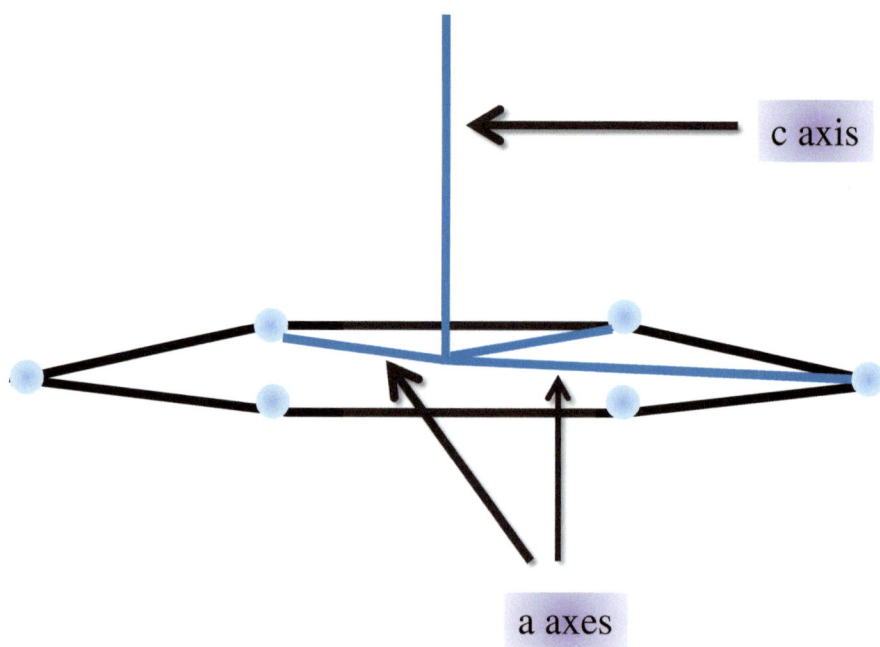

Fig. (7.11). Orientation of hexagonal ring in space. Three axes in plane known as "a axes" on the hexagonal plane, and an axis perpendicular to this molecular plane, c axis, define complete orientation of hexagonal ring in space.

In addition to these two types of ice phases, there exist also intermediate pressure variants (ice II, III, IV, V, VI & IX), and high pressure variants (VII, VIII). Ices III, V, VI and VII exhibit similar dielectric properties as hexagonal ice [72]. An interesting fact regarding these ice polymorphs is that they are found as pairs, the existence of which is based on proton order – disorder transitions [198]. These pairs, six in total found till date, are I_h–XI, III–IX, V–XIII, VI–XV, VII–VIII, and

XII–XIV. The first polymorph of each pair corresponds to proton disordered state, occurring at higher temperature. As temperature decreases, the protons in ice become more ordered. Occurrence of Bjerrum (D and L) or ionisation defects, however, limits the possible rotations.

Ice II, as in the case of hexagonal ice, has been built by the units of puckered hexagonal rings. In ice II, these puckered rings are moved up and down along c axis, which results in a rhombohedral stacking sequence (please see Fig. (**7.11**) for the definition of a and c axes). A 30 degree rotation around the c axis yields the structure of ice II.

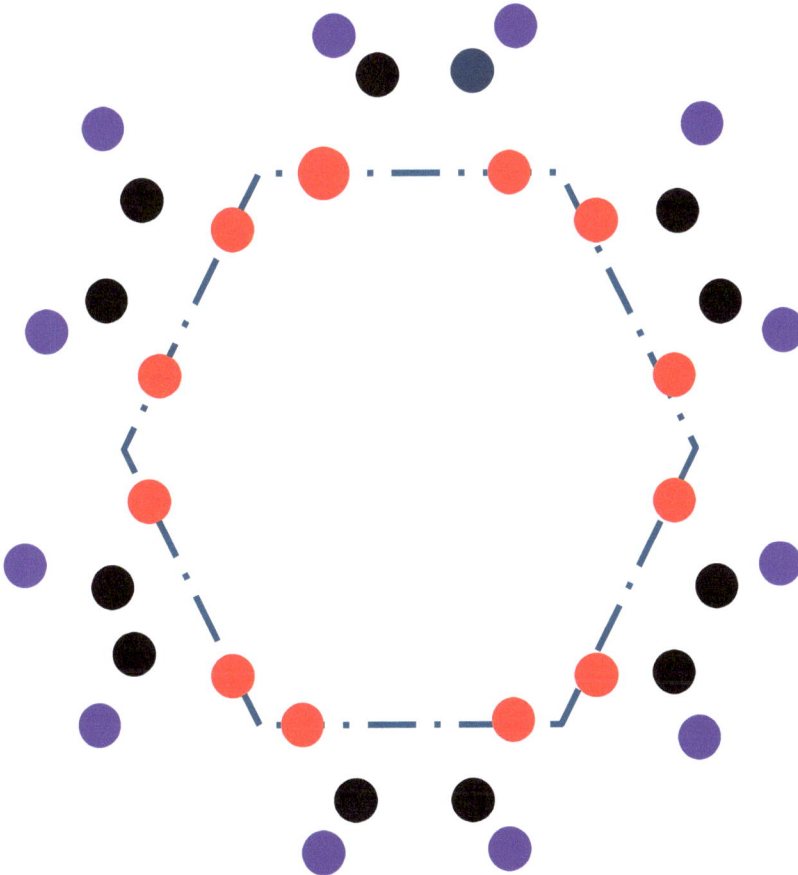

Fig. (7.12). Schematic diagram of the oxygen framework in ice II, viewed through the hexagonal c axis. Hydrogen atoms have not been shown for clarity.

The interatomic distance between two neighbouring oxygen atoms is estimated to be between 2.76 Å and 2.84 Å. Interestingly there are eighteen different oxygen–oxygen–oxygen angles in ice II, structure of which is shown in Fig. (**7.12**).

Ice III and ice IX (Fig. **7.13**) have the same symmetry, and the difference between them lies in whether the protons were ordered (ice IX) or disordered (ice III) [72].

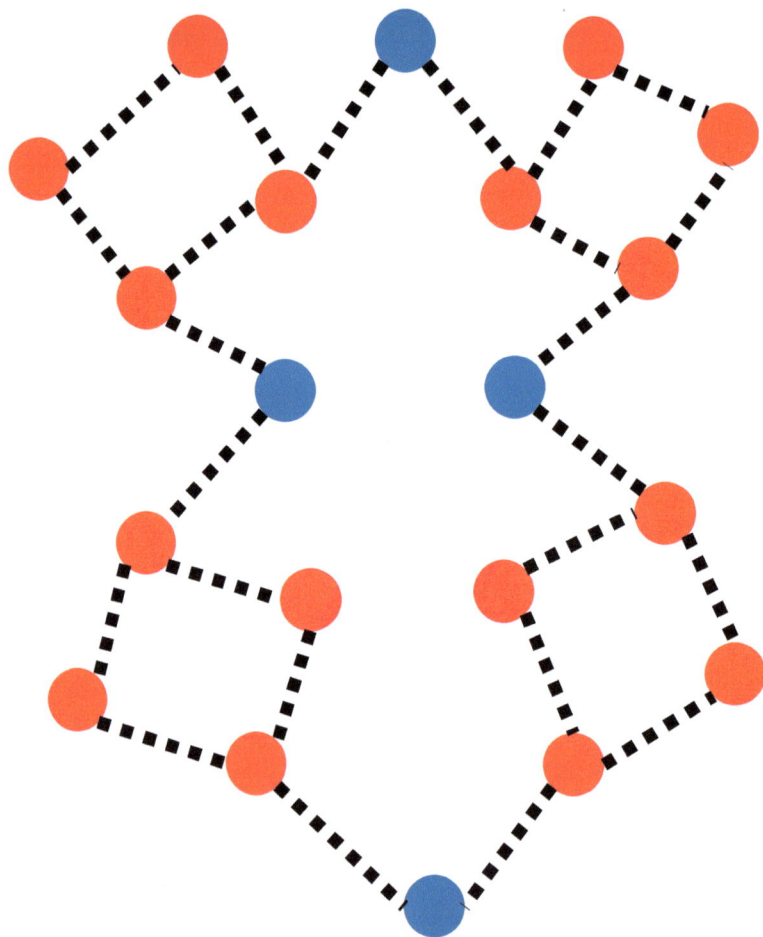

Fig. (7.13). Schematic diagram of Ice III and IX (viewed along c axis).Two types of oxygen atoms in the structure are shown as red and blue. Oxygen atoms shown in red are part of the principal network of water molecules forming helices, and the blue circles are the oxygen atoms that link these helices. Hydrogen atoms have not been shown for clarity.

The properties of ice VI (Fig. **7.14**) offer an interesting point for discussions on ring structures of water–based materials [202]. The variation of the hydrogen–oxygen–hydrogen angle is found to be the highest, form 98 to 115 degree, in this phase of ice: two fold increase compared to ice II and ice III [202]. Two kinds of hydrogen bonding patterns can be observed in ice VI: hydrogen bonds as part of a four membered ring and an eight membered ring with smaller and higher hydrogen–oxygen–hydrogen bond angles respectively.

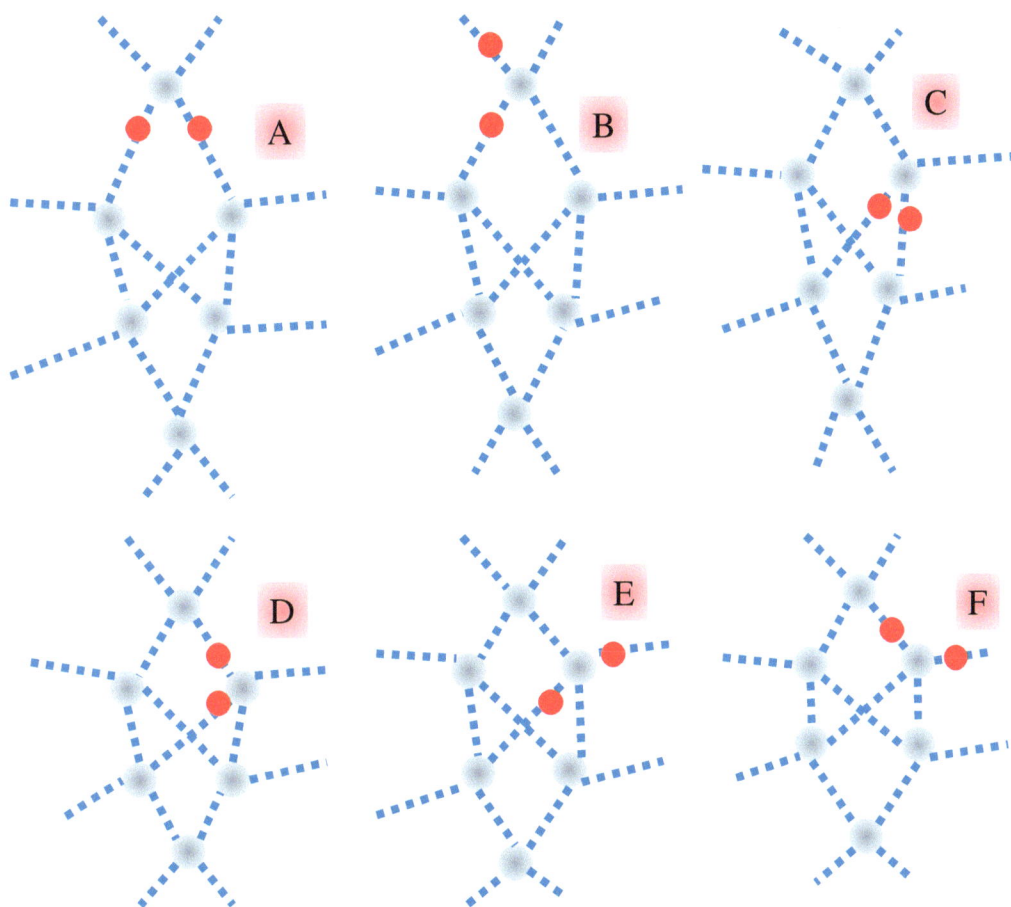

Fig. (7.14). Possible orientations of hydrogen atoms in ice VI (from a through f). Network of six oxygen atom is shown in the figures. If both hydrogen atoms, shown in red, are oriented towards any of the oxygen atoms shown in the figure as ash coloured circles, then the resulting configuration belongs to a four membered ring (**A, C & D**). Alternatively, one of the hydrogen atoms can orient towards oxygen atoms other than shown in the diagrams (**B, E & F**), resulting in an eight membered ring.

Computer simulation reveals that High Density Amorphous ice (HDA) has been transformed to a crystal that closely resembles to ice VII at high pressures [200]. On the contrary, experiments suggest that the resulting crystal is ice XII that has fourfold symmetry as shown in Fig. (**7.15**).

Fig. (7.15). Structure of Ice XII. Only oxygen atoms (shown as red circles) are shown for clarification.

The oxygen–oxygen pair correlation function reveals three distinct peaks within the distance of 6Å, indicating the sharp crystalline character of ice XII [210]. Oxygen–oxygen radial distribution functions of ice XIII and XIV, on the other hand, show four and five distinctive peaks within the distance of 6Å, suggesting that these phases are more rigid than ice XII [211]. Another salient feature of ice

XII (its density is closely matched with ice IV (1.4365 and 1.4361 g/cm^3 respectively), is that it is constituted entirely by five and seven membered rings of water molecules unlike other known phases of ice [77]. Occurrence of five, seven and eight member rings has also been reported for ice III, employing molecular dynamics simulations. As a result of deviation from ideal tetrahedral angle, the oxygen–oxygen–oxygen angle is found to be in the range of 80 to 140 degrees [212]. This suggests that the hydrogen bonding in the Ice III is weaker than the hexagonal ice. Distortion from the ordered tetrahedral structure has also been reported for ice III, V and VI in the literature [116]. Interestingly, Parinello *et al.* have noted that the formation of five and seven membered rings in ice crystal lattice is a characteristic of disorder, which leads to melting of ice [213].

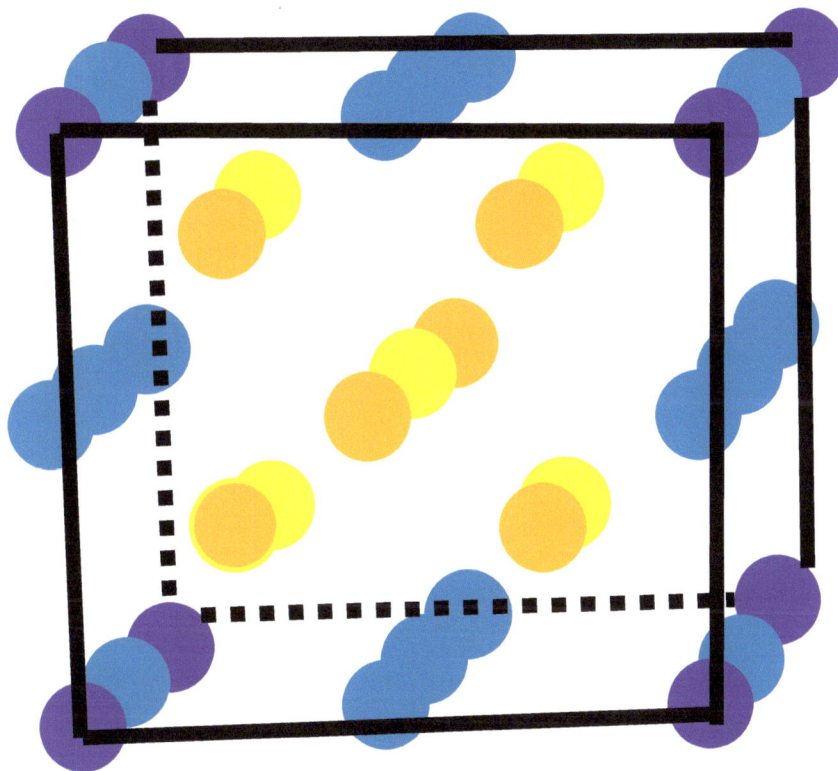

Fig. (7.16). The structure of ice VIII in a unit cell containing 16 water molecules. Oxygen atoms at the corners of the cell are shown violet and other oxygen atoms at the edges are indicated light blue. The other oxygen atoms within the cell are indicated by orange and yellow shaded circles. The hydrogen atoms have not been shown for clarity.

Ice VIII is considered as two fused cubic ice (I_c) structures, as clearly seen in the figure below, that more number of water molecules are squeezed in the given volume than in the cubic ice [214]. The schematic diagram of ice VIII is shown in Fig. (**7.16**).

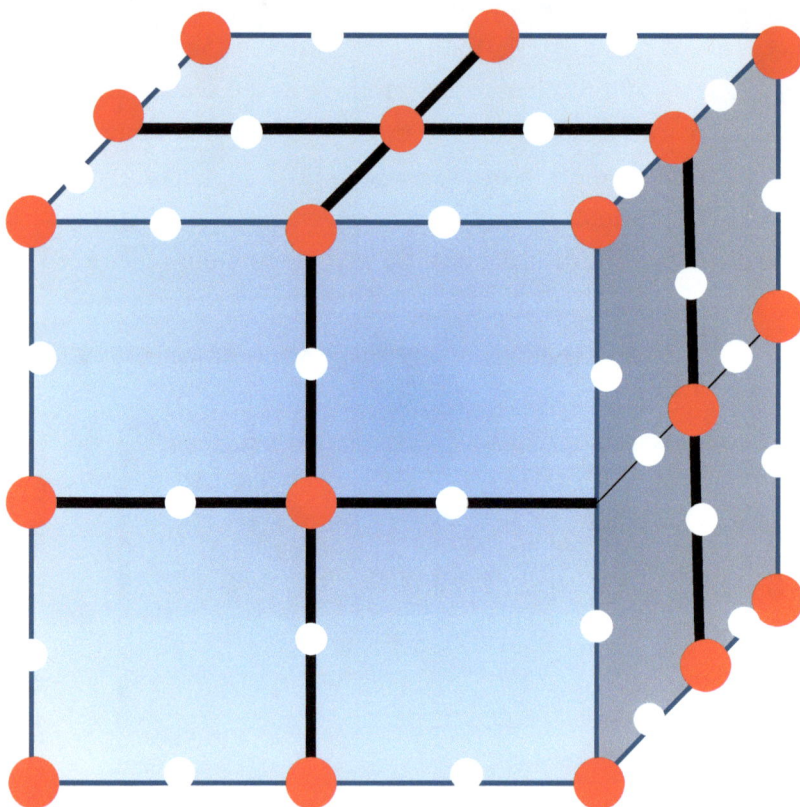

Fig. (7.17). Schematic diagram of ice X. Oxygen atoms are denoted by red circles, while hydrogen atoms are indicated by smaller white circles. The hydrogen atoms are positioned at exactly middle of the two neighbouring oxygen atoms.

Ice X is also known as symmetric ice [198]. Unlike other forms of ices, the hydrogen atoms in ice X (as shown in Fig. **7.17**) are positioned exactly in the middle of the distance between two neighbouring oxygen atoms [198].

The basic unit of ice XI is stacked with hexagonal rings of water molecules in three dimensional arrays as shown in the following diagram, (Fig. **7.18**) [215]. The major difference between ice XI and the hexagonal ice can be inferred from rotational spectroscopy [215]. The proton ordered ice XI produced four different modes of rotational motions (wagging, twisting and rocking), while rotational spectra of hexagonal ice yielded only two peaks. Configurations of water molecules in hexagonal ice are assorted, whereas in ice XI, since arrangement of hydrogen atoms are ordered, the "protonic" movements are restricted.

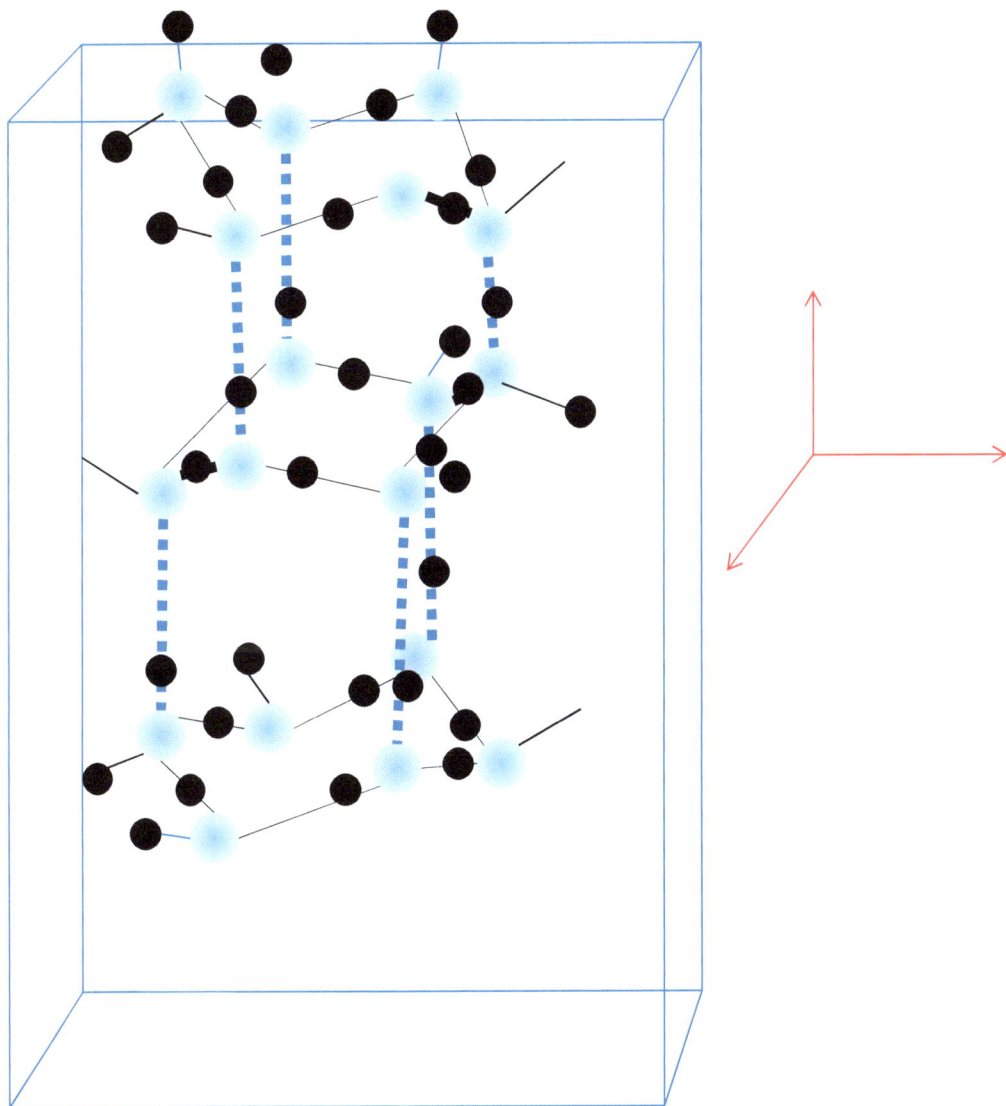

Fig. (7.18). Structure of ice XI. Oxygen and hydrogen atoms are denoted by sky blue and black circles respectively. Each layer in ice XI is constituted by puckered hexagonal rings.

It has been reported by several researchers that clusters, constituted by varying number of water monomers, are formed in all three major physical states of water [148]. Clusters with number of water molecules ranging between twelve and twenty six are considered as ice like since these water clusters contain hexagonal rings, important structural units of many of ice polymorphs that we have already seen in this chapter. Quantum chemical calculations suggest that in ice, large water clusters containing as many as 24 molecules can be formed, whereas in liquid water and water vapour, majority of water clusters contain lower number of water molecules [148]. However, clusters in supercooled water are of course much larger in size as well as number of participating water molecules than ice clusters as mentioned in the previous chapter.

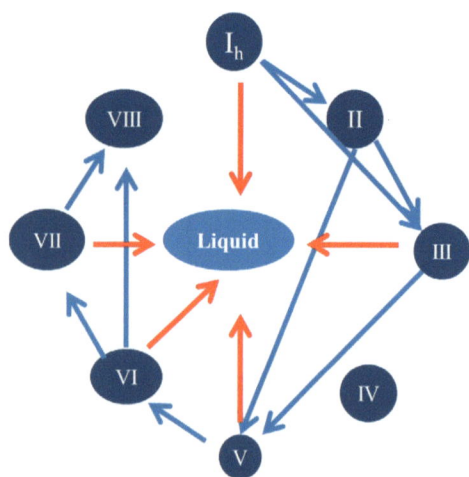

Transition	T (°C)	P(kbar)
$I_h \rightarrow II$	−35	2.13
$I_h \rightarrow III$	−22	2.08
	−35	2.13
II →III	−24	3.44
	−35	2.13
II → V	−24	3.44
III→ V	−17	3.46
	−24	3.44
V →VI	0.16	6.26
VI → VII	81.6	22
VI →VIII	~5	~21
II→VIII	~5	~21
$I_h \rightarrow$ Liquid	0.01	$6*10^{-6}$
III →Liquid	−17	3.46
V→Liquid	0.16	6.25
VI →Liquid	81.6	21.50
VII→Liquid	81.6	22.40

Fig. (7.19). Transformation of various forms of ice. The Interconversion between various crystalline ice forms is shown in the figure left. The red arrows indicate the transformations to liquid water and the blue arrows indicate transformations between various ice phases. On right, corresponding temperatures and pressures are shown.

Hydrogen bonding is the major driving force for the existence of intermediate ice forms, and the tetragonal coordination is a standard norm in ice structures. Much more interesting features can be observed in the case of high pressure variants: ice VII & ice VIII. In these high pressure forms of ice, each H_2O molecule is surrounded by eight other H_2O molecules, forming much stronger fused tetrahedral structures [216]. I_c, IV, IX, XII, XIII, and XIV forms are metastable while the rest are thermodynamically stable [197]. The transformation between these ice forms can be achieved by varying temperature and pressure. Various transformation routes have been shown in the picture given above, in Fig. (**7.19**). Interested readers are referred to the seminal book of Hobbs [72].

MODELS FOR ICE SIMULATION

As stated elsewhere, computer simulations have been emerged as one of the principal means to trace physical properties of materials in extreme conditions. Using the most commonly used computer models, Vega *et al.* have calculated the pressures and temperatures of its existence, which has been tabulated (Table **7.2**) [216].

Table 7.2. Experimental densities of various forms of ice obtained from computer simulations. Densities close to the experimental values have been shown in red. A detail analysis can be found in [216].

The form of Ice	Experimental Density (in g/cm³)	TIP4P/Ice	TIP4P/Ew	TIP4P	TIP5P	SPC/E
I_h	0.920	0.909	0.935	0.937	0.976	0.944
I_c	0.931	0.929	0.960	0.964	1.026	0.971
II	1.170	1.183	1.219	1.220	1.284	1.245
III	1.165	1.147	1.168	1.175	1.185	1.171
IV	1.272	1.276	1.308	1.314	1.371	1.324
V	1.283	1.255	1.289	1.294	1.331	1.294
VI	1.373	1.360	1.399	1.406	1.447	1.403
IX	1.194	1.174	1.202	1.210	1.231	1.219
XI	0.934	0.938	0.970	0.976	1.046	0.985
XII	1.292	1.282	1.312	1.314	1.340	1.313

As we can see from the Table **7.2**, TIP4P/Ice gives better estimate of density of the most of the ice phases.

Using various computer models, melting temperature of ice has also been tabulated for variety of water models, with which one can determine the stability of various forms of ice polymorphs. Abascal *et al.* have compared the stability of ice II and I_h using the 'modern water models (SPC, SPC/E, TIP3P, TIP4P, TIP4P/Ew & TIP5P) [217]. The calculated melting temperatures from computer simulations are found to be varied drastically from the experimental value. The melting temperatures of I_h are summarised in Table **7.3** [217, 218].

Table 7.3. Water models and melting temperatures of ice.

Water Model	Melting Temperatures of Ice
SPC	190 K
SPC/E	214 K
TIP3P	146 K
TIP4P	231 K
TIP4P/Ew	244 K
TIP4P/2005	250.5K
TIP4P/Ice	270 K
TIP5P	272 K
TIP5P/Ew	271 K
Experimental Value	273.15 K

Table **7.3** clearly demonstrates large differences among various water models in calculating physical properties such as melting temperature. Although SPC and TIP5P (so close to the experimental value) are two of the "efficient" water models that are widely used for molecular simulations today, the difference between them in this case is a huge 84 Kelvin. The discrepancy also exists in determining the stability of these two phases of ice, I_h and ice II [217]: from the comparative studies, except for TIP4P and TIP4P–Ew, all other models predicted the stability of ice II over I_h. We note that TIP4P–Ew is one of the two water models that provided better estimates of densities as we saw in Table **7.2**.

As mentioned in the chapter 4, the concerns over employing the most appropriate models in computer simulations are reflected here. Which is the best model that can accurately reproduce the physical quantities obtained from experiments for various forms of ice? This is an interesting question. If we have very precise answer, it will solve lots of questions that are still unanswered. As you can see, Vega *et al.* have extensively studied different modern water models that are popular nowadays in depth to find out the answer [216, 217]. Among these models, TIP4P variants gave the accurate estimates of various physical quantities: the most accurate values were found when TIP4P model was re–parameterised in the form of TIP4P/ice (please see Table **7.2** also). Several values have been obtained for the dipole moment of ice as in the case of water as well. Different genre of calculations yielded a broad distribution of results, ranging from 2.33 to 3.09 Debye [219].

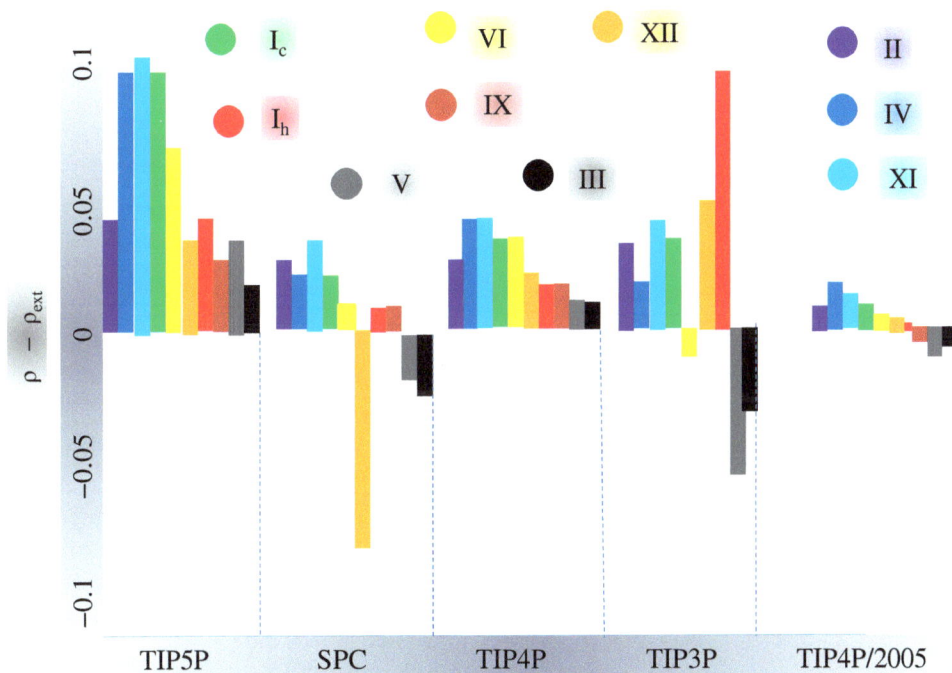

Fig. (7.20). Comparison of computationally calculated ice densities with experimental density for various contemporary popular water models. It can be seen that TIP4P models perform better than other contemporary water models.

Further to the studies conducted by Vega *et al.*, summarised in Table **7.2**, Aragones *et al.* have compared the performances of these water potentials (TIPnP

& SPC) for ice simulations [220]. They showed that another variant of TIP4P, TIP4P/2005, reproduced the experimental densities of ten ice polymorphs better than any other potentials (second best being TIP4P/ice with seven phases). Surprisingly, one of the latest water models belonging to TIPnP family, the five point TIP5P yielded poorer estimates of ice densities [220]. Please see Fig. (**7.20**) for a comparison between various water models.

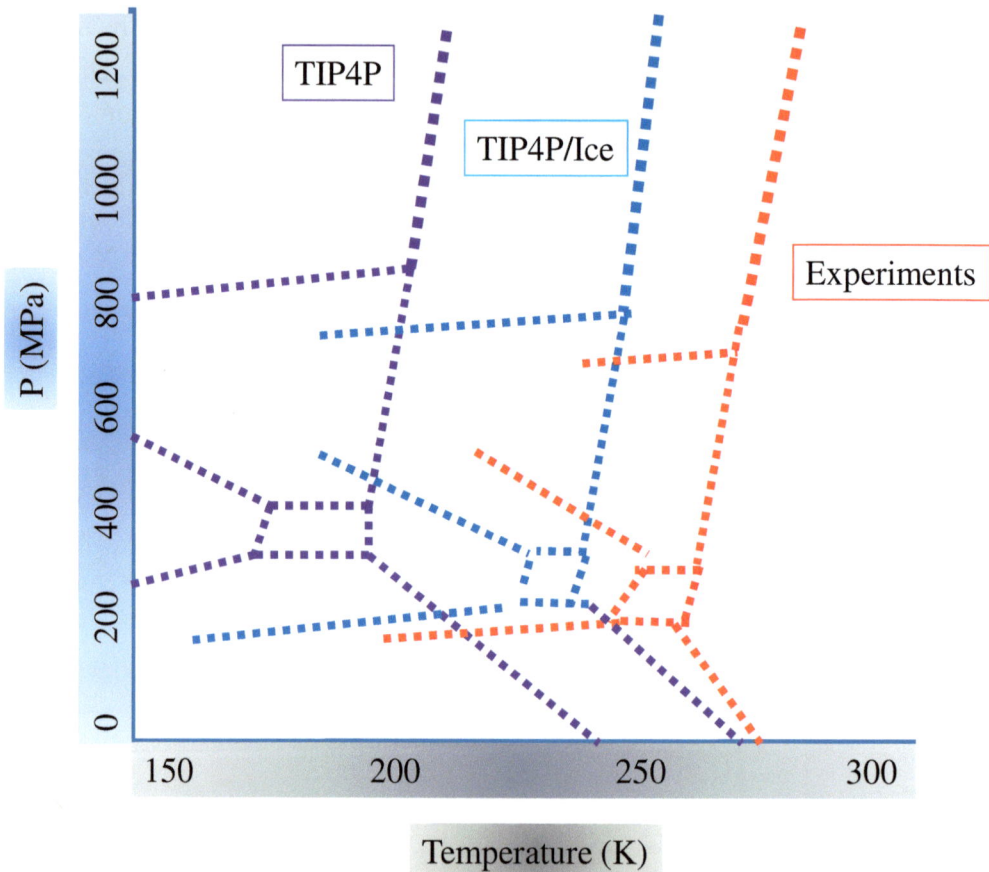

Fig. (7.21). Phase diagrams of ice obtained using TIP4P & TIP4P/Ice models. It is very clear from the figure that the phase diagrams obtained for both the TIP4P models resemble real ice. The idea has been adapted from [216].

The flexibility of computer simulations in mimicking experimental pressures and temperatures led to the computational phase diagram of the allotropes of ice. Vega *et al.* have obtained a phase diagram for ice (Fig. **7.21**) employing TIP4P and TIP4P/ice models [216]. The existence of various forms of ice with comparable accuracy has been predicted, suggesting that a refinement of the TIP4P/Ice model can reproduce the physical quantities very accurately.

The effectiveness of experimental investigations on cold water is often limited by crystallisation. On the contrary, since crystallisation is altogether absent due to large time scales required for nucleation, computer simulations can be employed for low temperature investigation on water and its crystalline and amorphous phases can be made by computer simulations. Because of this same reason, simulation of spontaneous crystallisation using most of the water models is difficult to achieve [221]. Sciortino *et al.* have, however, managed to investigate crystallisation, occurring at time scales ranging from nanoseconds to heptoseconds, using computational techniques. Crystallisation is characterised by sudden drop in potential energy and enhanced density fluctuations from supercooled state [221].

PROPERTIES OF ICE

Ice exhibits a wide range of interesting properties that can make significant impact on our lives. In this section, I venture into providing major properties and applications of ice known so far.

Ice exhibits thermoelectric effect, the ability to produce electrical potential difference due to the separation of opposite charges upon a temperature gradient [72]. This property is usually exhibited by metals and semiconductors. Another interesting property of ice is its stunning ability for storing charges (charge storage) [72].

Ice differs from water in many aspects due to the structural differences. Ice can be doped akin to semiconductors such as silicon (Si) and germanium (Ge), and in doing so a protonic p−n junction rectifier can be constructed [72]. Materials like hydrogen fluoride (HF) or polystyrenesulfonicacid can be employed as proton donors, whilst lithium hydroxide (LiOH) or poly 2−vinyl−N−n−butylpyridinium bromide (NBPB) acts as proton acceptors. The following diagram (Fig. **7.22**) shows a model protonic p−n rectifier.

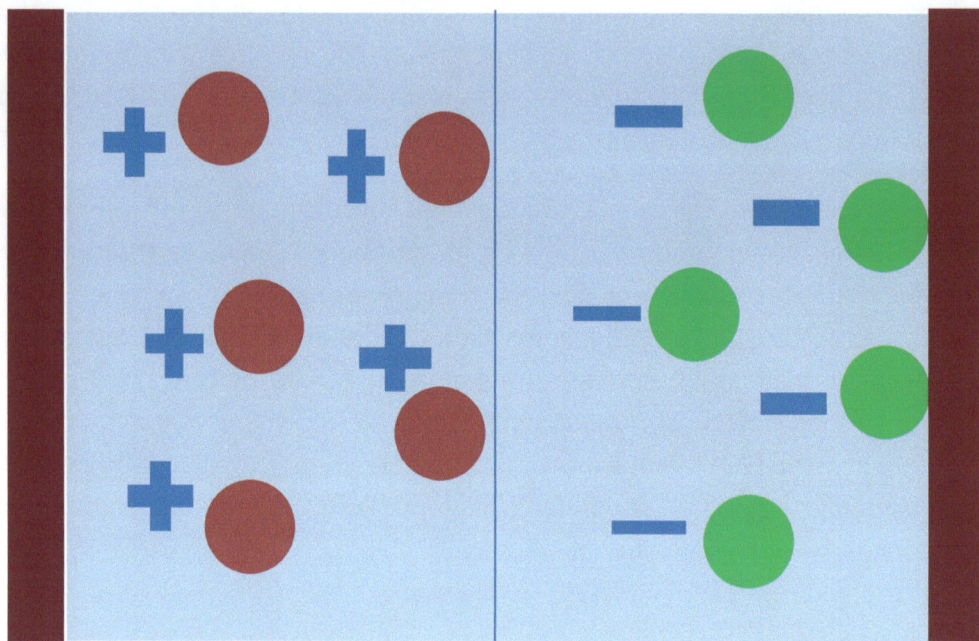

Fig. (7.22). Ice rectifier. A large block of ice is doped with negative ions (HF) and positive ions (LiOH) respectively, which is coated by two smaller compartments (shown in brown) which contain acid and alkali solutions promoting cationic and anionic exchange.

An interesting application of ice in relevance to chemical industry is its ability to catalyse certain biochemical reactions, for example hydrolysis of imidazole [72]. This property is quite unusual given the fact that decreasing temperature lowers the reaction rate. However, more investigations are required to explore the role of ice in catalysing major types of chemical reactions including oxidation and reduction, and dehydration.

Atmosphere acts as a buffer between the source (sun) and earth such that it regulates solar energy. Ice plays a vital role in dissipating excess solar energy by absorption and scattering, thereby keeping life on earth sustainable. These optical properties of ice have been employed in oceanography in the form of radio echo sounding techniques, which is used to identify the thickness of ice.

Ice has lower density than its liquid phase, normal water. This enables ice to float in water there by making a positive impact upon sustaining aquatic life. At lower temperatures, ice can mitigate the loss of heat from the liquid water underneath [72]. As temperature rises, ice crystals begin melting. What does cause the melting of ice? Kawamura *et al.* have provided a molecular level explanation, based on their Molecular Dynamics simulations [206]. At lower temperature, the rotational motion of the dangling oxygen−hydrogen bonds at the surface of ice is restricted by stronger hydrogen bond network of bulk ice. As temperature increases, the oxygen−hydrogen bonds gain rotational degree of freedom, which causes the distortion of structure of ice; this eventually leads to the melting of ice. Ice when heated by strong infrared or radio frequency radiation produces Tyndall flowers, a scintillating structure reminiscent of flowers as shown below, in Fig. (**7.23**).

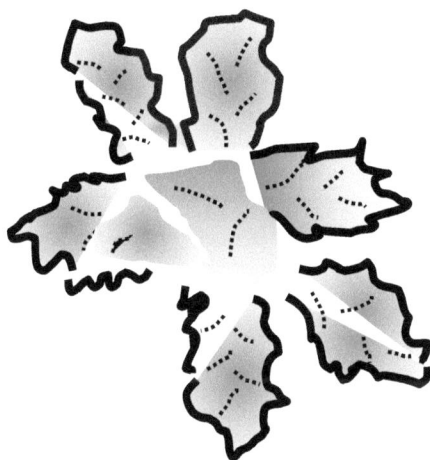

Fig. (7.23). A sketch of Tyndall flower, a structure obtained during metamorphosis of ice when it is heated by strong infrared radiation.

Ice also exhibits several other intriguing thermal properties, including thermoluminescence (emission of radiation due to heating) [72]. Several propositions have already been made by researchers in order to explain this thermal effect. A group of scientists believes that thermoluminescence occurs due to a liberated electron from the sample after warming. This electron is migrated to an impurity centre caused by radiation, while others believe that thermoluminescence occurs as a result of electron being trapped at structural defects within the ice crystal [72]. Ice can store heat, an interesting thermal property that can be made use for reducing the power cost for air conditioning [222]. Ice thermal storage technique is based on a simple concept: the ice that is formed overnight is stored in well−insulated storage. The coolness can be extracted from the ice for day time. It has been claimed, based on computer simulation investigations, that the electricity cost can be cut up to 45 percent of the cost of the traditional air−conditioning methods [222].

CONCLUDING REMARKS

Various polymorphic forms of ice were reviewed in this chapter. Numerous experiments and computer simulations have been performed in order to understand the transformation from one polymorphic form to the other. It has also been found out that at enhanced pressures, polyamorphous forms such as High Density Amorphous ice (HDA) can be converted to crystalline ice forms. In coming years, more and more inter−conversion routes are expected to be functional. Bernal−Fowler rules, devised decades ago, still serve as the fundamental guidelines regarding the possible orientations of protons in ice structures, and these orientations define the existence of various forms of ice. Both Bjerrum and ionisation defects play important roles in further restricting the number of orientations that water molecule can have. Despite the fact that various crystalline forms of ice possess plenty of excellent properties, an intense effort is required to convert them to industrial applications.

CONFLICT OF INTEREST

The author confirms that he has no conflict of interest to declare for this publication.

ACKNOWLEDGEMENTS

I express my gratitude to American Institute of Physics for granting me permission to use one figure under their copyright in this chapter.

CHAPTER 8

Water Above its Boiling Point

Abstract: Water beyond its boiling temperature has been investigated by numerous theoretical and experimental tools including classical and quantum simulations, Neutron Scattering, Nuclear Magnetic Resonance, Infra–Red, and a more recent Tetrahertz Vibrational–Rotational–Tunnelling spectroscopic technique. Water becomes Super Critical Water (SCW) when it reaches temperature 647 K with critical pressure and density at which vapour and liquid phases coexist. SCW has been found to have exceptional properties such that many chemical reactions can be efficiently carried out without the presence of catalysts, and hence is considered as a better alternative to many of the traditional reagents, which are currently being used in organic synthesis. It is also effective in biofuel production and in burying toxic waste products by converting them to water and CO_2. Simulations and experiments point into dramatic changes in the intermolecular structure of water at elevated temperatures signalled by depletion of tetrahedral hydrogen bonding network, which gives rise to higher population of water clusters than found in ambient water, with varying size and geometry. Cyclic water isomers are found to be stable up to clusters containing five water molecules, whereas three dimensional cage structures are found to be stable in higher analogues. Density along with temperature plays a vital role in determining diffusive properties of SCW. It has been demonstrated by numerous experiments and computer simulations that a proportional increase in hydrogen bonding is observed as density increases. On the contrary, diffusive motion of water molecules is retarded upon a hike in density.

Keywords: *Ab-initio*, Aldol, Benzamide, Cannizzaro, Density, Dimers, Friedel Craft, Hexamer, IR, NMR, Pentamer, SCW, Tetrahedral, Tetramer, Trimer, Water Clusters.

INTRODUCTION

Liquid water becomes water vapour beyond its boiling point, 373 K, under normal heating. The boiling temperature of water is inextricably linked to the upper limit of the temperature at which organisms can survive, 100°C (373 Kelvin): animals can survive temperature up to 51°C (324 Kelvin); Eucaryotic microorganisms can live bit longer (333 K), while photosynthetic procaryotes continue to exist up to temperatures around 350 K [223]. Certain bacteria can survive bit further, up to 373 Kelvin.

However, in analogy of supercooled water (below melting point of ice), liquid water can exist as superheated without being converted into vapour above its boiling point. Water beyond its boiling point becomes Super Critical Water (SCW) by attaining its critical temperature 647 K (374 °C), pressure 22.1 MPa and density 0.17 g/cm^3, and behaves like a gas–like fluid. Physical quantities such as dielectric constant (ε) and specific heat capacity (C_p) can take their anomalous values at these conditions, ~6 and 29.2 kJkg^{-1}K^{-1} respectively [224].

On the contrary, the role of Super Critical Water (SCW) in biological activities is almost negligible. However, it offers numerous unremitting applications, which have high economical, industrial and environmental impact. In this chapter, I venture into reviewing the applications and properties of water beyond its boiling point (in particular Super Critical Water) and major experimental and computational investigations on it. The chapter concludes with an analysis of small water clusters that are believed to exist at super critical regime.

APPLICATIONS OF SUPER CRITICAL WATER

The role of Super Critical Water (SCW) has been reported in the scientific literature principally as a reaction medium. Super Critical Water (SCW) can be useful in several ways: first it can act as solvent medium in several types of reactions including hydrolysis, dehydration, hydration and partial oxidation [224, 225]. Sub Critical Water (water heated beyond its boiling point without being boiled) is also found to be effective for better product conversion rate in certain aforementioned reactions [224, 226]. Enhanced reactivity in SCW can be exploited for chemical waste disposal since SCW can oxidise highly hazardous chemicals without any trace [227]. Hydrothermal technologies based on Super Critical Water (SCW) and Sub Critical Water can replace the traditional power generation methods to a greater extent [226].

Hydrolysis is one of the most common reactions in organic synthesis and the roles of SCW are remarkable in these types of reactions in terms of percentage yield and reaction kinetics. Hydrolysis of the ester 1,4–butanedioldiacetate is a typical example. Under normal conditions, hydrolysis of this ester gives only 38% yield. Amazingly with SCW, the yield can be increased to almost 100% [224]. Certain hydrolysis (for example benzamide hydrolysis) conducted in the presence of SCW occurs at faster rate. Similarly, in the dehydration reactions, for example ethanol to ethylene and glycerol to acrolein, SCW offers better yields than conventional free radical reactions. The exceptional catalytic activity of SCW is due to its ionic properties which spur on specific heterolytic bond cleavages [228]. Similar trends have also been observed in other two types of reactions just mentioned before. One of the exceptional properties of SCW is high feasibility of certain reactions without the presence of acidic or basic medium, which do not occur with normal water. Dehydration of 1,4–butanediol, a key industrial reaction to produce Tetra Hydro Furan (THF), is a prime example. There are several disadvantages, namely corrosion and generation of chemical waste, associated with the traditional synthetic routes of THF synthesis, due to the use of numerous chemicals including acids and salts. But with SCW the conversion from diols to THF can be achieved with high yield with minimal chemical waste [224]. Notable example for enhanced oxidation in the presence of SCW is catalytic reforming whereby methane, stable up to 723 K, is oxidised yielding CO_2 and H_2 with 90% conversion rate [226]. A large number of well–known reactions can be conducted with SCW without the support of catalysts: Cannizzaro reaction, Friedel–Craft reactions, Aldol condensation and to name a few [229]. This has a profound advantage in chemical industry if you consider the cost of chemicals and chemical industries polluting environment. Furthermore, as the world becomes more environment conscious, this "green" route is expected to replace traditional chemical methods polluting our healthy life. It is believed that this enhanced auto catalytic activity is due to the storage of vast amount of energy in SCW and to the fact that the concentrations of H^+ and OH^- ions in SCW is 30 times higher than ambient water [79, 229]. In addition, advanced computer simulation studies indicate that dissociation of acid and alkali increases in Super Critical Water [230].

Super Critical Water Oxidation (SCWO) can be highly useful for removing solid waste generated by large cities and industries. The waste can be converted (oxidised) to CO_2, water and molecules such as nitrous oxide (N_2O) and nitrogen (N_2) without being converted to ammonia (NH_3) and other toxic products including nitrogen dioxide (NO_2) [226]. Other elements containing in the waste, such as phosphorous, sulfur and chlorine, can be converted to corresponding mineral acids H_3PO_4, H_2SO_4 & HCl that can be used again in chemical reactions [226].

Traditional power generation methods, for example those driven by fossil fuels, pose threat to our environment by polluting natural resources by emitting relentlessly CO_2, one of the principal contributors to global warming, into atmosphere as mentioned in the introductory chapter. The degree of pollution caused by traditional power generating houses can be greatly reduced by adapting technologies that are effective in super critical conditions [226]. One of such enterprises is Ultra Super Critical Plants (USCP), where in pulverised coal in conjunction with Super Critical Water (SCW)/steam is used. This technology is found to minimise the drastic emissions and increase the efficiency by 10% than the traditional coal power plants [231].

PROPERTIES OF SUPER CRITICAL WATER

Super Critical Water (SCW) possesses very different properties with respect to normal water; for example, it forms less number of hydrogen bonds, and possess low dielectric constant than normal water. At the same time, its isothermal compressibility is higher than normal water [229]. The most exceptional behaviour of SCW is its ability to dissolve nonpolar substances such as hydrocarbons. The solubility of alkanes such as n−heptane, n−pentane and 2−methyl pentane in water is less than 10% of weight of solute in normal temperature and pressures. Beyond their critical solution temperatures (353°C, 351°C & 352 °C respectively) pressure has little effect on their solubility, and the solubility increases to 50% by the weight of the solute [232]. Benzene and water, immiscible at normal temperatures and pressures, are completely miscible at temperature beyond 570 K at 200 bar [232, 233]. Due to ion−ion interactions, on the contrary, ionic compounds do not dissolve in SCW, which is a reversal of solvation properties of water in the normal temperature domain. This feature of SCW is reflected in the decrease of dielectric constant of water: from approximately 80 at 298 Kelvin to 12 at 650 K and 300 bar [232, 233].

EXPERIMENTS AND SIMULATIONS ON SUPER CRITICAL WATER

The enigmatic properties of water at high temperatures (Sub and Super Critical Water) prompted a genuine interest in investigations of its structure and dynamics, which are too complex for facile summarisation. Besides classical computational approaches such as Molecular Dynamics (MD) and Monte Carlo Simulations (MC), experimental methods such as scattering (Neutron Diffraction experiments) and spectroscopic investigations (Nuclear Magnetic Resonance Spectroscopy (NMR), Infra−Red (IR) Spectroscopy and Far−Infra−Red (FIR) Vibration-Rotation−Tunnelling (VRT) spectroscopy) have been used in abundance to elucidate inter molecular structure of water at very high temperatures.

There is fundamental structural difference between supercritical water and normal water: tetrahedral network is one of the principal trademarks of water at ambient conditions (although there is a counter argument against this concept in the form of Mixture models as we saw in a previous chapter) and in lower temperature domain, whereas at higher temperatures this network tends to be broken. This comes about due to the reduction in number of hydrogen bonds available to complete the three dimensional framework at elevated temperatures, as revealed from early Neutron Diffraction investigations carried out by one of the pioneer experimentalists on water Alan Soper and co-workers, evidenced from the disappearance of peaks hydrogen − hydrogen and oxygen − hydrogen pair correlation functions [234, 73b]. In ambient conditions, oxygen − hydrogen radial distribution peak in water appears to be at 1.9 Å, which at 673 K disappears completely [73b]. However this trend has not been exactly reproduced in a later Neutron Scattering experiment: Guissani *et al.* observed that a weak peak is still retained at the supercritical state [235]. Notwithstanding the discrepancies between different sets of results, it is evident that the structure of ambient water is perturbed significantly at elevated temperatures. The alteration of temperature enfeebles orientation−correlations between neighbouring water molecules, as indicated by disappearance of two peaks centered at 2.4 Å and 3.8 Å in hydrogen

– hydrogen pair correlation functions [235]. Changes in oxygen – oxygen radial distribution functions are also observed, albeit less drastically: at higher temperatures, the characteristic peak of water at 3 Å moves towards higher distances and the subsequent peak at 4.5 Å disappears [73, 229]. The depletion of tetrahedral hydrogen bond network beyond a threshold temperature (> 400K) is also reported based on experimental and computational investigations [136, 236]. The peaks at 4.5 and 6.7 Å, corresponding to second and third shell neighbours are also broken at 416 K [236a]. The structural depletion has also been observed from the distribution of angles made by oxygen atoms of three neighbouring water molecules: at super critical temperature regime the characteristic peak in the distribution angle, corresponding to tetrahedral coordination, disappears [136]. This implies an increase in the molecular mobility of water molecules, as evident from the high diffusion coefficients of water at higher temperatures [237].

Notwithstanding a continuous decrease in the sharpness in the corresponding radial distribution functions is observed, first shell structure of water molecules, characterised by oxygen – oxygen pair distribution function for a large temperature range (from 220 K through 365 K) is retained [238]. At the same time, pair distribution plots derived from Empirical Potential Refinement Simulations (EPRS) reveals that the second peak of the oxygen–oxygen distribution functions is altered with higher temperatures [238a]. From Molecular Dynamics simulations, employed using SPC/E water model, the second peak was found to be totally disappeared while characteristic peak of water (the first peak) was retained at 2.85 Å [235]. The disappearance or weakening of peaks at elevated temperatures is attributed to the increase in kinetic energy that suppresses the attraction between water molecules due to hydrogen bonding [230].

From the above discussions it is turned out to be that as temperature increases one observes a lack of correlation in all the three water pair distribution functions [230]. It is interesting to note that the hydronium oxygen – water oxygen, as well as hydroxide oxygen – water oxygen pair distribution functions show a greater level of structuring than that of pure water even above critical temperature; however, the structuring also in this case diminishes generally as temperature increases. This property has however both pros and cons in relation to the application of Super Critical Water (SCW) in chemical industries. The advantage is that higher ionisation constant facilitates acid and base catalysed reactions with better yield. On the other hand it can be seen that higher ordering of hydronium (H_3O^+) ions and hydroxide ions (OH^-) is a result of higher degree of dissociation, as more hydroxide or hydronium ions are produced, which can result in corrosion reduces the life span of reaction vessels.

As mentioned in chapter 3 at length, Neutron Scattering techniques are so helpful in elucidating water dynamics. A comparison of water dynamics in ambient water and Super Critical Water has been made by Bellisent–Funel *et al.* [80c]. They identified three types of movements of water molecules (in terms of three different bands) corresponding to intermolecular bending and stretching motions of hydrogen atoms, and liberation motion of water molecules in the bulk. On the contrary, in the super critical state, only one band was present which indicates the reduction of hydrogen bond population [80c].

Despite the depletion of hydrogen bond network for SCW, it is really surprising to see that even at temperatures beyond 673K, three hundred Kelvin beyond the boiling temperature of water (373 K), approximately 40% of hydrogen bonds is retained [229]. This means that every water molecules can make at least 1.5 hydrogen bonds per molecule. Kalinichev *et al.* argue based on IR and diffraction experiments that hydrogen bonds even exist at temperature as high as 800 K [86b]. They compared the intensities of oxygen – hydrogen vibration band in water across a wide range of temperatures. The intensity, indicative of the strength of hydrogen bonding, decreases with respect to temperature but is found to be higher than water monomer in gaseous phase at 823 K. Bellisent–Funel *et al.* using SPC/E water model have provided a quantitative estimation of average number of hydrogen bonds (2.2), higher than the experimentally measured 1.8, existing at 653 K, above the critical temperature of water [235]. Nuclear Magnetic Resonance Spectroscopy (NMR) experiments further reveal a proportional increase in the number of hydrogen bonds in Super Critical Water (SCW) with respect to the density.

"Generally" one can say that as temperature increases, the average number of hydrogen bonds per water molecule decreases [239]. As a result, the hydrogen bond network existing at normal and sub–zero conditions is broken at high temperatures [240]. Similar results have also been obtained from Empirical Potential Structure Refinement (EPSR) simulations by Soper and *ab−initio* simulations performed by Parinello *et al.* [73a, 241]. From a Molecular Dynamics simulation study, Vallauri *et al.* have found out that the average number of water molecules in a cluster or water network is reduced dramatically from 500 (298 K) to a mere 6 (673 K) [136]. But this value depends upon how hydrogen bonding angle is defined: when hydrogen bond angle was changed from 30° to 45° and 90° they obtained 249 and 403 respectively at 673 Kelvin [136]. In addition, various other computer simulation investigations employing different methodical protocols suggest the weakening of hydrogen bond structure, albeit quantitatively they differ with each other (with *ab initio* findings showing a large variation). Contrary to two other simulation findings, *ab−initio* radial distribution functions show a complete collapse, which might be due to the fact that the simulation length was too short in the order of several pico seconds (ps). A better and conclusive picture can be obtained from Dio's approach. Obtaining accumulative radial distribution functions (grund radial distribution function (grdf)) Dio has showed that the hydrogen bond length increases upon the increase in temperature, a trend indicative of the weakening of hydrogen bond at elevated temperature [242]. Further, it shows the absence of oxygen atoms and hence water molecules, from the corners of tetrahedral structure, strongly indicating the rupture of three dimensional water frameworks. However, there is in fact a slight possibility, as Kalinichev *et al.* argues, for the existence of the tetrahedral structure, albeit with short lifetime, even around supercritical conditions because of the freely moving water monomers, which may cluster together albeit for very short time scales [86b].

Is the thermal energy supplied in excess in high temperatures only responsible for the breaking hydrogen bonds in Super Critical Water? Precisely not since one must also consider the variations in density while accounting for the number of hydrogen bonds because of the formation of wide range of density regions, a characteristic phenomenon observed in Super Critical Water. Jin and Ichi Ikawa demonstrated from their IR spectroscopy experiments that the existence of hydrogen bonding depends primarily upon density by monitoring molar absorption intensity over concentration [237]. Also, it has been shown by

Nakahara *et al.* employing Molecular Dynamics simulations that variation in density significantly alters the hydrogen bonding environment, corroborating the experimental observations of Ikawa *et al.* [243]. It has to be noted that the energy cost for the formation of a hydrogen bond is 20 kJ /mol, whereas average thermal energy is only 5−6 kJ/mol in the temperature range around the supercritical temperature (573 K to 673 K). What we can infer from these is that breaking and making of hydrogen bonds is primarily influenced by constantly changing density map of Super Critical Water.

NMR spectroscopy can equally be employed to monitor the transport properties such as diffusion in liquids. Nakahara *et al.* have measured diffusive properties of water for a wide range of temperatures [244]. It can be seen that at a particular density threshold the diffusion is reduced drastically, in sub and super critical region for water. Density affects more than temperature does on the diffusive properties of Super Critical Water [79]. Diffusion is found to be directly proportional to temperature, while an inverse relationship holds in the case of density [244]. Density is also influential in governing the interactions existing in water: at higher density intramolecular interactions are preferred over inter molecular interactions [80b]. With IR spectroscopy also, one can investigate the bonding environment. The dynamics of water at high temperature was investigated Jin *et al.* by employing Infra−Red (IR) spectroscopy [237]. Fragmentation of hydrogen bond network has been correlated with the free movement of water molecules, which they inferred from the fast decay of absorption of hydroxyl group (OH) overtone band at around 6500 cm^{-1}, attributed to hydrogen bond free water molecules [237].

Although the physical properties extracted from simulations primarily depend upon the choice of water models, a conscience on the shape of phase diagram of water at higher temperatures has been generally achieved, with a density maximum within the range of 600 K to 650 K. A rough sketch of a phase diagram of water derived from a computer simulation investigation is shown in the following figure (Fig. **8.1**) [240].

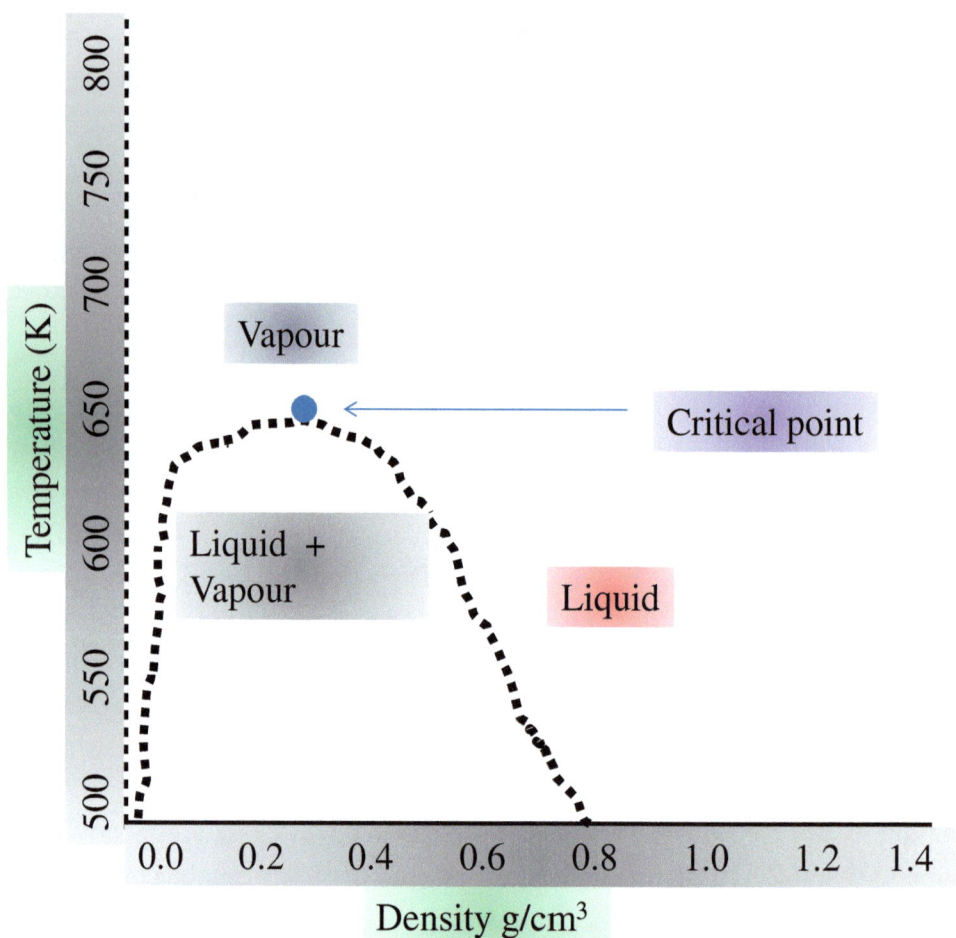

Fig. (8.1). Temperature − density profile of water around its super critical temperature (647 K), obtained from computer simulations using a flexible water model (BJH). The critical point is marked with a blue dot where the density is recorded maximum (0.3 g/cm³) at 647 K. Inside the density contour, liquid and vapour forms of H_2O coexist.

WATER CLUSTERS

The depletion of hydrogen bonding network has an important consequence on the dynamics of a system: decoupling of translational, rotational and vibrational degrees of freedom [80b]. This 'structural disorder' however does not preclude the association of water molecules because they in fact can associate themselves as clusters. Employing far Infra−Red Vibrational−Rotational−Tunnelling (VRT) spectroscopy, the internal dynamics of water clusters of varying sizes has been characterized in depth and reported by Saykally and co–workers in numerous articles [90, 245]. There are principally two intermolecular vibrational bands of water can be monitored using VRT experiments: one band corresponds to the translational motion signifying hydrogen bond stretching vibration at 180 cm^{-1} and the other corresponding to liberation motion, which spurs on breaking and making of hydrogen bonds, in the range of 300 cm^{-1} to 1000 cm^{-1} [246].

Small water clusters, the water aggregates containing from three to twelve water monomers, can be broadly classified into three: small quasi–planar water clusters with only three to five water monomers per structure belong to the first category, hexamers, the smallest three dimensional cluster belongs to the second category, and larger three dimensional structures ranging from seven to twelve water monomers belong to the third category [148]. The presence of smaller water clusters have been noted in super–critical temperature regime by several researchers. Ludwig *et al.* have theoretically predicted the abundance of water pentamers [148], while using a flexible water model. Kalinichev and Churakov proposed the existence of small water clusters even above critical temperature of water, with maximum number of water molecules per cluster equal to ten, and with a mix of high proportion of linear chains and relatively low proportion of closed chains [240].

We saw in previous chapters that how cluster size is affected by change in temperatures [139]. X–ray inelastic scattering experiments also reveal the same fact [81]. At 473 K, clusters containing six or more water molecules are abundant in the bulk water with very small percentage of monomers. At 573 K, the size and number of clusters decreases while number of water monomers increases. Further rise in temperature results in dominance of water monomers with very small number of water clusters with five $((H_2O)_5)$ or six $((H_2O)_6)$ members [81]. Thus it can be said that smaller clusters, in particular five and six water monomers are likely to be abundant in supercritical conditions [247]. This can be attributed to the fact that the angle between oxygen atoms of these monomers (O–O–O) is very close to the tetrahedral value minimising hydrogen bond interactions [245a]. These clusters are believed to undergo rearrangements (for example from pentamers to hexamers and *vice versa*), which influences many properties of water such as high melting and boiling points [248].

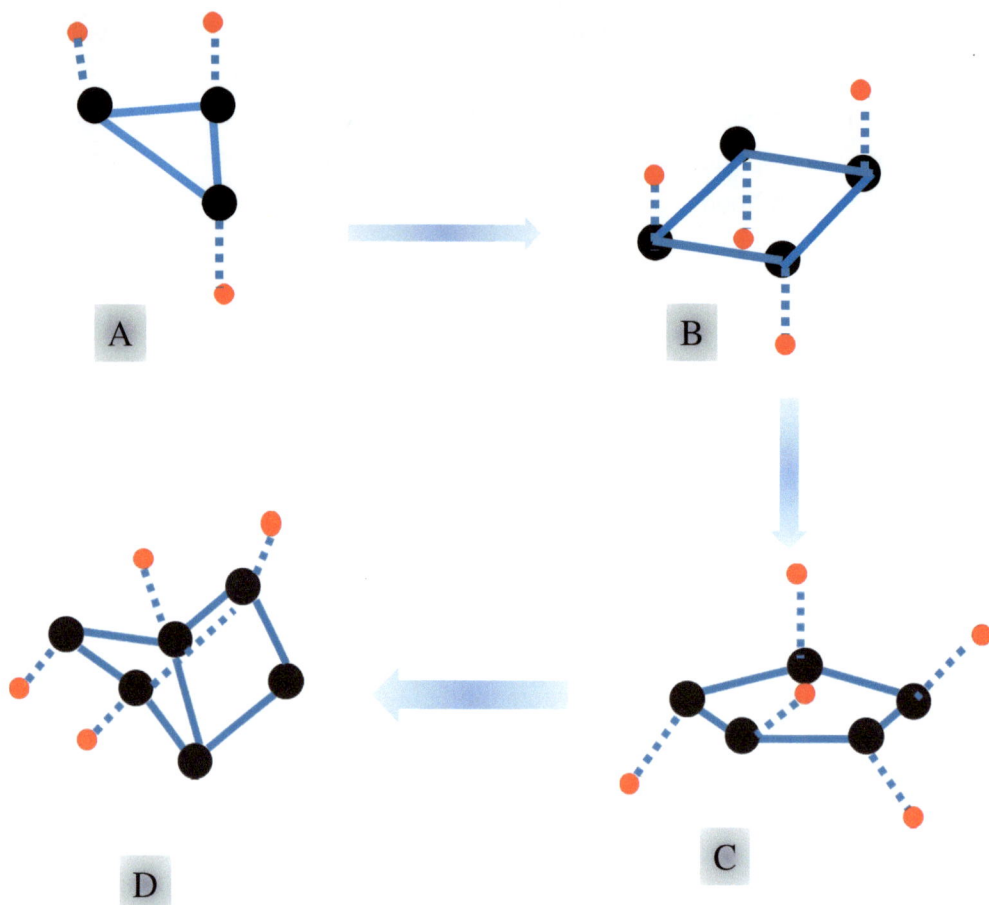

Fig. (8.2). The lowest energy structures of water to trimers (**A**), tetramers (**B**), pentamers (**C**) & hexamers (cage) (**D**) are shown in the figure. It must be noted however that the hexamer cluster having the lowest energy has not been agreed upon universally [255].The blue lines show hydrogen bonding network, whereas the red spheres are hydrogen atoms covalently attached to oxygen atoms (shown black), which are responsible for vibrational motions detected from experiments and simulations alike. Some of the hydrogen atoms in hexamer cluster have been omitted for clarity. The idea has been adapted from [249a].

Ab−initio calculations, which can take account of complex electronic effects, have reinvigorated interests in the investigations of water clusters, with which micro level analysis of the topology of water clusters of varying sizes formed under various thermodynamic conditions can be made [249]. Electron delocalisation play a non negligible role in the dynamics of small clusters containing three or four water molecules [250]. Quantum chemical calculations reveal that monocyclic arrangement in trimers through pentamers are found to be having minimum binding energy (global minimum), the energy required to break the system into its constituents [251]. These structures are characterised by oxygen framework of water monomers set up by hydrogen bonding between each of them (each water monomer can act as a single donor and a single acceptor of hydrogen bonds), with remaining hydrogen atoms projecting out above and below of the oxygen frame. This skeleton of the oxygen atoms (O–O–O…..) defining the molecular plane determines the shape of the water clusters, coordination number of individual monomers and average number of hydrogen bonds [252].

The oxygen molecular frame in pentamers undergoes distortion than in trimers and tetramers known as puckering [248, 253]. These monocyclic structures are found to be more stable than their nearest counterparts by 2–3 kcal/mol [251]. In the case of hexamers, however, theoretical and experimental findings diverge drastically. Several theoretical works have produced different conclusions assigning prism, cyclic, book and cage structures as the most stable hexamer clusters [254]. Fig. (**8.2**) shows energetically favourable water clusters for trimers through hexamers.

Higher configurations of water clusters have also been studied well primarily by theoretical tools. Using high level quantum chemical calculations, Kirchner has identified the cubic isomer is the most stable among octamers due to possibility to form more hydrogen bonds [256]. Kazimirski *et al.* have observed that as the size of the clusters increases, they become highly symmetrical and compact [252]. Two competing effects, reduction of geometrical strain and maximising the number of hydrogen bonds, are suggested to be the deciding factors regarding the stability of clusters [90]. This observation is supported by high level *ab–initio* calculations performed by Gadre *et al.* They found that high symmetric & non–steric structures for nonamers $((H_2O)_9)$ and decamers $((H_2O)_{10})$ are found to be energetically more stable than structures which are made up by more number of hydrogen bonds [257]. One must note that the stability of structures obtained from simulations depend heavily on the parameters used in the simulations as can be inferred from the comparative studies of water clusters using different water models [258]. As stated in the chapter 4, it is one of the drawbacks of computer simulations. In Fig. (**8.3**) shown is the most "stable" structure of water clusters, whose size is ranging from seven to ten water molecules per cluster, calculated using computer simulations.

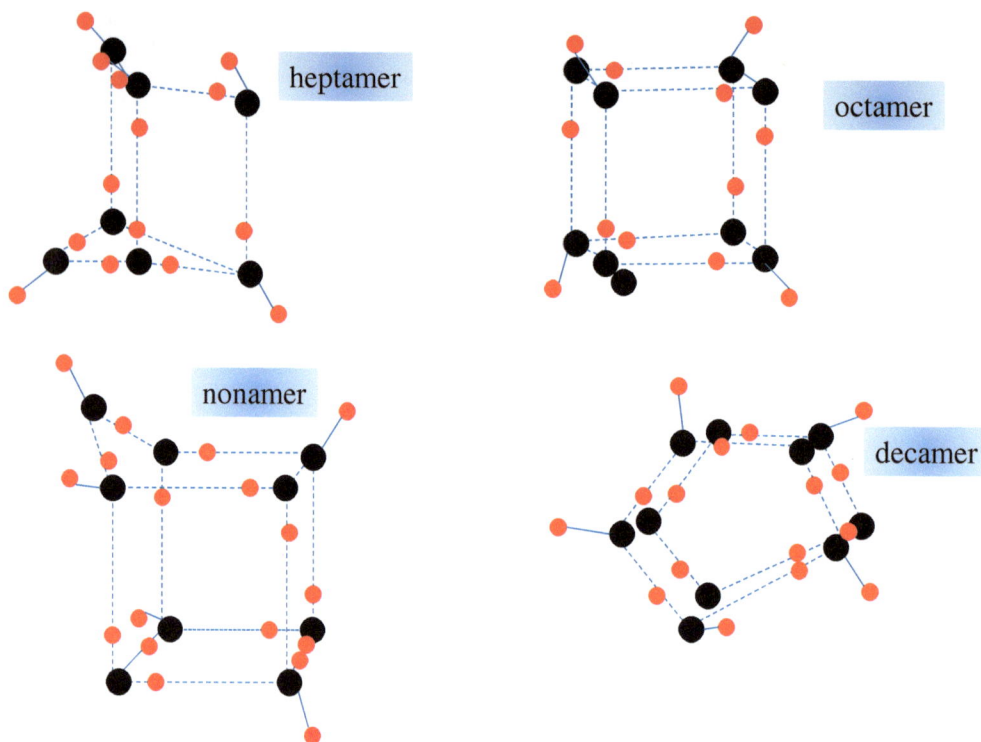

Fig. (8.3). Stability of water clusters from heptamer through decamer obtained from classical simulations. Black and red circles denote oxygen and hydrogen atoms respectively. Compactness and steric factors are found to be the major contributors to the structural stability of these structures. This set of structures serves as mere guidance to the reader since different authors (employing various methods and potentials) have obtained structures with different geometries as the most stable structure representing each clusters, see for example [257] and [258]. This "structural conundrum" exists beyond hexamers. The structures have been taken from [258] with permission from publishers.

Internal Dynamics of Water Clusters

The torsional motion (flipping out of the molecular plane) leads into the inter conversion of structures with characteristic energies [255]. Flipping also gives rise to induced dipole moment and exceptional vibrational properties in water clusters, which can be characterised by experiments and simulations [259]. In addition, concerted motions of the monomers in a cluster results in Hydrogen Bond Network Rearrangement (HBNR) by swapping donor and acceptor oxygen atoms, which is energetically favourable costing only small amount energy. A water monomer participant in the hydrogen network can act as either double proton donor, or double proton acceptor or a single donor and single acceptor as indicted in Fig. (**8.4**).

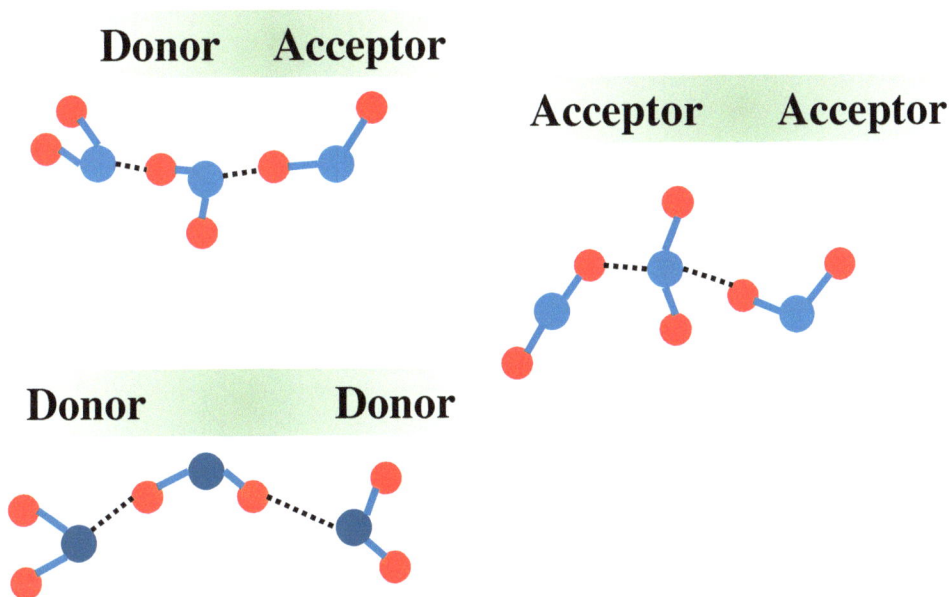

Fig. (8.4). Orientation of water monomer in water clusters. Water monomer can either act as single acceptor and single donor (top), or double acceptor (centre right) or double donor (bottom). Oxygen–hydrogen covalent bonds are indicated by light blue lines, whereas the black dotted lines denote hydrogen bonds between adjacent water molecules. Blue and red circles indicate oxygen and hydrogen atoms respectively. These ideas have been adapted from [90, 251].

The flipping motion is otherwise known as hydrogen–bond tunnelling motion, (suggestive of its quantum mechanical origin) are of different types. One of such motions, whose energy cost is estimated to be too low at about 0.12 kcal/mol, is the rotation of each water molecules in the clusters about their hydrogen bond axes [245b, 260]. Quantum effects are considered to be the principal reason for the flipping, which can result in generation of several configurations (in particular in smaller clusters such as dimers and trimers) with energy difference between them negligible, called degenerate states [261].

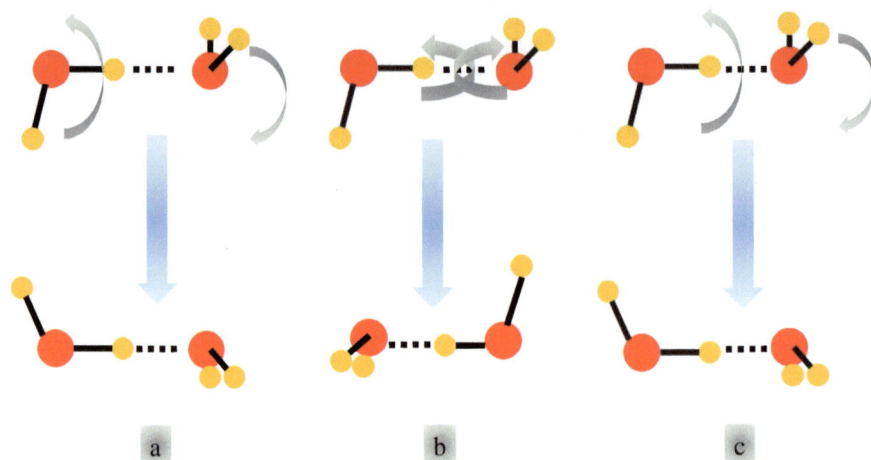

Fig. (8.5). Structural rearrangement in water dimers. In water dimer, three types of hydrogen bonding rearrangement occur as shown here, from **a** through **c**. In Acceptor Switching pattern (as shown in diagram **a**), which is energetically the most feasible of the three, the water monomer, donating hydrogen atom shown in the left, completes a 180° rotation, while the accepting monomer flips (indicated by an arrow downward). In the Interchanging Tunnelling (IT) rearrangement (diagram **b**), both monomers swap the roles of donor and acceptor, after several rotations. The third rearrangement shown in the right (**c**), energetically the most expensive of the three, is known as Bifurcation Tunnelling. In this pattern the acceptor monomer (shown in extreme right) flips towards a configuration in which both hydrogen atoms point downward, and then the acceptor monomer exchanges its hydrogen atom after rotation through the molecular plane. Oxygen and hydrogen atoms have been shown yellow and red respectively. Covalent bonds have been shown by continuous black lines, while dotted lines indicate hydrogen bonding. These ideas have been adapted from [262].

The rearrangement of hydrogen bonding pattern in dimers, the simplest of all water clusters in which two water molecules held together by a hydrogen bond, occur in three distinct ways: Acceptor Switching (AS), Interchange Tunnelling (IT) and Bifurcation Tunnelling (BT) [262]. In dimers, two water monomers are connected by a single hydrogen bond, with one monomer acting as a donor and the other as an acceptor. In Acceptor Switching (AS) motion (the lowest in terms of energy), a flipping motion of acceptor monomer occurs first, followed by a rotation of donor monomer. In Interchanging Tunnelling (IS), the donor and acceptor monomers swap their roles after several molecular rotations. The Bifurcation Rearrangement, with the highest energy barrier among the three, is the rotation about the axis that bisects water molecules (in group theoretical terms this axis is known as C$_2$ axis), resulting in the breaking of hydrogen bonds and their reforming during the process [91]. The energy barrier for the whole process is approximately 2 kilo calories/mol. Fig. (**8.5**) illustrates the three principal hydrogen bond rearrangements occurring in water dimer.

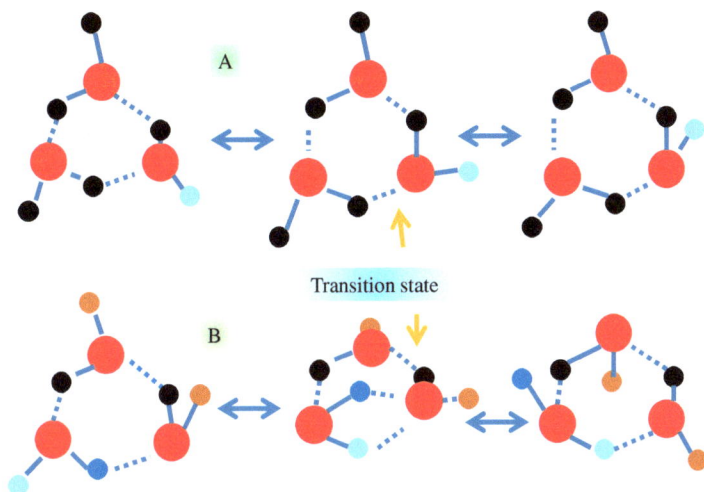

Fig. (8.6). Structural rearrangement in trimers. The hydrogen bond rearrangement occurs in trimers principally in two ways: Flipping motion (**A**) and Bifurcation Tunnelling motion (**B**). In Flipping motion, also known as torsional motion, the hydrogen atom (shown in magenta), which is initially on the lower side of the hydrogen bonded plane, flips over. Bifurcation Tunnelling motion results in exchange of hydrogen atoms participating in hydrogen bonding in one water monomer, and flipping of free hydrogen atoms of the other two water monomers. In the transition state, all atoms except two hydrogen atoms of the rotating monomer are out of plane (see the positions of hydrogen atoms represented by blue and magenta circles). Oxygen atoms have been shown by large red spheres. These ideas have been adapted from [262].

The hydrogen bonding rearrangement in trimers occurs principally in two ways: Torsional motion and Bifurcation Tunnelling motion [262]. In the first type, hydrogen atom situated in one side of the molecular plane, which is highly planar and defined by the oxygen atoms, is flipped to its opposite side. The flipping angles in trimers, constituted by the dangling oxygen–hydrogen bonds with the hydrogen bonded molecular plane, are estimated to be within the range of 50°-65° [263]. Water trimers, compared to their high symmetrical higher analogue tetramer, can easily undergo torsional motions due to the lack of symmetry of free hydrogen atoms connected to the molecular plane. In Bifurcation Tunnelling motion, exchange of the free and the bound hydrogen atom in the hydrogen bond network occurs in concomitant with the flipping motions of monomer subunits of the cluster [262]. The rearrangement is completed *via* a transition complex structure in which one water monomer acts as double hydrogen bond donor and one of its neighbour acts as double hydrogen bond acceptor. Two hydrogen atoms of this water molecule align with the two free orbitals of the oxygen atom of the neighbour monomer acting as double hydrogen bond acceptor [264]. Fig. (**8.6**) illustrates the hydrogen bonding rearrangement pattern in trimers.

The tetramer on the other hand is highly symmetrical (belonging to group S_4 in group theoretical terms) than its smaller analogues, dimers and trimers, and is found to be rigid than trimers and pentamers [265]. High symmetry of water tetramer is exemplified by the fact that its inertia defect, which is the linear combination of principal inertia moments, has a positive value, indicative of quasi–planarity. It has to be noted that in a perfect planar structure, inertia defect is exactly zero. Due to its high symmetry with four oxygen atoms lying at the four corners of a square, structural evolutions after several hydrogen bond rearrangements (including bifurcation rearrangement) are very much constrained in tetramers [262, 265]. Two equilibrium tetramer structures are shown in Fig. (**8.7**).

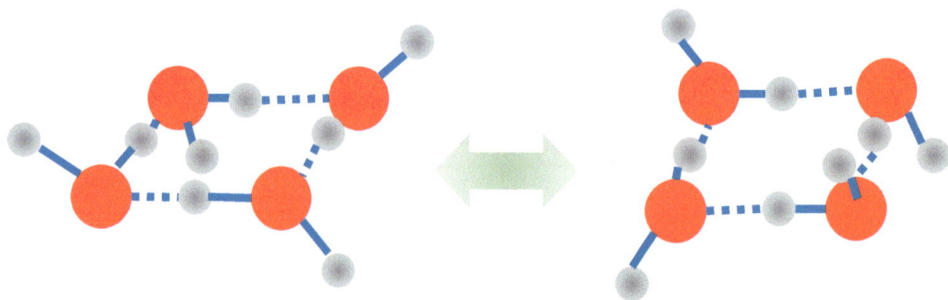

Fig. (8.7). Two equilibrium tetramer structures. It is believed that these two tetramer structures are inter–convertible *via* torsional flipping. All four oxygen atoms (shown red) act as donor and acceptor in this highly symmetrical near planar structure. The free hydrogen atoms, shown as ash circles, position alternatively above and below the hydrogen bonded plane. The idea has been taken from [265].

The two principal modes of structural rearrangement, flipping and bifurcation rearrangement, as in the case of trimers are also observed in pentamers. The free hydrogen atoms are found to occupy alternatively above and below the hydrogen bonded plane, which is unlike trimers and tetramers puckered by 13° [248]. The interoxygen angle (O−O−O) is found to be about 108°, very close to the tetrahedral angle. This might be a reason as to why the cyclic structure in pentamers remains intact for a wide range of temperature as revealed from Molecular Dynamics simulations [262, 266]. Fig. (**8.8**) illustrates the hydrogen bonding rearrangement occurring in pentamers due to flipping of a water monomer and subsequent puckering of the plane containing oxygen atoms of each monomer [267].

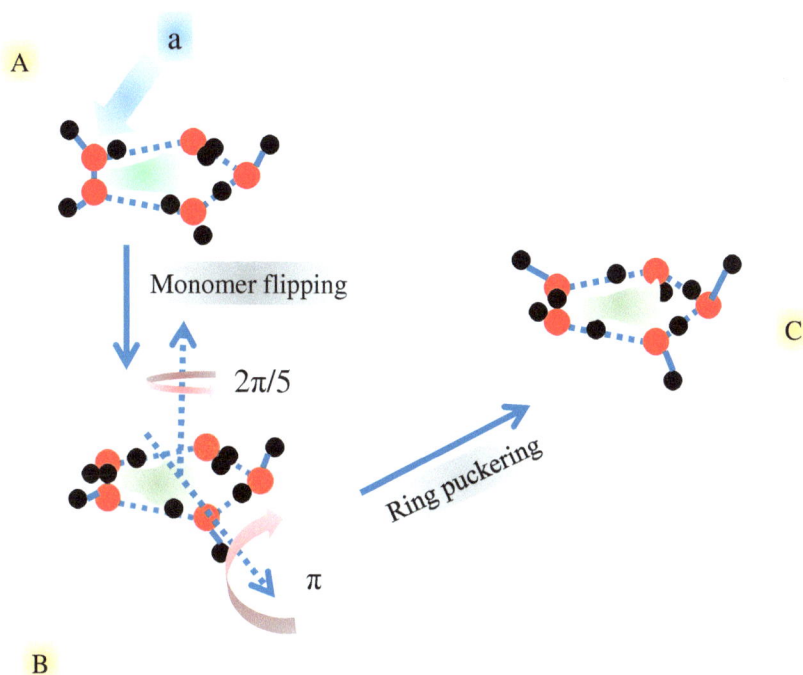

Fig. (8.8). Structural rearrangement in pentamers. The hydrogen atom bonded to monomer a in structure (**A**) is flipped to generate structure (**B**). Successive rotations and ensuing ring puckering on the transition structure (**B**) generate a non degenerate structure (**C**). Oxygen and hydrogen atoms have been shown red and black circles respectively. The idea has been adapted from [267].

In hexamers, three dimensional cage structures (held together by eight hydrogen bonds) are found to be stable than other structural isomers [259]. Four of the constituent water monomers, set up O−O−O−O plane, participate in three hydrogen bonds (with two water monomers act as double donors and single acceptors and the other two act as double acceptors and single donors), while the remaining two water molecules, positioned on the above and below of the oxygen plane, indulge in two hydrogen bonds each (single donor and single acceptor) [91,262]. Although the exact mechanism of transformation between various hexamer configurations (cage, prism, boat, and so on) has not been identified yet, a tentative mechanism has been suggested by Saykally *et al.* as shown in the following diagram, Fig. (**8.9**) [261].

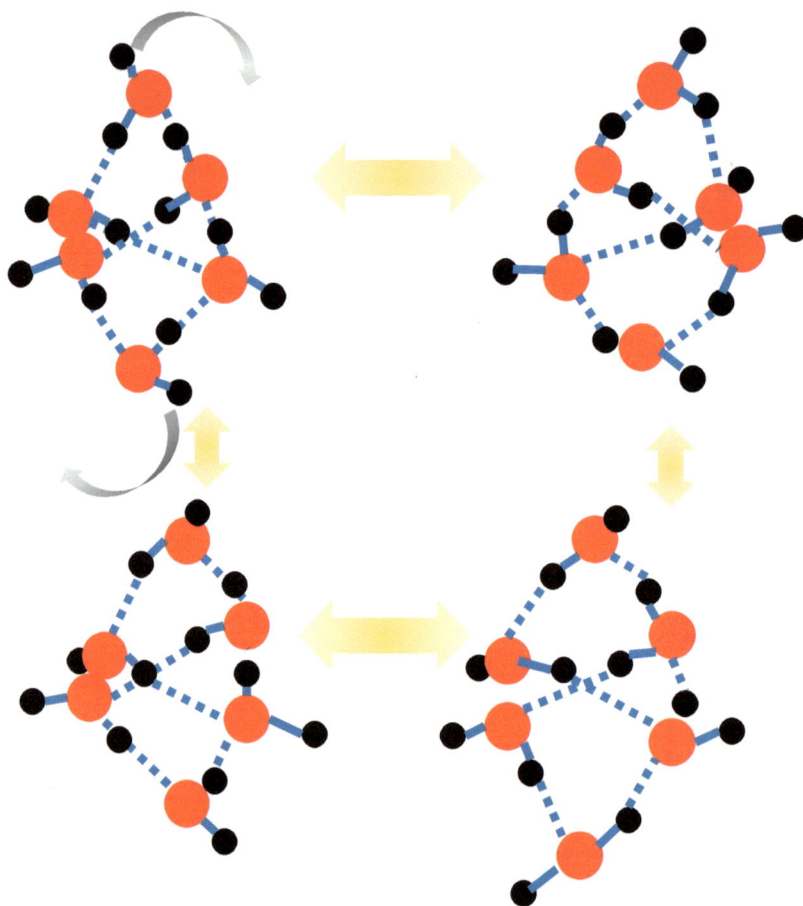

Fig. (8.9). Structural rearrangements in water hexamers. Structural rearrangements in water hexamers (cage) occur principally *via* flipping motions of free oxygen − hydrogen bonds, leading to four non−degenerate structures. The structure shown in bottom right has the largest dipole moment (3.45 D) and the structure shown in the top left has the lowest (2.05 D). Oxygen and hydrogen atoms have been shown black and red circles. The idea has been adapted from [261].

Interatomic Distance and Angle

We have already seen in the previous chapters that various conditions such as temperature influence the size of water clusters. Laser experiments indicate that the variation of oxygen–oxygen distance in water clusters is inversely proportional to the cluster size [90]. This trend has also been reproduced by *ab−initio* simulations. The simulations yields the following interatomic distance 2.976 Å, 2.90 Å, 2.76 Å & 2.74 Å respectively for dimer through pentamer [268]. However it must be noted that a variation of 0.2 Å can be found in the reported bond lengths based on various levels of theories and experiments [245a, 266]. The decreased bond length in larger clusters is attributed to the increased hydrogen bonding strength owing to multiple body interactions (for example non−pair wise three body interactions) known as cooperative effects [251,262]. From Monte Carlo computer simulations, it has been estimated that the oxygen–oxygen distance for cage hexamer be 2.85 Å, which matches exactly with that of ambient water, and interoxygen distance for cyclic hexamer 2.76 which is very close to the reported oxygen–oxygen distance (2.759 Å) for Ice, I_h [259]. Spectroscopic experiments performed by Saykally and co−workers reveal that the interoxygen distance for cyclic hexamer be 2.82 Å, which is the reported oxygen–oxygen bond length in ice IV [269]. Notwithstanding the differences, Saykally *et al.* argue that the close proximity of these values (inter−oxygen distance and dipole moment (discussed next)) with those of ambient water suggests the fact that cage and cyclic hexamers may share similar properties with ambient water and ice (ice I_h or ice IV) respectively. Using all atom polarizable model parameterised by *ab−initio* simulations Leslie *et al.* have calculated the oxygen – oxygen bond length for larger clusters beyond hexamers [270]. They found that the separation is converged to a value of 2.8 Å. This is found to be in good agreement with classical Molecular Dynamics simulations using SPC/E, TIP3P and TIP4P water models [258]. Density Functional Theory (DFT) calculations indicate that as the size of water clusters increases up to clusters with six water molecules, the hydrogen bond length decreases (from 1.85 Å (trimer) through 1.67 Å (hexamer)) [250]. This trend however seems to be absent with the calculated distances using the three of the most popular water models for higher structural analogues (SPC/E, TIP3P and TIP4P). It has been found that the hydrogen bond distance converges to 1.8 Å from $(H_2O)_6$ through $(H_2O)_{25}$ [258]. As cluster size increases, a proportional increase has been recorded in the hydrogen bonding angle (the angle O....H−O): in trimers the hydrogen bonding angle reads 148.5° and in hexamers they are found to be 176° [250]. It must be noted that this trend is prominent in clusters with three to six water molecules.

The angles between oxygen atoms in various clusters also show characteristic deviations. The angles between neighbouring oxygen−oxygen bonds (oxygen−oxygen−oxygen) deviate largely from an ideal angle 109.28′ of the tetrahedral hydrogen bonding network. For example, based on highly sophisticated spectroscopic experiments, Saykally and co−workers have proposed a model cage hexamer with interatomic angles ranging between oxygen atoms 77.24° and 88.1° [261].

The dipole moment for isolated water monomer is 1.86 D. It is interesting to note that the enhancement of dipole moment to values in the range 2.4 D − 3.0 D for average monomers in the clusters is very close to that of ambient water [269]. This is very important observation which accentuates the proposition of the existence of clusters in water. Dipole moment primarily depends upon the geometrical construct of the system, and this explains its variation in different clusters. Unlike the hydrogen bond distance and inter oxygen separation, convergence of dipole moment does rarely occur as the cluster size increases [270]. In small clusters (up to tetramer), it has been shown (*via ab−initio* simulations) that as the number of water molecules in a cluster becomes larger, dipole moment of constituent monomers increases. This is due to large electric field generation by neighbouring water molecules in the cluster [269]. However in cyclic hexamers and pentamers the dipole moment vanishes due to their symmetry of higher order [269].

Impact of Density on Water Clusters

We saw in the previous chapters and also earlier in this chapter how the alteration in density affects the hydrogen bonding in water, supercooled water and Super Critical Water (SCW). Density plays a key role in determining the size and shape of water clusters, and number of hydrogen bonds as Monte Carlo computer simulations indicate [247]. The formation of number of hydrogen bonds is commensurate with increase in density [247]. In fact, density variations alone can change the number of hydrogen bonds at a fixed temperature. Computer simulations reveal that at very low density as 0.01 g/cm^3 most of the water molecules (around 90%) do not engage in hydrogen bonding, and this percentage decreases considerably upon tenfold increase in density (at 0.1 g/cm^3) [243]. An important observation Kalinichev *et al.* have made is the existence of "infinitely long" hydrogen bond network in the Super Critical Water (SCW), the presence of which is observed when the density of water reaches around two times that of critical density (0.60 g/cm^3). They have estimated that the clusters can on average contain 150–200 water molecules when density becomes 2–3 times higher than the critical density beyond the critical temperature, similar to supercooled water, as we saw in chapter 6 [247]. At much lower density around 0.04 g/cm^3 on the other hand a very high population of monomers, over 60%, is observed [243]. This high proportion of monomers is expected to influence the diffusive properties of Super Critical Water (SCW) at low densities as the NMR investigations indicates [244]. In the high temperature region the balance between the rotational motion of water molecule and its attractive interaction with surrounding molecules turns into the favour of the former [243]. This effect has been corroborated by the enhancement of peaks pertaining to dipole time correlation at temperature as high as 1500 K than observed in much lower temperatures [243].

CONCLUDING REMARKS

The primary aim of this chapter is to provide the reader a brief account of the behaviour of water beyond its boiling point, in particular Super Critical Water (SCW). Numerous investigations unearthed its amazing catalysing properties; further development of these properties is expected to reduce the environmental pollution and production cost of certain chemicals. A brief account of the Super Critical Water (SCW) is also given in this chapter based on experimental and computational investigations reported in the literature. Albeit studies in general agree upon the depletion of the three dimensional network that is so characteristic of ambient water, the ability of water molecules to associate themselves towards forming aggregates, water clusters, is not constrained completely at high temperatures. Clusters with up to ten water molecules can exist in Super Critical Water (SCW). Using state–of–the–art spectroscopic techniques and computationally demanding *ab–initio* calculations, the geometry and internal dynamics have been studied detail. These studies, in particular VRT experiments primarily applied to clusters from dimers to hexamers, reveal the structural rearrangements that can occur among various water clusters. Further investigations are warranted in order to investigate existence of clusters with higher water molecules and their probable rearrangement mechanism.

CONFLICT OF INTEREST

The author confirms that he has no conflict of interest to declare for this publication.

ACKNOWLEDGEMENTS

I express my gratitude to the following organisations for granting me permission to use figures under their copyrights in this chapter: National Academy of Sciences (U.S.A.), American Institute of Physics and American Chemical Society.

CHAPTER 9

A Brief Review of Water Anomalies

Abstract: Numerous anomalies of water have been reported in the literature. Anomalous behaviour of liquid water is so striking when it is supercooled below the melting temperature of ice, T_m. Several physical properties have been found to be diverging in the supercooled liquid phase, including isobaric heat capacity, isothermal compressibility, relaxation time and thermal expansion coefficient. Interestingly hydrogen bond life times show a divergence at this temperature, indicating its connection to these singularities. Liquid water exhibits both density maximum and minimum, the latter has been discovered by a recent Small Angle Neutron Scattering (SANS) experiments, considered to be two of its most notable thermodynamic anomalies. Unlike other liquids, translational and diffusive motions in water exhibit contrasting behaviour and product of these two diffusive constants is found to be insensitive to temperature and density. Formation of water clusters of varying sizes dictates the nature of diffusion in supercooled water. Several propositions have been made in order to account for water's anomalies, which include Liquid–Liquid Critical Point theory, Singularity Free hypothesis, Critical Point–Free hypothesis and Stability Limit conjecture. In bulk phase, water shows its most of the anomalies. In addition, it exhibits several other anomalous characters when confined to nanoscale geometries and is near to macromolecular surface. It has to be noted that in the vicinity of non–polar solutes the strength and lifetimes of water network increases.

Keywords: Boltzmann constant, Critical point, Entropy, Hydrophobic hydration, Isochore, Isotherm, Life times, Short range, Singularity–free, Spinodal, TMD, Well depth, Widom line.

INTRODUCTION

Anomalies of water were known to the scientific community for centuries. There are numerous anomalies that have been reported for water, several of them have been studied in much depth by experimental and computational methods.

Jestin Baby Mandumpal

As mentioned in chapter 5, water is a very small molecule, containing three atoms with molar mass just over 18 g per mole. However, there is no other material that possesses large number of anomalous properties as water does. Most of the anomalous behaviour exhibited by water is in the supercooling range, between the melting point of ice, T_m, and the temperature of homogenous nucleation, T_H, and therefore much of the efforts have been made to understand the nature of water in this region. In the remaining portion of this chapter, I wish to discuss about notable anomalies of water as a liquid, which are thermodynamic, kinetic and structural in nature. The chapter is concluded with four major interpretations that have been suggested to interpret anomalous behaviour of water.

WATER AS LIQUID

The most notable unusual property of water is its very existence as a liquid at normal temperature domain [121]. The interaction that dominates water is undoubtedly hydrogen bonding, whose interaction strength (20 kJ/mol or 5 kcal/mol approximately) is higher than van der Waals forces and lower than ionic bonding. The hydrogen bonds can overpower thermal fluctuations (one tenth of the former) in water, and provides vital strength to water to be remained as liquid [121].

HIGH ENERGY OF DISSOCIATION

I briefly mentioned in chapter 2 about the role of pair interaction functions in modelling liquids in general. Water under normal thermodynamic conditions is a classic example of matter in condensed phase. Water has an amazing pair potential well depth. This means that there exist forces that bind the molecules together. Classically these forces can be categorised into two: long range electrostatic forces and short range van der Waals forces (non– bonded interactions) (I have mentioned about these forces briefly in chapter 2). The short range forces heavily depend upon the distance between the molecules, and are attractive at larger distances but very repulsive at shorter distances. This suggests that there exists an effective distance which corresponds to a minimum in energy, lower than zero. The potential well depth (ε) is the distance between the energy minimum and the point at which energy is zero, which is also the approximate energy required to separate a pair molecules that are close to each other in a condensed liquid such as water. Values of potential well depth for some smaller molecules and water are given (as $(\varepsilon/k_B)/K$, where K_B is the Boltzmann constant) in the following Table **9.1**.

Table 9.1. Pair potential well depth.

Atom/Molecule	$(\varepsilon/k_B)/K$	Boiling Point/K
He	11	4.2
Ar	142	87
Xe	281	166
CH$_4$	180-300	111.5
H$_2$O	2400	373.2

It can be seen from Table **9.1** that water possesses high energy of dissociation (the energy required to break its condensed phase into constituent molecules), almost 8 times than that of methane (CH$_4$). The boiling point also increases dramatically. The high pair potential well depth of water explains another anomalous behaviour of water: high boiling point.

DENSER LIQUID PHASE

Most of the solids are denser than their corresponding liquids and therefore these solids cannot float in their corresponding liquid phases. On the contrary, normal water is denser than its solid form, ice, and the latter floats in liquid water; this is a signature of water's volumetric anomaly [271]. Enthusiastic readers can verify themselves this fact by comparing how ethanol cubes behave in liquid ethanol [272]. Like ethanol, other hydrogen–bonded liquids, for example, dinitrogen tetroxide (N$_2$O$_4$) or hydrogen peroxide (H$_2$O$_2$) too do not behave like water does, so this points to the fact that water's exceptional properties do not solely lie in its ability to form hydrogen bonds, rather how it forms. In liquid state, water can have numerous configurations, forming clusters with varying number of water molecules (for example, hexamer and decamer) as we saw before. Nevertheless, as temperature drops by, in particular below the melting point of ice, nearly all of the water monomers prefer to have four nearest neighbours with linear oxygen – hydrogen – oxygen angles.

Density of water has been a focal point of research for quite long time. Earlier measurements were reported as early as 1837 [97a]. This can be clearly understood if we compare densities of water and ice, as shown in the figure below (Fig. **9.1**). Between 273 K and 283 K, the density of water shows a distinct maximum, whereas the density of ice is found to be very stable in this range. The following figure demonstrates the density anomaly of water, which in other words is known as Temperature of Maximum Density (TMD).

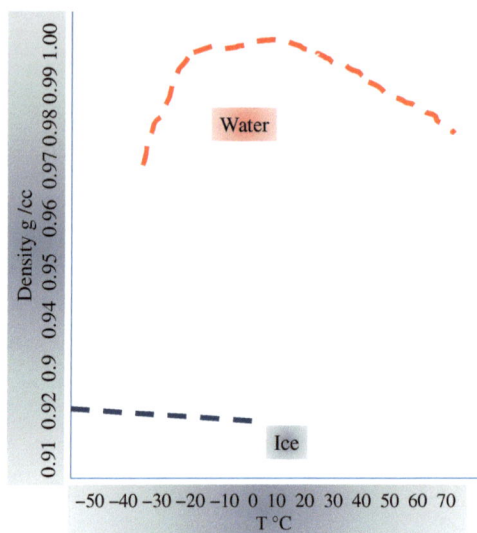

Fig. (9.1). The experimental density of water and ice. The densities of water (shown in red) and ice (shown in blue) demonstrate two contrasting behaviours. The water density shows a maximum between 0 and 10 degree Celsius, as opposed to a constant change exhibited by ice. The Temperature of Maximum Density (TMD) at −4∘C is one of the anomalous properties of water. The idea has been adapted from [97a].

THERMODYNAMIC SINGULARITIES

In addition to the density anomaly, Temperature of Maximum Density (TMD), occur other thermodynamic anomalies which include water's expansion upon cooling, a minimum in isothermal compressibility and large heat capacity [180]. Water has high heat capacity at constant pressure (C_p), which the derivative of enthalpy with respect to temperature. This demonstrates the capacity of water to store thermal energy at the expense of hydrogen bonds and van der Waals interactions [271].

Experimental studies show that isothermal compressibility of water under 273 K rises very sharply [69], and it can be easily seen that at −45°C, the compressibility diverges, and upon increasing pressure, this temperature (−45°C) is shifted to lower temperatures [173]. Several other physical properties, including density (as you saw in the previous section), diffusion coefficient, viscosity, heat capacity, and dielectric relaxation time, are also divergent at this temperature. With the support of some calculations, a tentative explanation had already been given as early as 1976 by Speedy and Angel that this trend is caused by the changes in orientation of water molecules with respect to each other (in particular between the O−H axes of one of the water molecules and the oxygen atom of the second water molecule culminating in hydrogen bonding). Alteration of this bond angle results in fluctuations in both energy and density [69], which is reflected in the anomalous increases in aforementioned quantities. Fig. (**9.2**) shows thermodynamic singularities in density, thermal expansion coefficient, heat capacity, and isothermal compressibility. We must note that all of these "singularities" occur in supercooled water.

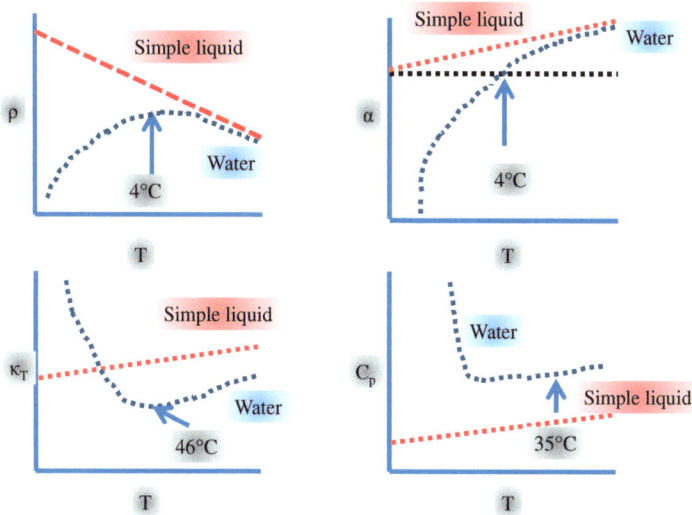

Fig. (9.2). Thermodynamic anomalies of water in the supercooling range. Contrasting behaviour of water in its supercooling regime with respect to a normal liquid in density (top left), thermal expansion coefficient (top right), isothermal compressibility (bottom left) and heat capacity (bottom right) shown in the figure. The idea has been adapted from [173].

These anomalous properties can be explained by the concept of metastability curve, as shown by the figure below (Fig. **9.3**). As can be seen from the figure, the thermodynamic singularities appear to be 'permanent' in the supercooled water as changes in pressure does not diminish these singularities.

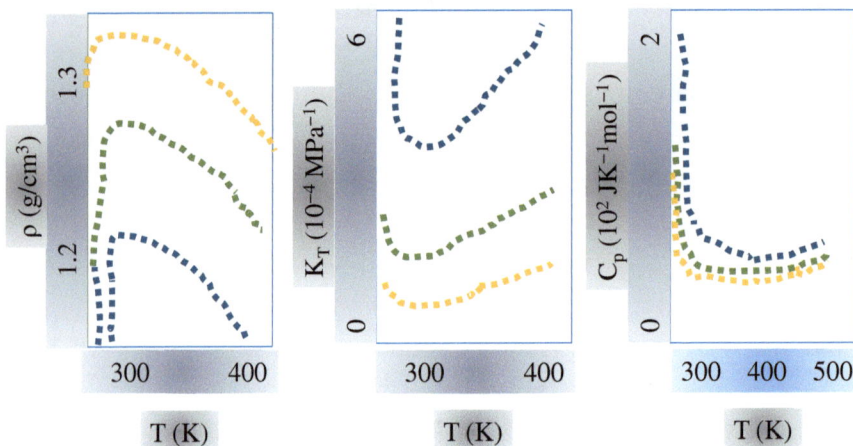

Fig. (9.3). The limit of metastability existing in water. The thermodynamic singularities explained by the concept of metastability. In all three thermodynamic quantities, density (shown in the left panel), isothermal compressibility (shown in the middle panel) and the heat capacity (shown in the right panel,) there exist lines of thermodynamic singularities, which do not diminish upon changes in pressure. The curves for three different pressures have been indicated in the diagram: blue line corresponds to pressure equal to zero MPa, green line at 200MPa and the orange lines correspond to pressure equal to 300 MPa. The idea has been adapted from [273].

The Temperature of Maximum Density (TMD) is one of the exceptional characteristics that water exhibits. TMD is the locus of temperature at which density is the maximum at constant pressure. The significance of TMD lies in the fact that it divides the entire pressure–temperature phase diagram into two regions which have two different properties [164]. For example, the coefficient of thermal expansion, defined as the relative increase in volume (proportional to the product of fluctuations in volume and entropy) is negative in the low temperature side of the phase diagram, while on the higher temperature side, this quantity takes positive value. A plausible explanation can be given based on the quality of hydrogen bond network that can be formed at lower temperature. Quality refers to the strength, average number of neighbouring water molecules to which the central water molecule bonded to (four), and orientation of the hydrogen bonds, which confer higher stability and lifetimes than that of those formed in higher temperatures. These stable, near perfect hydrogen bonded clusters naturally have higher volume per molecules (*i.e.* change in volume takes a positive value) at the expense of decrease in entropy (hence the product of fluctuations of volume and entropy, (proportional to the coefficient of thermal expansion), takes a negative value).

TMD can also be obtained from pressure–temperature diagram at constant density. TMD is the locus of temperature at which pressure takes a minimum value at constant density, as Fig. (**9.4**) demonstrates [274]. From the figure, one can observe that TMD changes its slope while passing from positive pressures to negative pressures, and this change occurs at T= 4°C (277 K) at density equal to approximately 1g/cm³ [274]. It has also to be noted that according to Stability Limit conjecture (which will be discussed in length later in this chapter), TMD does not intersect with the spinodal line, where the derivative of pressure with respect to the molar volume at constant temperature becomes zero, beyond which liquid phase is no longer mechanically stable.

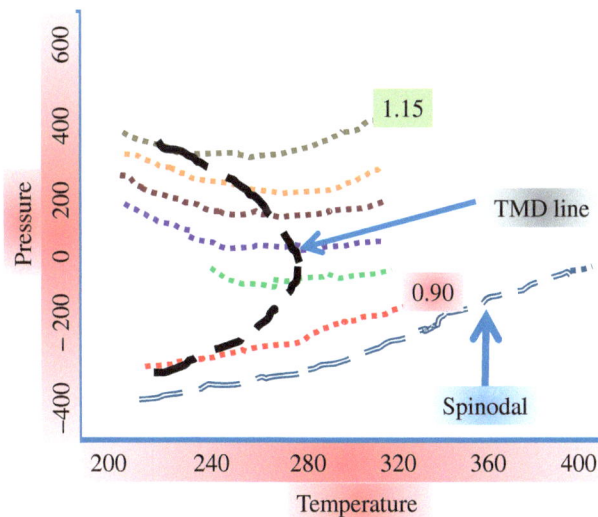

Fig. (9.4). Pressure dependence of Temperature of Maximum Density (TMD). Pressures have been recorded in MPa, whereas density and temperature in g/cm³ and Kelvin respectively. Pressures have been computed for wide range of temperatures for constant density from 0.90 g/cm³ to 1.15 g/cm³ with the interval of 0.05 g/cm³. The black line drawn across the pressure-temperature lines (isochores) indicates the Temperature of Maximum Density (TMD). The spinodal line of water is shown at the bottom.

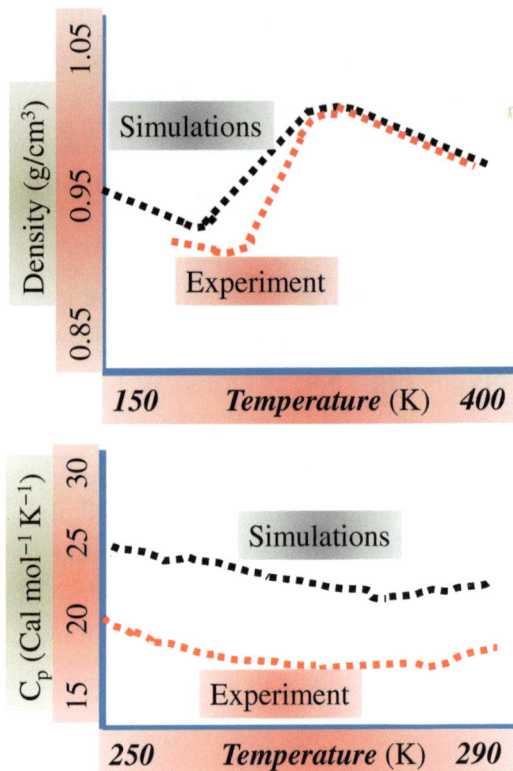

Fig. (9.5). A rough sketch of the comparison of two physical properties obtained from computer simulations and experiments. The comparison of density maximum and minimum (top, and isobaric heat capacity (bottom) are shown. The black line indicates the simulation results using TIP4P/2005, and the red line indicates the experimentally measured values. In general, simulated values are in agreement with the experimental values. The idea has been adapted from [275b].

TMD has been a favourite area of researchers, and it has been mainly used, apart from theoretical view point, for the validation of water models [275]. McBride *et al.* has been shown that TIP4P/2005, a variant of TIP4P water model, better reproduce most of the thermodynamic anomalies discussed above than models such as SPC, TIP4P and TIP5P. Fig. (**9.5**) provides a comparison of several physical properties of water such as isobaric heat capacity and density maximum obtained from computer simulations and experiments. It can be seen that simulations employed using TIP4P/2005 are in good agreement with experiments. This is in fact promising considering the fact that large number of anomalous properties of water so far unearthed is in this temperature range.

Density Minimum

In a recent experimental advancement, Small Angle Neutron Scattering (SANS) experiments, Hsin–Chen *et al.* have reported density minimum in supercooled water [74]. Density minimum has not been reported to be exhibited by many materials, and hence by this reason it can be regarded as another anomalous behaviour by water. The following Fig. **(9.6)** demonstrates the maximum and minimum in density of supercooled D_2O (due to scattering properties D_2O is preferred over H_2O in actual experiment) [74].

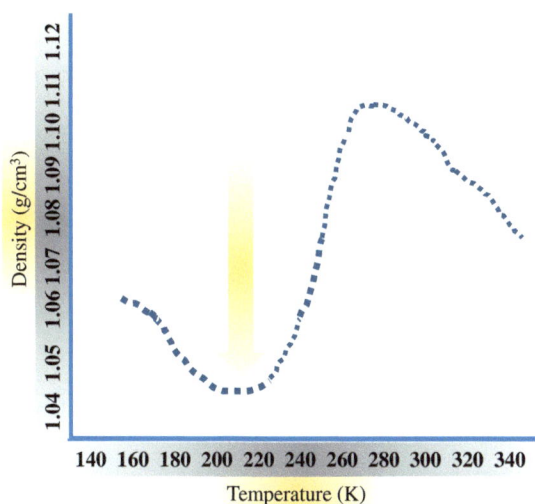

Fig. (9.6). Density minimum in supercooled D_2O. Density maximum (1.105 g/cm^3) and minimum (1.041 g/cm^3) exhibited by supercooled D_2O. The density minimum has been shown by an arrow pointing downwards. The idea has been adapted from [74].

Thermal Conductivity Maximum

Thermal energy is not evenly distributed in the system: some particles possess higher energies, while some other have lower energies. According to one of the basic thermodynamic principles, heat flows from higher energy states to lower energy states, and the resulting heat transfer (temperature gradient) is material specific and can be expressed as thermal conductivity. Thermal conductivity of water is of particular interest not only because of its significance in many applications of water in high temperature, for example as a medium in organic synthesis, but also because of its complex nature. Thermal conductivity of water is affected thermodynamically by pressure, temperature and density, and structurally by water's peculiar hydrogen bonding pattern. Anomalous increase of thermal conductivity of water has been reported from experiments, and by means of computer simulations. It has been found out, using models such as SPC/E and TIP4P/2005, that there exists a maximum at around 500 K [276].

Thermal Conductivity Minimum

Minimum in thermal conductivity and thermal diffusivity (the ratio of heat flux and temperature gradient) in supercooled water is a recent addition to the family of anomalies in water [277]. Using TIP5P water model, Kumar *et al.* have found this minimum temperature, 250 K, and surprisingly this temperature corresponds to the maximum for isobaric heat capacity, as shown in Fig. (**9.7**) below.

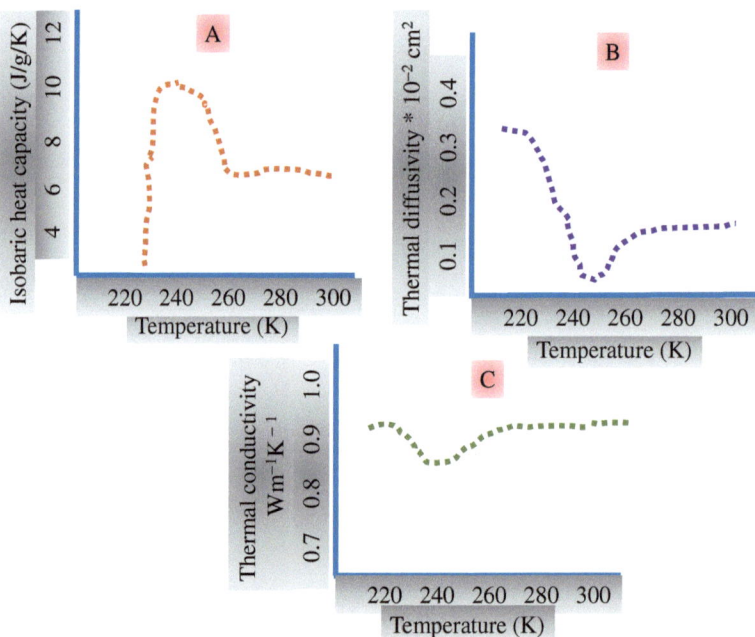

Fig. (9.7). Isobaric heat capacity, thermal diffusivity and thermal conductivity *versus* temperature. In the panel **A**, it can be seen that isobaric heat capacity shows a distinctive maximum at 250 K which corresponds to the temperature of Widom line. Thermal diffusivity and thermal conductivity are shown in the panels, **B** and **C** respectively. It can be clearly seen that the minimum in these thermodynamic quantities correspond to the maximum in heat capacity. The idea has been adapted from [277].

Pressure Anomaly

Simple liquids become solid upon increasing pressure, but for water, the opposite is true. This special characteristic can be explained by simple phase diagram showing solid–liquid–gas coexistence (pressure – temperature diagram). In the Fig. (**9.8**) (on right), it can be seen that simple materials have a positive solid–liquid phase boundary slope, whereas in the case of water we observe a negative slope.

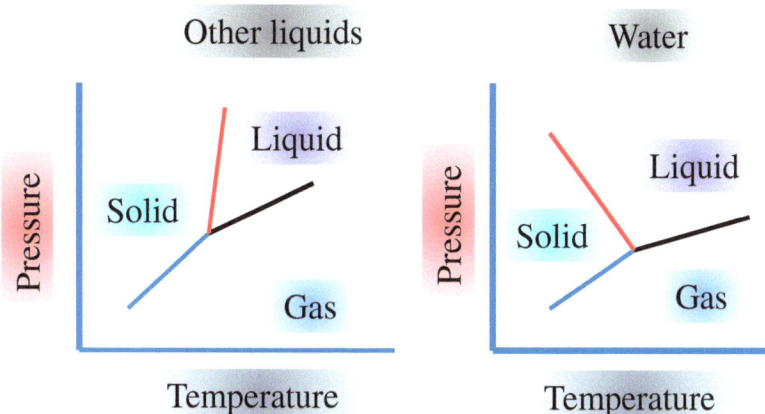

Fig. (9.8). Pressure anomaly of water. A comparison of simple face diagrams of water and other liquids is shown. In other liquids (left) as pressure increases, the liquid is transformed to solid. On the contrary, for water, upon increasing pressure more solid is formed from liquid phase (right).

DIFFUSION

Both experiments and computer simulations offer interesting insights into the diffusive motions of water molecules. These investigations reveal that rotational and translational movements of water molecules in supercooled domain show contrasting behaviour [173]. The rotational motion of water molecules is of shorter scale, at around 2 pico seconds (ps), replicating Arrhenius behaviour. Translational movements, on the other hand, exhibit non–Arrhenius behaviour, with a higher order of magnitude. Moreover, these diffusive movements increase in supercooled water dramatically upon hike in pressure [278]. This anomaly has been explained on the basis of Stability–Limit conjecture. In addition to this hypothesis, a statistical explanation has also been given for the anomalous increase of rotational motion (about 250% in the range from 0.1 through 250

MPa). The energy barrier between different energy states becomes narrower at high pressures, which increases rotational motions of H_2O molecules dramatically (see chapter 6) [278].

In Fig. (**9.9**), (the upper and lower panel provide results from computer simulation and Nuclear Magnetic Resonance (NMR) spectroscopic investigations respectively), it can be seen that there is a distinct maximum especially lower temperatures, whereas the profile becomes flat upon increase in temperatures [279].

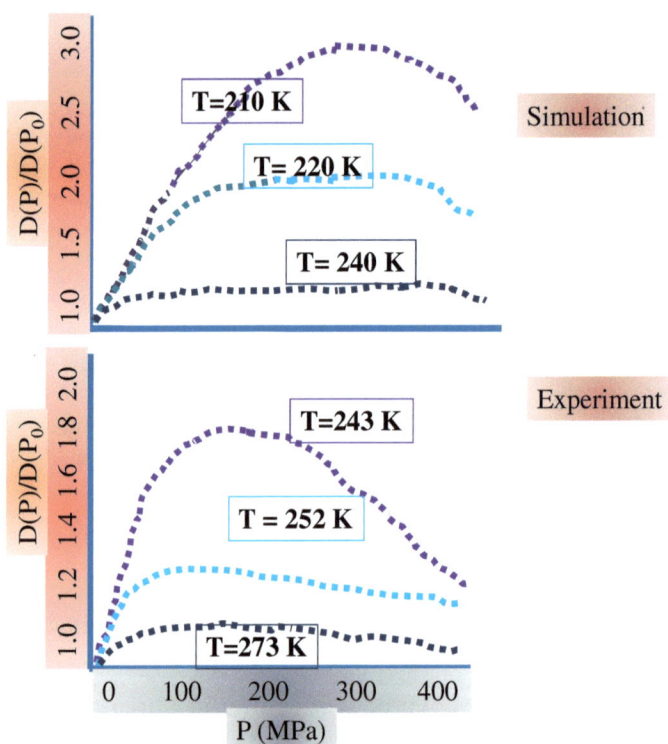

Fig. (9.9). Anomalous increase in diffusion coefficient upon variation in pressure. In the upper panel shown is the computer simulation results using SPC/E water model, and in the bottom panel shown is NMR results. On the x axis plotted is the ratio of diffusion coefficient corresponding to various pressures with respect to that of zero MPa pressure. The idea has been adapted from [279].

In fact, the occurrence of diffusion anomaly is due to the interplay between several thermodynamic variables: density, pressure and temperature. For a wide range of temperatures, from 210 K to 240 K, the density corresponding to maximum diffusion coefficient is found to be 1.15g/cm^3 [167] (readers are referred to see relevant discussions in chapter 6). The decrease of translational motion (diffusion coefficient) of water molecules can be observed for higher densities beyond this threshold density. This is expected, and can be explained by steric factors because increasing density implies that more molecules are confined to less available space. However, at lower densities than the threshold value (1.15 g/cm^3) the mobility of water molecules decreases contrary to the expectations [280]. Further decrease in density (beyond the density corresponding to the minimum in diffusion coefficient) disrupts hydrogen bonds, which leads to the preponderance of small water clusters and water monomers. The formation of smaller water units at the expense of large clusters leads into the enhancement of mobility of water molecules.

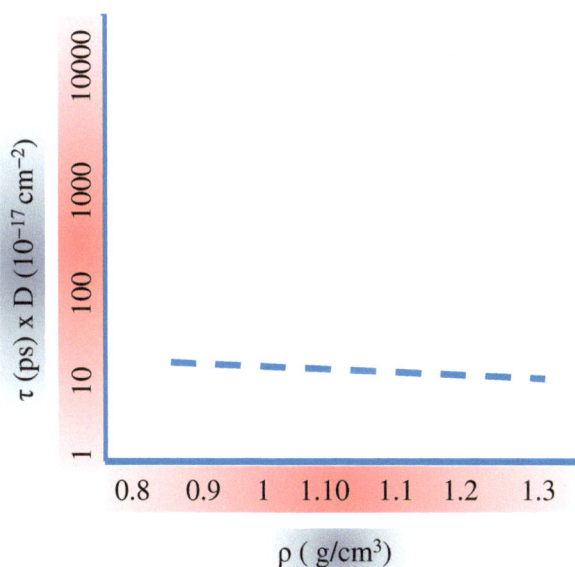

Fig. (9.10). The product of diffusion coefficient and rotational relaxation time. The idea has been adapted from [280].

It is interesting to note that the product of translation diffusion coefficient and rotational relaxation time is found to be independent of temperature and density, as shown in Fig. (**9.10**). Stanley *et al.* has interpreted this abnormal trend using computer simulation of SPC/E water model [280]. However, Netz *et al.* argue that there is indeed an indirect influence of density in this behaviour since the changes in the size of water clusters solely depends upon density. The changes in density have a drastic implication in altering the coordination of water molecules: as density increases, water molecules can have higher coordination number than 'perfect' tetrahedral coordination. This situation however does not alter the number of hydrogen bonds in the system, but stronger tetrahedral hydrogen bonds are replaced by much feeble hydrogen bonds due to the presence of extra molecules around each water monomer. As a result, water molecules are not only free to rotate about their axes, but also to move to a considerable extent. Although density influences the coordination numbers of water molecules, the overall movement of water molecules (considering translational and rotational motions together) is not affected by it since their product is constant for a wide range of densities (for details of the calculations and results, please see [280]).

Breakdown of Stoke–Einstein Relationship

Stoke–Einstein (SE) relationship, which connects diffusion, temperature and viscosity to hydrodynamic radius of a molecule, does not valid in supercooled water [281]. Most of the supercooled anomalies of water occur at the temperature roughly equal to 1.8 times that of glass transition temperature (T_g), where Stoke–Einstein (SE) equation fails. This has been computationally found to be around Widom line passing through the Liquid–Liquid Critical Point. The disentanglement of diffusion with viscosity in the supercooled water is the result of heterogeneity in water due to mobile water clusters. These clusters show interesting thermodynamic behaviour: at higher temperatures (*i.e.*, temperatures above 260 K) it is independent of pressure and only exhibit slight temperature dependence. As temperature and pressure decreases below the nucleation temperature of water (240 K), cluster size of water molecules increases dramatically.

THEORETICAL INTERPRETATION FOR THE ANOMALIES IN SUPERCOOLED WATER

Four important theories have been widely circulated in the scientific circles to treat anomalous behaviour of water in the supercooling regime, namely

Stability–Limit conjecture, Liquid–Liquid critical point theory, Singularity Free scenario, and Critical Point Free model.

Stability–Limit Conjecture

Stability–Limit conjecture primarily provides the limit of mechanical stability of water (spinodal curve) for a wide range of temperature and pressure [69]. The location of the spinodal curve (thick black dotted shown in Fig. (**9.11**)) connects the thermodynamic anomalies and the nucleation temperature, beyond which water under normal conditions crystalizes. Crystallisation can be explained on the basis of thermodynamics as a result of fluctuations in energy, entropy and density (please refer to [69] for an in depth discussion). Later it was observed that these fluctuations in water, unlike other liquids, increase at low temperatures, are attributed to the variation in the angle of hydrogen bonds (O…..H–O), as mentioned in the previous chapter [180]. This is indeed an important development concerning the fact that the Stability–Limit hypothesis conjectured over thirty years ago was verified by sophisticated experiments.

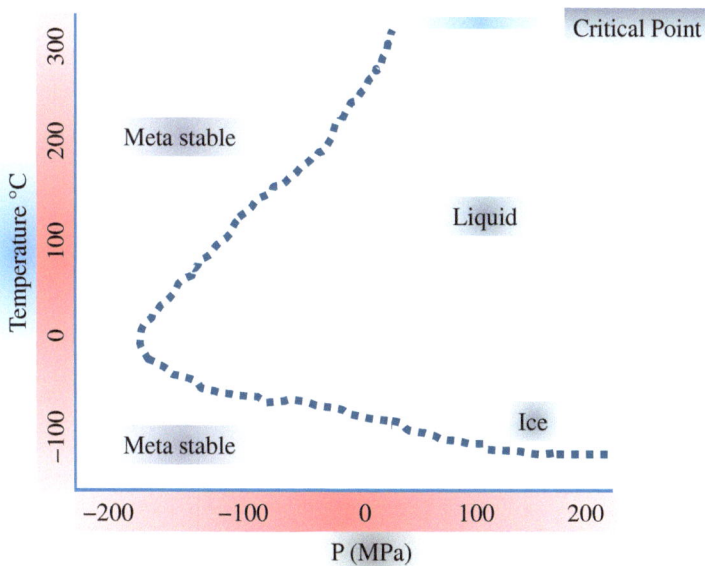

Fig. (9.11). Schematic diagram of Stability Limit conjecture. To the left of the line of mechanical stability, denoted by the line of blue dots, water is metastable with respect to its more stable phases, liquid and ice. The idea has been adapted from [278].

Liquid–Liquid Critical Point Theory

Liquid–Liquid Critical Point theory advocates that there exits an additional critical point (temperature 200K, pressure 100Mpa, and density 1 g/cm^3) besides the well–known critical point of water at temperature 600K, pressure 20 MPa and density 0.3 g/cm^3 where the liquid and gaseous phases become indistinguishable [175, 282]. At this additional critical point, the separation of low density and high density liquid phases terminates [186]. The liquid water isotherms obtained from computer simulation employing TIP5P water model (Fig. **9.12**) below clearly manifest the temperature dependence of the Liquid–Liquid phase transition. An inflection in the isotherms can be observed from 250 K downwards which culminates in two flat regions at 215 K, which is attributed to the segregation of two liquid phases [274].

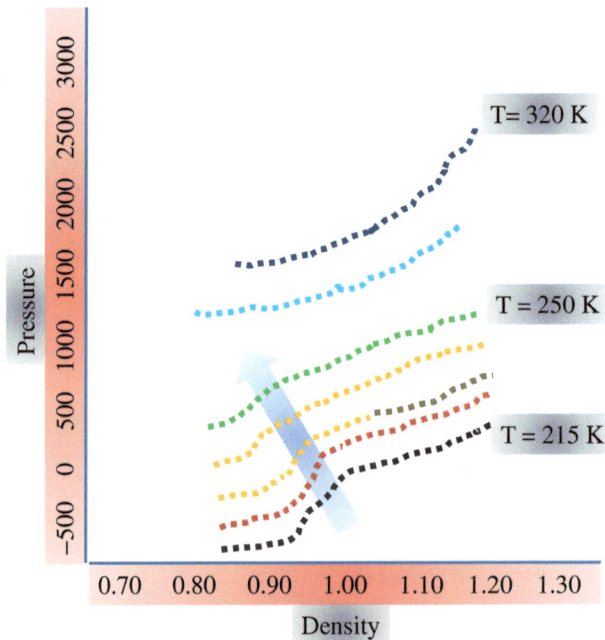

Fig. (**9.12**). Liquid water isotherms. Pressure-density plot of water for a wide range of temperatures, computed using TIP5P water model is shown. An inflection develops, as shown by blue line drawn across low temperature isotherms, from 250 K downwards, which is absent in higher temperatures. The inflection culminates into two separate flat regions corresponding to two different liquid phases, according to the Liquid–Liquid Phase Transition theory. Pressure (in MPa) has been rescaled for clarity. The density has been measured in g/cm^3. The idea has been adapted from [274].

The liquid–liquid and liquid–gas coexistence curves can be extended beyond the critical points, and these extension curves are known as Widom lines, which lead into one–phase region [277]. It is very clear from the following Fig. (**9.13**) that two Widom lines correspond to two critical points in water, C and C'. The notion of Widom line is so significant in the case of water that anomalies such as Temperature of Maximum Density (TMD) originates from one of the two Widom lines (Liquid–Gas Widom line) as indicated by the diagram below (Fig. **9.13**).

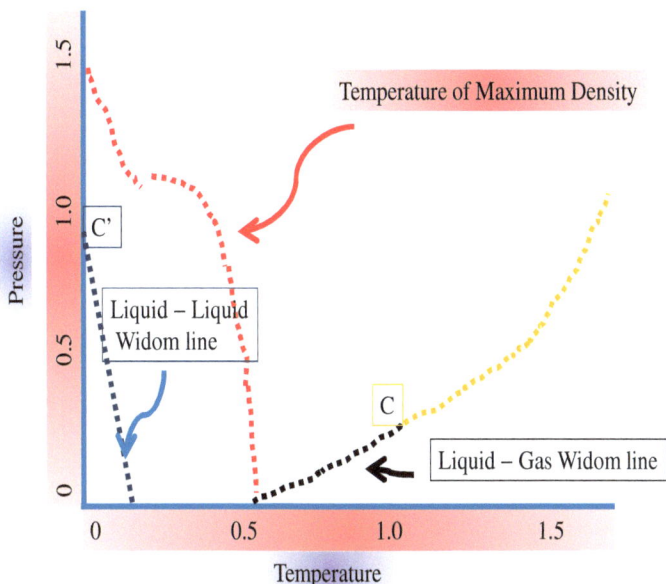

Fig. (9.13). Widom line and critical points. Widom lines are shown in the diagram, as extensions to two known critical points in water, (Liquid-Liquid (denoted by C') & Liquid-Gas (denoted by C)). As can be seen from figure, Temperature of Maximum Density (TMD, shown in thick red dotted line) and thermal expansion coefficient (thick blue dotted line) are emanating from Liquid-Gas Widom line. The Liquid-Liquid critical point is appeared to be lower, at around 0.5, reported in another paper co-authored by Stanley [182]. Both pressure and temperature have been scaled for clarity. The idea has been adapted from [155].

The significance of this second critical point is that at this point high density and low density liquids (HDL & LDL) (two phases of supercooled water) become identical, that is to a single phase as mentioned earlier. Liquid–Liquid Critical point theory (a schematic diagram is shown as Fig. (**9.14**)) was originally proposed by computational physicists and is well supported by experiments as well [74,163]. One of such notable experiments was conducted by Suzuki *et al.*, whereby they demonstrated the discontinuous transformation between the two liquid phases (HDL an LDL) through very short range of temperature (111 K to 114.5 K) [181].

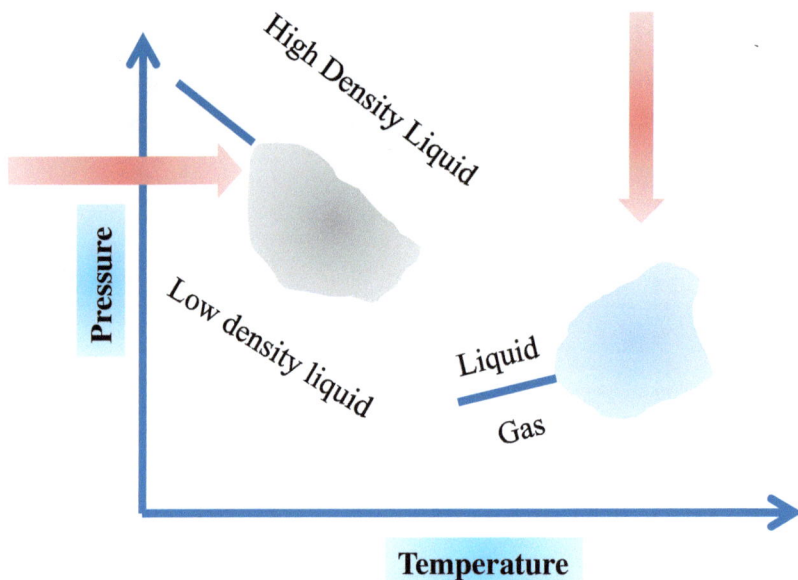

Fig. (9.14). Schematic diagram depicting Liquid–Liquid Critical Point theory. In addition to the well-established critical region, which separates liquid and gas, LLCP theory posits an additional critical point in the low temperature region separating two fluid phases, Low Density Liquid (LDL) and High Density Liquid (HDL). The arrows point into the critical regions in the diagram. The idea has been adapted from [282].

According to the Stability Limit conjecture, there exists a boundary in phase diagram (known as spinodal curve) originating from the liquid–gas critical point that connects superheated, supercooled and stretched states (negative pressure) of water. Existence of this continuous line rules out of any thermodynamic singularity in the supercooled water. In the following Fig. (**9.15**), the difference between the Stability Limit conjecture and Liquid–Liquid Critical Point theory is demonstrated. As can clearly be seen, in Stability–Limit conjecture (depicted in panel a) the spinodal curve extends from liquid–gas critical point, through negative pressures into supercooled temperature domain (spinodal curve retracing). On the contrary, in Liquid–Liquid critical point scenario (as shown in panel b), the spinodal line does not re–enter to the positive pressures. Another notable point is that the Temperature of Maximum Density (TMD), which runs parallel to the spinodal curve in the Stability–Limit conjecture (panel a), whereas in the LLCT, only the lower part coincides with the spinodal curve.

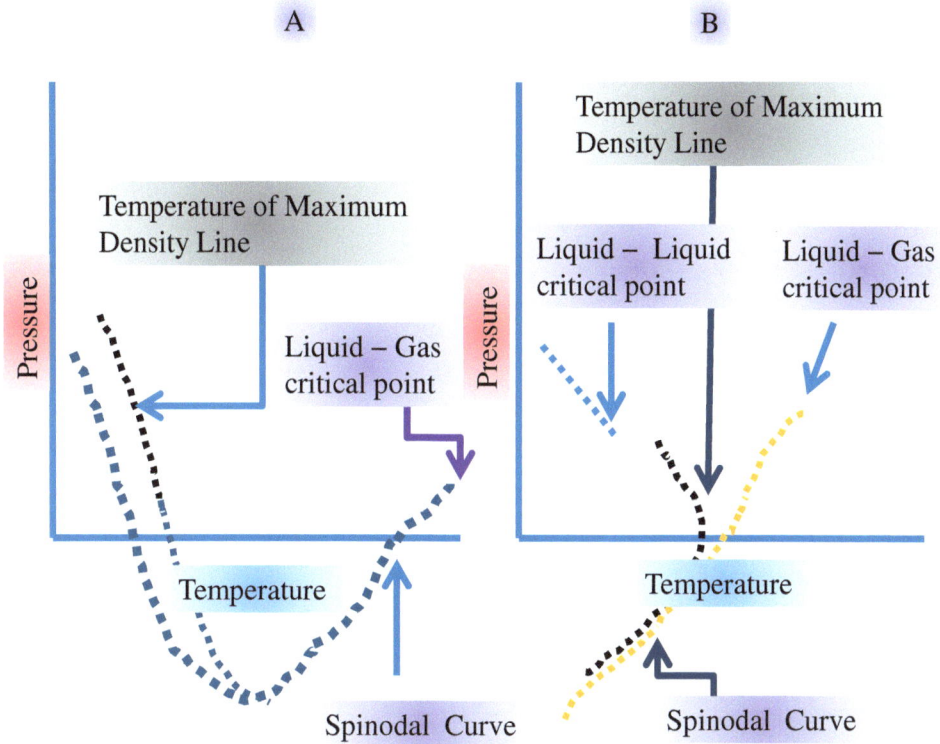

Fig. (9.15). Comparison of Stability Limit conjecture and Liquid–Liquid Critical Point (LLCP) theory. In the Stability Limit Conjecture shown in the panel **A**, the spinodal curve retraces back to the positive pressure region in contrast to LLCP theory shown in the panel **B**.

Singularity Free Hypothesis

Singularity Free (SF) hypothesis was suggested by Stanley *et al.* in order to explain "thermodynamic ambiguity" in supercooled water [283]. As shown before, the slope of Temperature of Maximum Density (TMD) curve is appeared to be only negative in the pressure–temperature diagram according to Stability Limit conjecture, whereas this requirement is not necessitated in the Liquid–Liquid Critical Point scenario. This indicates that the shape of Temperature of Maximum Density (TMD) curve plays a key role in determining whether the spinodal curve traces back or not. It can also be shown that the nature of slope (negative) of TMD can be linked to increase in isothermal compressibility (it has to be noted that "anomalous" increase in this thermodynamic function is one of the thermodynamic anomalies observed in supercooled water). TMD changes its slope at the point of its intersection with the locus of Temperature of External Compressibility (TEC). Three cases, as shown in the following Fig. (**9.16**), arise at the point of intersection between TMD and TEC: whether this point of intersection corresponds to compressibility maximum, minimum or the point of inflection between maximum and minimum. Furthermore, it has been shown by the proponents of the SF hypothesis that it is indeed compressibility maximum existing in lower temperatures, which negates any thermodynamic singularities (Fig. **9.16**).

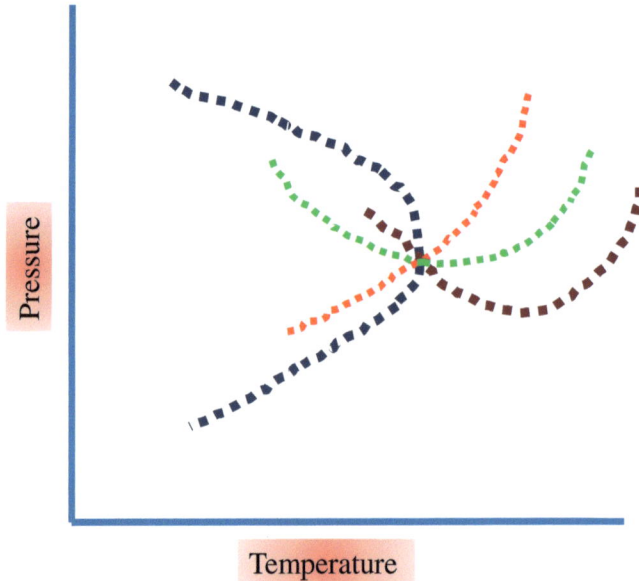

Fig. (9.16). Singularity Free Hypothesis. The change of slope of TMD curve (blue dotted) precludes the necessity of spinodal retracing. For negatively sloped TMD, the compressibility (red, green and maroon dotted lines indicating three different scenarios of compressibility extrema) indeed increases at positive pressures, suggesting that anomalous increase in compressibility is a thermodynamic requirement.

Succinctly speaking, the Singularity Free (SF) hypothesis posits that the anomalies occurring in the supercooled water are due to thermodynamic requirement related to Temperature of Maximum Density and Isothermal Compressibility, rather than any singular behaviour.

Critical Point Free Hypothesis

A recent addition to the aforementioned hypotheses is Critical Point–Free hypothesis proposed by Angel [110]. This theory primarily challenges the concept of LLCP theory, as shown in the following Fig. (**9.17**). The major distinction between Critical Point Free scenario and Liquid–Liquid Critical Point theory is that the liquid–liquid critical point is blended with the metastability contour in the former case, whereas according to LLCP theory, the critical point is clearly distinct (as evident in the phase diagram), that accounts for the thermodynamic anomalies observed in the supercooled water.

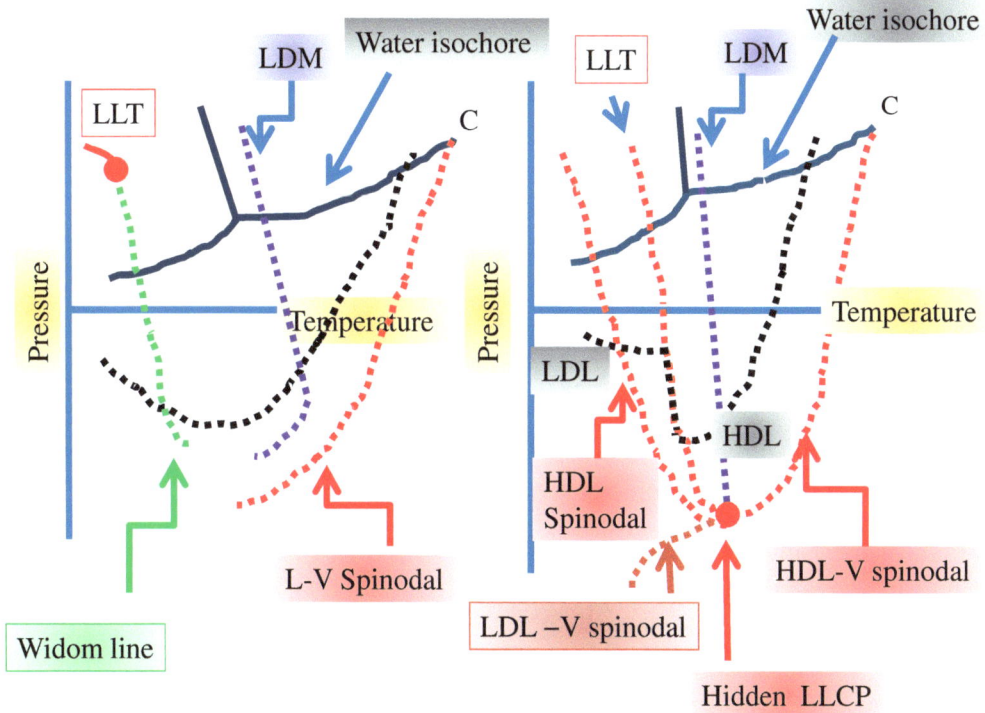

Fig. (9.17). Comparison of Liquid-Liquid Critical Point theory (left) and Critical Point Scenario (right). The primary distinction between two scenarios can be noted in the disappearance of Liquid-Liquid Critical Point (LLCP) in the Critical Point Scenario (note that the thick red dot that appears at positive pressures in the left panel merged with the metastability limits (HDL-Vapour & HDL spinodal lines) in the right panel). The line of maximum density is indicated by LDM. The idea has been adapted from [284].

DIVERGENCE OF HYDROGEN BOND LIFETIMES

The anomalous behaviour observed at −45°C (known as Angell Temperature, T_A) points out into another interesting phenomenon: strengthening of tetrahedral structure. Using computer simulations, using ST2 water model, Stanley *et al.* have studied the hydrogen bond structural variation in supercooled water across a wide range of temperatures [157]. They observed that the hydrogen bond lifetimes in water are found to be diverged at T_A, and concluded that many of the anomalous properties of water at this temperature are attributed to this diverging lifetime, as demonstrated in the following Fig. (**9.18**).

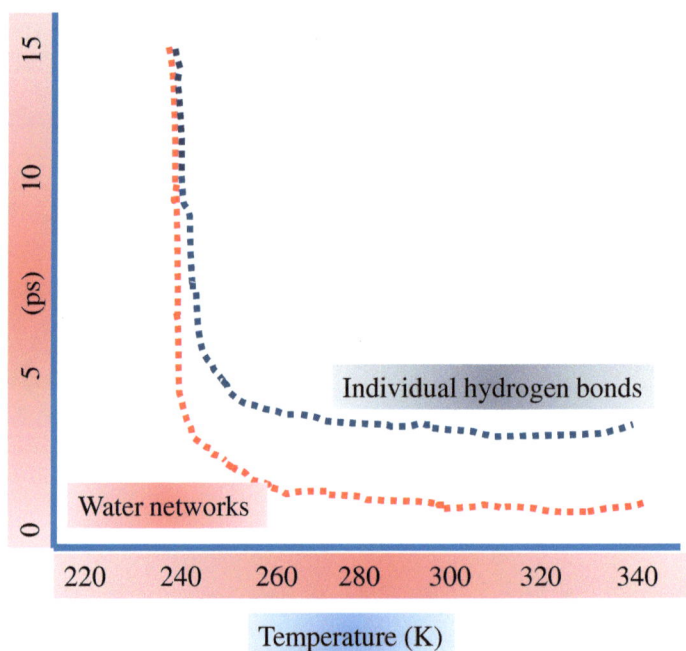

Fig. (9.18). Hydrogen bond lifetimes diverge at temperature −46°C. Hydrogen bond lifetimes of individual hydrogen bonds (shown in blue) and water networks (shown in red) are shown. The idea has been adapted from [157].

It can clearly be seen (Fig. **9.18**) that the lifetime of a single hydrogen bond is always higher than that of water network, except at the diverging temperature, T_A, at which the hydrogen bond lifetimes converge.

CORRELATION BETWEEN STRUCTURAL, KINETIC & THERMODYNAMIC ANOMALIES

The foregoing discussions on water anomalies were centered on several structural, kinetic and thermodynamic anomalous behaviours, and they are correlated to each other. Naturally, one would rather be interested as to how they are inter−related. In an excellent article, Debenedetti *et al.* have answered this question by applying order parameter concept within the framework of density and temperature as shown in the following, Fig. (**9.19**) [285]. Structural anomalies in liquid water, as shown by the outer green contour region (Fig. **9.19**), occur at a wide range of density (0.9 −1.2 g/cm^3) and temperature (~200 K to 325 K). The structural anomalous region is defined by two order parameters, translational and orientational [285]. Translational order denotes the optimum separation of two neighbouring molecules in a system, whereas the orientational order parameter provides a quantitative estimation of structural compactness of molecules (tetrahedral orientation in the case of water). The structural anomalies are bound by extremes of these two order parameters: lower densities and higher densities corresponding to maxima of orientational order and minima of translational order respectively. On the other hand, kinetic anomalies exhibited by water occur in narrower region, as shown by blue shaded region in the following Fig. (**9.19**), which is defined by extreme values in diffusion coefficients. The thermodynamic anomalous region is defined by the Temperature of Maximum Density (TMD), shown by the inner region shown in red. TMD is characterised by a hike in density upon increasing temperature at constant pressures.

HYDROPHOBIC HYDRATION

Another interesting unusual behaviour that H_2O possess is hydrophobic hydration, a process by which water molecules associate themselves around non polar solute in solution thereby minimising the degree of contact with neighbouring water molecules [286]. The hydrophobic hydration is an influential factor in water−macromolecule interactions [286]. Non polar solutes such as noble gases, organic compounds such as methane, and sulfur hexafluoride (SF_6) are completely soluble in water despite being incapable for making hydrogen bonds. From thermodynamic point of view, this process is favourable due to increasing entropy and negative free energy. When hydrophobic hydration occurs, non–polar solutes cluster together, pushing aside water molecules (hydrophobic interaction). However, the water molecules that are more ordered around the non–polar solutes, and Van der Waals type interactions existing between non–polar solutes and them, leads to a decrease of order in water molecules which is accompanied by an increase in entropy [286].

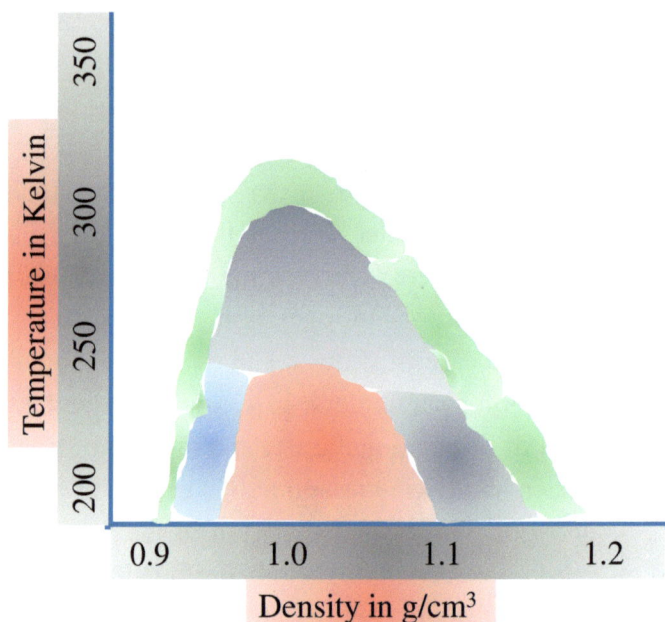

Fig. (9.19). Relationship between structural, dynamic and thermodynamic anomalies. The region where thermodynamic anomalies are exhibited by water is shown in red–shaded area; the blue–shaded region indicates the region in which diffusive anomalies are dominant, and the outer green–shaded region denotes the region of structural anomalies of water.

At molecular level, it can be interpreted as a structure making effect by which these non–polar solutes safely accommodate the structural network of water, without breaking it altogether [138b]. Water prefers to form hydrogen bonded network as we have seen in earlier chapters and it has been proved that in the presence of non–polar solutes the stability of these hydrogen bonded network increases [286]. The following Fig. (**9.20**) shows how water molecules orients towards nonpolar solutes.

As a result of the interaction between polar water and non−polar solutes *via* Van der Waals type interactions, the lifetime of the water network increases. Enhanced solubility of hydrophobic gases in water upon decreasing temperature can be considered as a related anomaly [161b].

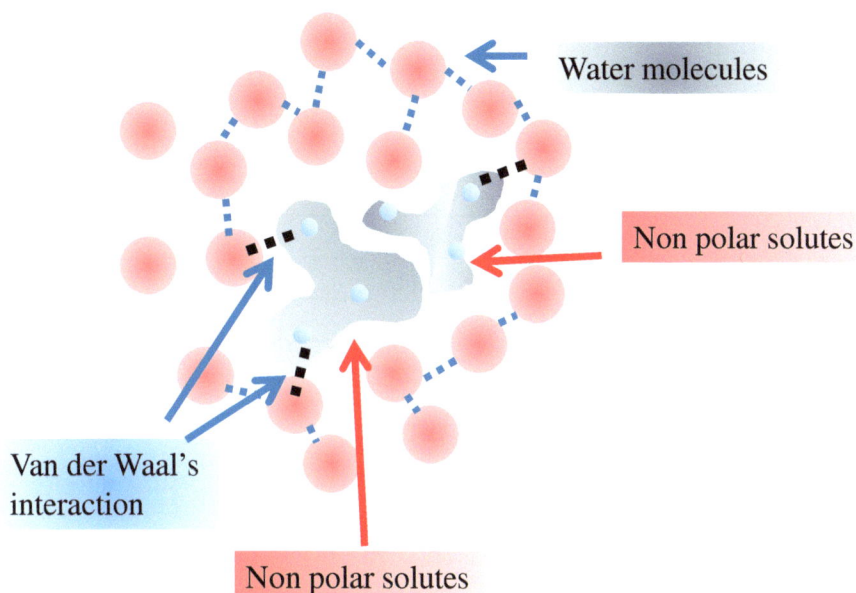

Fig. (9.20). Hydrophobic hydration in water. In the presence of non-polar solutes water (shown by red circles) cluster together around them in a much more ordered fashion, but the order is disrupted due to the van der Waals interaction between solute and water particles contributing to the spontaneity of the hydrophobic hydration. Hydrogen atoms of water molecules have been omitted for clarity.

CONCLUDING REMARKS

The following is the list of notable anomalous properties of water in bulk phase, which is the essence of this chapter, presented as a quick reference for a reader. I have described the anomalies of water in this chapter, which are studied extensively by computational and experimental means. There are several other anomalies, which I haven't described in this chapter, in particular the properties of aqueous solutions [287], and no effort was made into review the anomalies that water exhibit when it is confined to nanoscale, and near the macromolecular surface (for example protein) since I fear that this will digress the subject. Nevertheless, it is interesting to note the connection between the theories developed for bulk water and water in macromolecular environment. Kumar *et al.*

have extensively employed computer simulations in order to understand complex mechanism of interaction of water with biomolecules, and their findings reveal clearly a link between Liquid–Liquid Critical Theory (LLCT) and protein dynamics. More computational and experimental efforts in the coming years could unearth further anomalous behaviour of water under specific conditions [277]. Below (Table **9.2**) given are the list anomalies that have been investigated in this chapter.

Table 9.2. Notable Water anomalies discussed in this chapter.

Number	Anomalous Character	Thermodynamic Conditions
1	Existence of water as a liquid	273 K – 373 K
2	High boiling point	373 K
3	Density maximum	Supercooled
4	Density minimum	Supercooled
5	Diffusion coefficient	Supercooled
6	Product of diffusion coefficient and rotational relaxation time (Stokes-Einstein Relationship)	1.8 * Tg
7	Heat capacity maximum	Supercooled
8	Thermal conductivity maximum	Supercooled
9	Hydrophobic hydration	Normal temperature
10	Hydrogen bond lifetime	Supercooled
11	Liquid-Liquid phase transition in supercooled state	Supercooled

With increasing sophistication and accuracy of the experiments and theoretical tools, it is expected that the list will grow considerably in coming years.

CONFLICT OF INTEREST

The author confirms that he has no conflict of interest to declare for this publication.

ACKNOWLEDGEMENTS

I express my gratitude to the following person and organisations for granting me permission to use figures under their copyrights in this chapter: Professor Carlos Vega, American Physical Society, and Nature Publishing Group.

CHAPTER 10

A Journey Through Water - A Review

Abstract: Recent advances in computing and development of sophisticated experimental techniques have enabled us to make giant leaps in understanding the microstructure of water. The structure of water in normal temperature range is still shrouded in Continuum-Mixture model controversy while new characterisation methods are reported on yearly basis, leaving water an interesting and controversial theme for ever. A multifaceted approach, amalgamating social, economic, political, geographical and technological aspects, is required to alleviate the issues related to the scarcity of water to considerable extent. Centres have been established for performing cutting edge research on water across the globe in order to develop efficient technologies in response to chronic water scarcity. Molecular scientists can greatly contribute to the technological advances that could allay the problems related to fresh water, and influence the policy makers at various organisation levels.

Keywords: Gel, Moor's law, Supramolecular chemistry, UNESCO, Water models.

INTRODUCTION

Our journey ends with this chapter. The motive of writing this chapter is twofold: to summarise the major landmarks throughout this journey and to provide an outlook on future actions for combating the problems related to the availability of fresh water. In the first part of this chapter (from 2 through 4), a review of theoretical and experimental background of molecular investigations and theories on water is provided. The remaining portion of this part is devoted to important research findings on various forms of H_2O, described in chapters from 5 through 9.

In the second part of this chapter, pragmatic approaches, from a molecular scientist's point of view, are discussed in relation to tackling water crisis. This

Jestin Baby Mandumpal

includes application of confinement techniques within the framework of nanotechnology. The chapter is concluded by two important recommendations for increasing the participation of molecular scientists in international stage, and for offering much more effective solutions to water–based problems the world faces.

VARIOUS STAGES OF THE JOURNEY

In this section, I present the essence of what have been discussed in the nine previous chapters. This will help the reader quickly rewind the whole book and understand "water" in a wider perspective.

Water, the Centre of Life

In the first chapter, at the beginning of our journey, the central role of water plays in our life was discussed. I presented water in social, political, economic and technological landscapes, the dimensions of which provide rather bleak picture unless a serious and focussed approach is taken. It advocates the importance of considering water in a broader perspective in order to solve one of the chronic problem our world faces, the scarcity of clean drinking water. This approach combines political consensus among different groups standing for solving water crisis. Although water's growing demand surpasses what any solutions can offer, the crisis will become worse without the development and implementation of efficient technology comprising all aspects of water, and molecular scientists can provide a strong support in achieving this aim in the different parts of the world.

A Snapshot of Liquid State

Water is used the most when it is in liquid state, and therefore in the second leg of the journey, the discussion was centered on liquids, starting from their elementary properties. The major portion of this journey is devoted to intermolecular forces, and it would be appropriate here to paraphrase what professor Anthony Stone opined about the applicability of some crude potential forms such as Lennard Jones model in molecular simulations in his seminal book "The theory of intermolecular forces" [47]: having made a great progress in diagnosing what the intermolecular forces are, it is high time to replace such a model, though still useful, for betterment of estimation of various physical properties. The remaining section in this part of the journey was centered on supercooled water, which we would again visit later in the book. Theoretical tools of glass physics have been applied to investigate the glassy water, and a considerable level of progress has been made to understand the behaviour of the puzzling supercooled water. These brilliant theoretical works include Adam – Gibbs theory, Mode Coupling Theory (MCT) and Energy Landscape theory, and most of them are delicate mix of kinetics and thermodynamics.

Experimental Tools for Microanalysis of Water

Third leg of our journey offers us a tour on a wide range of experimental techniques that have already been employed for the microstructural elucidation of water. A closer look at the experiments on water reveals that three major techniques have been widelyemployed in the investigation of various phases of water: scattering, spectroscopicand calorimetric techniques.

Although computer simulations have several advantages over real experiments as we would see in chapter 4, we must not underestimate the role of experiments as the "final word" on any scientific matter. Stanley and his co–workers have summarised this fact in one of their review papers: "one experiment is enough to kill a hypothesis" [182]. This statement was based on several recent experimental observations, which refute several already existing theoretical propositions. For example, with the experimental report that claims the observation of the third form of supercooled water, the existence of Liquid–Liquid Critical Point (LLCP) theory that advocates only two phases in supercooled water has been challenged. Further, recent findings on the structure of water reveal that water is structured up to five hydration shells. It is interesting to note that the first and fourth shells are almost without any significant change upon alteration in temperatures [180]. This is indeed very opposite to the traditional view point that the "overall" water structure is disrupted.

Development in nanotechnology opened up new horizons of empiricism, hybridisation of various experimental tools. The recent discovery of the 18th stable phases of ice, by exploiting the properties of graphenes, is a perfect example for this approach [104]. A more recent method to investigate the formation of hydrogen bond is another example of hybridisation: by this method (called as molecular surgical method) which involves creating an opening of C_{70} molecule that allows a single water molecule enter inside, the dynamics of monomers or dimers can effectively be investigated [288]. Hybridisation encourages researchers to apply hybrid experimental techniques in unearthing physical and chemical properties of materials.

Fundamentals of Molecular Simulations

Computer simulations were a step change in scientific investigations when they were introduced in the latter half of twentieth century. The formative period of computing is over and few would be sceptical about the role of computer simulations in increasing our understanding of the world, and science is too not spared from this. As a result, many popular misconceptions regarding the hitherto known scientific facts have been come to the lime light [289]. In this phase of our journey, various aspects of computer modelling as a powerful enterprise for the investigation of matter at atomic and molecular scale are reviewed.

The history of molecular simulations begins with two forcefield based methods, namely Monte Carlo and Molecular Dynamics. These techniques were firmly built on the principles of statistical mechanics using the concept of ensemble, which provides a theoretical basis for deriving physical properties by significantly reducing number of particles in the system of interest. The errors can also be induced from various approximations made during several stages of the simulation protocol to reduce computational cost. First drastic approximation we encounter in traditional molecular dynamics simulation is the truncation of long range non−bonded interaction, which in the case of associated liquid like water can lead to erroneous outcomes. Other types of errors are model based. For example, a three point water model can save enormous amount of computer time compared to a five point model at the expense of accuracy. Exploiting the development of robust computer simulation methods and water models, performing computer simulations on water is now a days a routine task. Over the years, water models of various classes, developed by the pioneers like Stillinger, have played a non−negligible role in enhancing and bettering our understanding of water. One must, however, note that the usefulness of these methods is limited by the choice of parameters that define the intra and inter molecular interactions defined within the forcefields.

By the introduction of *ab−initio* Molecular Dynamics simulations (combination of classical molecular dynamics simulations and electronic structure calculations), these drawbacks can be overcome to a greater extent. In addition, one can investigate the dynamics of condensed phase such as water at very high resolution, at electronic level. Nowadays, various protocols of *ab−initio* methods are commercially and non−commercially available to academics. Approaches based on Density Functional Theory (DFT) are the pick of *ab−initio* theory since the information (electron density) they provide can directly be validated by experimental methods. Combined electronic and classical approaches (QM−MM for example) are also increasingly becoming popular among molecular scientists. However, notable disadvantages of these methods include longer simulation time in order to achieve a reliable estimate, which force one to restrict to sample the system of interest for relatively short time scales than traditional approaches.

For every use of computer simulation methods for the investigation of materials such as water, there are several detractors too. The most notable disadvantage is the selectivity of water models for attesting a theory. For example, the verification of Liquid−Liquid Critical Point (LLCP) theory, which is accessible with a Stillinger Type potential (ST2), is inaccessible with a much modern water model like SPC/E [186]. Other typical example is the variation of the Temperature of Maximum Density (TMD) estimated by different water models.

New research articles come to light claiming better representation of a particular physical property of water with one particular model than others, often referring to experimentally measured values [275b]. This would mean that assumptions based on other tested models in previous research endeavours became obscure. The situation becomes very complex when this very 'successful' model fails to reproduce some other physical property, raising concerns over the reliability of the model. This dilemma is a growing matter of concerns for experimentalists too: which simulation findings can be considered as authentic in order to compare their experimental work?

Another uncertainty in the use of computer experiments lies in the fact that there lacks strict norms regarding the length of simulation, which sometimes makes comparison of results obtained by different models, methods and research groups very unreasonable. Despite the advancement of computing speed the "normal simulation time" attained is only nanosecond/microsecond level, which is far lower than the time scales of real life. If Moor's law can be trusted, simulations at second's level will become a norm within the span of ten years.

The construction of a robust model (ideal water model) for mapping "complete" phase diagram of water is very essential, and the estimation of physical properties using this model is expected to be accurate under any thermodynamic conditions (temperature, pressure and density). Together with sophisticated experimental techniques the properties of water can be unravelled further. This is expected to have tremendous impact upon the development of water based technologies since most of them are fundamentally of chemical nature.

Water Between Its Boiling Point and Freezing Point

In the fifth phase of our journey, water in the normal temperature range (between 273 K and 373 K) was discussed. The nature of hydrogen bond is dynamic in nature, spurring sporadic changes in its local structure, which can be effectively probed by various spectroscopic and scattering techniques. Local structure of water molecules is influenced by thermodynamic effects. Water undergoes a cascade of morphological changes upon alteration in temperature across wide range, which is still a fascinating subject for many researchers.

Several propositions have been made in order to account for water structure, Tetrahedral Network theory and Mixture models. One school supports the notion of extensive three dimensional networks of hydrogen bonds in water (Uniformist model). On the other hand, there are growing number of researchers who argue that the idea of homogenous structure in water is outdated, and advocate that there exist preponderance of networks of various sizes and dimensions in water. It is worthy of attention the experimentally and theoretically verified clusters of varying sizes are present in water, and hence it is vital to diagnose the extent to which water forms clusters and its mechanism. Molecular Dynamics simulations have been extensively used for studying internal dynamics of water clusters, from dimers to clusters with over hundred water molecules [290]. Quantum chemical calculations have bettered our micro level understanding of water. Quantum mechanical calculations are very fruitful in several aspects in the context of microanalysis of water. For example, one can compare different configurations of water clusters with fixed number of water monomers to the utmost accuracy. Further, one can investigate the electronic effects in the solvation [250].

The Mixture model, with obvious heterogeneity in its character, had been gaining popularity among scholars until the publication of article on X−Ray Diffraction (XRD) investigations on water [149]. In this article worthy of attention, the authors claim that they have provided experimental evidence for the homogeneous character of water structure. Thus, the debate still continues over the nature of networks in the form of hydrogen bonds in water. While discussing about water network theories, it is important to mention Percolation model, which provides a quantitative picture of hydrogen bonding in liquids.

While there is still confusion regarding the microstructure of water whether it can accurately be represented by Mixture or Continuum models, it is "convenient" to perceive water as locally structured transient gel. Water has a local structure, as one can clearly infer from radial distribution functions, extending from 4 to 8 Angstroms [291], which has been resolved up to 12 Å by a recent X−ray scattering experiment [180]. At the same time, water has similar features of gel, thanks to its enigmatic hydrogen bond network by which individual molecules are connected to each other. However, due to very short life times of the connecting bonds, this gel is "transient" (constantly reformed).

One may be tempting to think that these debates are of little interest beyond a pool of chemists and physicists who are working on water, since our lives are not directly affected by theories on hydrogen bonding that connects water molecules together. However, we learn that it is the basis of understanding water itself, and hence a solid theoretical development on this subject may contribute to the development of new technologies, which I will certainly mention latter in this chapter. We must note the fact that theories developed hitherto on hydrogen bonding played non−negligible roles in emerging fields like supramolecular chemistry [292].

Supercooled and Glassy Water

Much of the understanding on cold water has come from scholarly analyses based on numerous computer simulations over the years. The sixth phase of the journey was devoted to explore two low temperature and non−crystalline forms of water, supercooled and glassy water. These two states, though very metastable with respect to the stable crystalline form of water, ice, are very pivotal in supporting the existence of several microorganisms in low temperatures. The geometries of water clusters formed in supercooled water range from two dimensional chains to three dimensional protein−like structures. The growth of clusters from a water molecule to a three dimensional structure can give way to crystals. We note that diffusion in supercooled water occurs in two different ways: heterogeneous and homogenous, analogous to heterogeneous and homogenous nucleation processes. Another important question discussed in this session was how long supercooling can be postponed. The lowest recorded temperature at which water can be in supercooled state is 200 K, but a large pool of scientists has not taken this claim seriously.

Despite the fact that performing experiments on supercooled water is cumbersome, scientists have managed to get a major experimental breakthrough by identifying the transformation between High Density Amorphous (HDA) and Low Density Amorphous (LDA). On the contrary, the determination of accurate glass transition temperature at a given cooling rate still shrouded in mystery even after numerous experiments and computer simulations. An important theoretical achievement obtained so far is the notion of strong and fragile glasses, a qualitative tool for glass identification. At the same time controversies are around on either sides of a particular region in the phase diagram, between supercooled and glassy/crystalline states, which is still known as no−man's land due to lack of certainty of the properties of water.

Ice, the Crystalline Phase of Water

In the seventh phase of the journey, the crystalline forms of water were explored. When we talk about ice, the most important factor that must come into our mind is that unlike water ice is a polytype, and number of forms that ice possesses has not been fixed yet as in every six years a new form is discovered. Some of the properties of these polytypes make their investigations cumbersome: for example pairing between the various ice structures, the transition between the ordered and disordered structures. Due to the existence of ice as numerous poly types, studies on ice crystals provide wonderful and exciting opportunities for those who are interested in symmetry.

Dynamic character of hydrogen bonds plays important roles in the order–disorder transition. The movements of protons in ice, restricted by Bernal–Fowler rules, result in Bjerrum and ionisation defects, which are responsible for their dielectric properties. In addition, what make ice studies interesting are its excellent electrical, optical, mechanical, thermal & surface properties. Its promising catalytic properties are relevant for environmentally friendly chemical industry. More theoretical and experimental efforts are required to tap these advantages, and it is believed that this would open up a new door in investigations related to environmental damages such as the depletion of ice layers in Antarctic.

Water above Its Boiling Point

In the eighth phase of our journey, water above its boiling point was investigated. Water is transformed to Super Critical Water (SCW) at critical pressure, temperature and volume. SCW has been found to have exceptional properties such that many chemical reactions can be efficiently carried out without the presence of catalysts, and hence is considered as a better alternative to many of the traditional reagents, as in the case of ice. It seems that Super and Sub critical water are suggested to be more effective than ice in combating pollution due to chemical industries. Another interesting fact is that depletion of tetrahedral hydrogen bonding network occurs at elevated temperatures, leading to the preponderance of water clusters than found in ambient water, with varying size and geometry. Number of hydrogen bonds shows a proportional increase as density increases. The presence of hydrogen bonds even beyond the maximum temperature at which organisms can survive is truly significant and it shows how fundamental this type of intermolecular force is. Super Critical Water (SCW) has also been found to have exceptional properties to catalyse several types of reactions, which is environmentally, kinetically and economically more viable than the traditional methods employed in the current chemical industry. Every year billions of dollars are spent on chemical reagents, which, in turn, cause environmental damages to our surroundings. It is clear that the replacement of these costly reagents by Super Critical Water can reduce environmental damages to a greater extent and immeasurable costs inflicted upon research laboratories across the world, troubled by dwindling financial support.

A Brief Review of Water Anomalies

The penultimate phase of our journey is devoted to what water is known for, its mysterious anomalous properties, some of which have already been known to the scientific community for centuries, and have been studied in much depth by experimental and computational methods. Most of the notable anomalies of water have been reported in the low temperature regime; anomalous behaviour of liquid water is so striking when it is supercooled below the melting temperature of ice, T_m. Singularities in several physical properties (including isobaric heat capacity, isothermal compressibility, relaxation time and thermal expansion coefficient) are signatures of anomalies in supercooled water. Interestingly hydrogen bond life times show a divergence at this temperature, indicating its strong connection to the anomalous behaviour.

A brisk survey was done on several propositions that have been made in order to account for water's anomalies, including Liquid–Liquid Critical Point theory, Singularity Free hypothesis, Critical Point–Free hypothesis and Stability Limit Conjecture. Although most of the anomalous characters that are exhibited by water and supercooled water are in the bulk phase, several others when confined to nanoscale geometries and near to macromolecular surface have also been reported. While nanoscience breaks the boundaries of various disciplines, "new" anomalous properties of water is very likely to be reported on regular basis in near future.

This was indeed a short and quick journey, passing through very extremes: from low icy temperatures to very high steamy temperatures and from solid phases to vapour states. It is very clear now, due to the monumental efforts of numerous researchers who worked diligently, that these conditions imprinted very different properties on the nature of water. The most important landmark on this journey has definitely been the discussions based on hydrogen bonding, the network that underpins characteristic properties of water across the whole thermodynamic map. It is therefore so crucial to procure a complete understanding of these elusive hydrogen bonding interactions that lead to mysterious properties of water, and aqueous solutions in which water is a principal ingredient.

A Quest for a Unified Theory of Water

Enormous volume of work with bewilderingly varied results prompts a profusion of hypotheses on several aspects of water (for example, networks in water or hydrogen bonding patterns). We note that despite the hitherto advancement in scientific understanding of water drafted in great number of articles year by year, a theory that unifies all of structural peculiarities of water is still long way off. One would expect that with this unified theory on water all of its known properties including anomalies be explained. The immediate consequences this theory is that one can simply explain the behaviour of water at all conditions.

I wind up this section with a reference to one of the prominent theoretical physicists of our era Stephen Hawking on the logical requirements for a "good" theory:

1. The theory must accurately describe a large class of observations based on a given mathematical construct (model) containing only a limited number of arbitrary elements.
2. The theory should be useful for making definite predictions about the results of future observations.

Is such a unified theory possible in the case of water? I am not confident of an affirmative answer due to the following reasons: firstly, the list of properties of water is still growing, and scientists are still struggling to explain these individually let alone collectively; and secondly, new forms of water framework, in particular its solid form, (ice), are identified on regular basis. Meanwhile, a new state of water has been reported by a group of American researchers at the time of the release of this book, adding spice to the already existing controversies related to water based materials [293].

MOLECULAR INVESTIGATIONS IN WIDER PERSPECTIVE

We just reviewed glimpses of knowledge (at molecular level) produced so far on water. A genuine question follows is that what molecular scientists can offer as solutions to water related issues, in particular water pollution and water scarcity. How can they deal with water at micro level that brings concrete solutions to the problems related to water? Before discussing this, we must think about the global platforms that are available for scientists to engage in water related issues.

A continental initiative combining various aspects of nanoscience can help resolve several problems related to water. One of such initiatives is **nano4water**, exclusively created for the application of nanotechnology towards obtaining "affordable solution" to the growing water crisis [294]. The developers of "nano4water" have so far managed to combine various themes under one roof. This initiative is the demonstration of European researchers working in different disciplines, from nearly all parts in the European Union (EU), that they have realised the importance of joint approach in combating water related issues. However, it is unfortunate to say that the participation of molecular scientists in bigger science platforms and scientific initiatives are still mediocre. The biggest of such initiatives has been taken by United Nations Educational, Scientific, and Cultural Organisation (UNESCO) by founding water related centres across the globe. As a result, about twenty five institutions targeting specific water studies have been erected so far. The focus of research includes ground water management (Latin America, East Africa), river basin management (Nigeria), ecohydrology (Indonesia, Portugal), water resource management (USA), water hazard and risk management (Japan), and urban water management (Iran).

In order to address the issue of water scarcity, at national level too, several research centres have been set up to tackle water scarcity. The federal government in the United States (U.S) has taken initiative in establishing numerous water research institutes. Most notable of them is The Institute of Molecular Engineering at The University of Chicago. Membrane technology and the development of novel catalysts are among the principal area of research at this centre. Another important institute worth to mention is Water Research Institute (WRI) under the auspices of Council for Scientific and Industrial Research. This research centre is a conglomerate of twelve research subdivisions including Hydrogeology, environmental chemistry and microbiology.

In Asia, several countries have taken measures to set up centres for performing research on water. In Israel, several famous water research institutes have already been set up. Stephen and Nancy Grand Water Research Institute (GWRI) works in private–public partnership, receiving generous amount of research funds. Zuckerberg Institute for Water Research (ZIWR) in Israel adopts both theoretical and experimental methods in order to perform cutting edge research in water. The centre has even developed, probably the first outside Europe and America of its kind, an undergraduate course in water studies, "hydrology and water quality". Surely, it is an important step towards creating awareness in younger generation, by promoting education extensively on water related subjects. Islamic Republic of Iran has established Water Research Institute under its Ministry of Energy, which coordinates all research activities of centres affiliated to it. Kuwait has also instituted a state–of–the–art research centre, under Kuwait Institute of Scientific Research, which focusses more on wastewater treatment. Masdar Institute in the United Arab Emirates (UAE) has set up a sophisticated water research centre (iWater). The centre specialises in several potential research themes including climate change, water resource management, and desalinisation.

In other developing countries too, research initiatives has been taken in the form of institutes, for example International Water Management Institute (IWMI) in Colombo. The institute, focussed on efficient use of water in ensuring food safety in developing world, is working along with the governments in the developing world, in particular Asia and Africa. In India, a Water Technology Center has been functioning under Indian Agricultural Research Institute (IARI) since 1969. IARI aims at developing water based technologies by performing fundamental and applied research on water management. Indian Institute of Water Management has also been instituted for developing water management technologies. National Water Academy is an Indian Government enterprise set up for coordinating all water related research in the country, which coordinates different science and engineering departments.

In the Far East, Japan has initiated its effort to face water crisis by founding Japan Water Research Centre. China Institute of Water Resources and Hydropower Research is a major organisation, which aims to develop water based technologies. Several research centres function in Australia, aiming to perform various research activities on water. These research centres include Water Research Centre (WRC) in New South Wales, Centre for Water Management and Reuse at University of South Australia, Advanced Water Management Centre (AWMC) at the University of Queensland and International Water Centre at the University of Western Australia (UWA).

There are several interesting ideas are being developed at the aforementioned research hubs as part of their research initiatives. The extraordinary development of nanotechnology has brought molecular scientists a great opportunity for solving problems related to water. Nanotechnology is based on the fact that exploitation of matter can be achieved at molecular level, controlling atoms and molecules individually, with high degree of precision. Although applications of nanotechnology on water have not been explored yet completely, researchers have realised its potential as an answer to water scarcity.

One of such projects is the use of sun light with the doped TiO_2 nanomaterials. As this technology is based on the molecular level manipulation of materials, molecular scientists can contribute to this industry–academia joint venture. Another recent development that can be economically viable and technologically efficient from molecular perspective is development of hydrogels, polymer networks that can absorb large amount of water. This type of materials is promising due to its ability to store water in large amounts [295]. Specially tailored hydrogels can be used to purify water due its high absorption capacity in high salt content water.

FINAL WORD

The author calls for better cooperation between molecular scientists, being one among them, and other researchers, and policy makers towards finding effective solutions to water related problems the world faces. Molecular scientists are in good position than ever to exert pressure on governments and organisations such as United Nations Organisation (UNO) to allow generous financial support for molecular level investigations, thereby tuning water at very molecular level to quench the technological needs. At the same time, the author feels that there should be more cooperation within the community of molecular scientists, its theorists and experimentalists, towards understanding water better. This will aid understanding each other's methodology better, which would help them optimise their working tools.

CONFLICT OF INTEREST

The author confirms that he has no conflict of interest to declare for this publication.

ACKNOWLEDGEMENTS

Declared none.

REFERENCES

[1]   Molle, F.; Mollinga, P.P.; Meinzen-Dick, R. Water, politics, and development: introducing water alternatives. *Water Altern.,* **2008**, *1*, 1-6.

[2]   Gleick, P.H. Water in crisis: paths to sustainable water use. *Ecol. Appl.,* **1998**, *8*, 571-579.
[http://dx.doi.org/10.1890/1051-0761(1998)008[0571:WICPTS]2.0.CO;2]

[3]   Sowers, J.; Vengosh, A.; Weinthal, E. Climate change, water resources, and the policies of adaptation in the Middle East and North Africa. *Clim. Change,* **2011**, *104*, 599-627.
[http://dx.doi.org/10.1007/s10584-010-9835-4]

[4]   Lazarova, V.; Levine, B.; Sack, J.; Cirelli, G.; Jeffrey, P.; Muntau, H.; Salgot, M.; Brissaud, F. Role of water reuse for enhancing integrated water management in Europe and Mediterranean countries. *Water Sci. Technol.,* **2001**, *43*(10), 25-33.
[PMID: 11436789]

[5]   Available from: www.water.org.

[6]   Franks, F. Water a matrix of life. In: *The Royal Society of Chemistry*, 2nd ed; **2000**, 1-225.

[7]   Pimentel, D.; Houser, J.; Preiss, E.; White, O.; Fang, H.; Mesnick, L.; Barsky, T.; Tariche, S.; Schreck, J.; Alpert, S. Water resources: Agriculture, The Environment, and Society. *Bioscience,* **1997**, *47*, 97-106.
[http://dx.doi.org/10.2307/1313020]

[8]   Berman, I.; Wihbey, P.M. *The new water politics of the Middle East*; Strategic Review, **1999**.

[9]   Zeitoun, M. *Power and Water in the Middle East. The hidden politics of the Palestinian - Israeli Water conflict*; I.B. Tauris and Co.: London, New York, **2009**.

[10] Fisher, F.M. The Economics of Water dispute resolution, project evaluation and management: An application to the Middle East. *Water Resources Development,* **1995**, *11*, 377-390.
[http://dx.doi.org/10.1080/07900629550042092]

[11] Arnell, N.W. Climate Change and global water resources: SRES emissions and socio-economic scenarios. *Glob. Environ. Change,* **2004**, *14*, 31-52.
[http://dx.doi.org/10.1016/j.gloenvcha.2003.10.006]

[12] Murphy, R.P. Did we miss the boat? The clean water act and sustainability. *Univ. Richmond Law Rev.,* **2013**, *47*, 1267-1300.

[13] Mejla, A.; Hubner, M.N.; Sanchez, E.R.; Doria, M. *Water and sustainability UNESCO*; France, **2012**, p. 52.

[14] Kaika, M.; Page, B. The EU water framework directive: part 1. European policy making and the changing topography of lobbying. *Eur. Environ.,* **2003**, *13*, 314-327.
[http://dx.doi.org/10.1002/eet.331]

[15] Andreen, W.L. Water Quality Today - Has the Clean Water Act been a success? *Ala. Law Rev.,* **2004**, *55*, 537-593.

[16]   Page, B.; Kaika, M. The EU water framework directive: Part 2. Policy innovation and the shifting choreography of governance. *Eur. Environ.,* **2003**, *13*, 1-17.

[17]   Gupta, J.; van der Zaag, P. Interbasin water transfers and integrated water resources management: Where engineering, science and politics interlock. *Phys. Chem. Earth,* **2008**, *33*, 28-40. [http://dx.doi.org/10.1016/j.pce.2007.04.003]

[18]   Houweling, E.V.; Hall, R.P.; Diop, A.S.; Davis, J.; Seiss, M. The role of productive water use in women's livelihoods: evidence from rural Senegal. *Water Altern.,* **2012**, *5*, 658-677.

[19]   Arthington, A.H.; Naiman, R.J.; McClain, M.E.; Nilsson, C. Preserving the biodiversity and ecological services of rivers: new challenges and research opportunities. *Freshw. Biol.,* **2010**, *55*, 1-16. [http://dx.doi.org/10.1111/j.1365-2427.2009.02340.x]

[20]   Shannon, M.A.; Bohn, P.W.; Elimelech, M.; Georgiadis, J.G.; MariA as, B.J.; Mayes, A.M. Science and technology for water purification in the coming decades. *Nature,* **2008**, *452*(7185), 301-310. [http://dx.doi.org/10.1038/nature06599] [PMID: 18354474]

[21]   Shiklomanov, I.A. *World fresh water resources*; Oxford University Press: the U.K., **1993**.

[22]   a) Saleth, R.M.; Dinar, A. *The Institutional Economics of Water: A Cross Country Analysis of Institutions and Performence*; Edward Elgar Publishing Inc.: U.S., **2004**. [http://dx.doi.org/10.1596/0-8213-5656-9] b) Gleick, P. H. *The world's water: The Biennial report on Fresh Water Resources*; island press: Washington DC, **1998**.

[23]   Harou, J.J.; Paulido-Velazquez, M.; Rosenberg, D.E.; Medellin-Azuara, J.; Lund, J.R.; Howitt, R.E. Hydro-economic models: Concepts, design, applications, and future prospects. *J. Hydrol. (Amst.),* **2009**, *375*, 627-643. [http://dx.doi.org/10.1016/j.jhydrol.2009.06.037]

[24]   Barker, R.; Dawe, D.; Inocencio, A. Economics of Water Productivity in Managing Water for Agriculture. In: *Water Productivity in Agriculture: Limits and Opportunities for Improvement, J.W*; Kijne, R.B.; Molden, D., Eds.; , **2003**; pp. 19-35. [http://dx.doi.org/10.1079/9780851996691.0019]

[25]   a) Karr, J.R. Biological integrity: a long neglected aspect of water resource management. *Ecol. Appl.,* **1991**, *1*(1), 66-84. [http://dx.doi.org/10.2307/1941848] [PMID: 27755684] b) Arnold, C.A. Clean-water land use: connecting scale and function. *Pace Envtl. L. Rev.,* **2006**, *23*(2), 291-350.

[26]   Magesh, R. In *OTEC Technology - A world of clean energy and water Proceedings of the world congress on Engineering London,* **2010**.London

[27]   White, D.C. Clean Water hardly anywhere and that not safe to drink. In: *Clark Lecture,* **1995**.

[28]   Bosch, A. Human enteric viruses in the water environment: a minireview. *Int. Microbiol.,* **1998**, *1*(3), 191-196. [PMID: 10943359]

[29]   Public Service and Outreach, Clean Water Act. The University of Georgia http://outreach.uga.edu/

[30] Ball, P. Water as an active constituent in cell biology. *Chem. Rev.,* **2008**, *108*(1), 74-108.
[http://dx.doi.org/10.1021/cr068037a] [PMID: 18095715]

[31] Smolin, N.; Oleinikova, A.; Brovchenko, I.; Geiger, A.; Winter, R. Properties of spanning water networks at protein surfaces. *J. Phys. Chem. B,* **2005**, *109*(21), 10995-11005.
[http://dx.doi.org/10.1021/jp050153e] [PMID: 16852340]

[32] Ermler, U.; Fritzsch, G.; Buchanan, S.K.; Michel, H. Structure of the photosynthetic reaction centre from Rhodobacter sphaeroides at 2.65 A resolution: cofactors and protein-cofactor interactions. *Structure,* **1994**, *2*(10), 925-936.
[http://dx.doi.org/10.1016/S0969-2126(94)00094-8] [PMID: 7866744]

[33] Ohno, K.; Kamiya, N.; Asakawa, N.; Inoue, Y.; Sakurai, M. Effects of hydration on the electronic structure of an enzyme: implications for the catalytic function. *J. Am. Chem. Soc.,* **2001**, *123*(33), 8161-8162.
[http://dx.doi.org/10.1021/ja015589w] [PMID: 11506591]

[34] a) Krauss, M.; Gilson, H.S.; Gresh, N. Structure of the first shell active site in Metallolactamase: Effect of water ligands. *J. Phys. Chem. B,* **2001**, *105*, 8040-8049.
[http://dx.doi.org/10.1021/jp012099h]
b) Erhardt, S.; Jaime, E.; Weston, J. A water sluice is generated in the active site of bovine lens leucine aminopeptidase. *J. Am. Chem. Soc.,* **2005**, *127*(11), 3654-3655.
[http://dx.doi.org/10.1021/ja042797q] [PMID: 15771473]
c) Wang, L.; Yu, X.; Hu, P.; Broyde, S.; Zhang, Y. A water-mediated and substrate-assisted catalytic mechanism for Sulfolobus solfataricus DNA polymerase IV. *J. Am. Chem. Soc.,* **2007**, *129*(15), 4731-4737.
[http://dx.doi.org/10.1021/ja068821c] [PMID: 17375926]

[35] Derat, E.; Shaik, S.; Rovira, C.; Vidossich, P.; Alfonso-Prieto, M. The effect of a water molecule on the mechanism of formation of compound 0 in horseradish peroxidase. *J. Am. Chem. Soc.,* **2007**, *129*(20), 6346-6347.
[http://dx.doi.org/10.1021/ja0676861] [PMID: 17472375]

[36] a) Bizzarri, A.R.; Cannistraro, S. Molecular Dynamics of water at protein-solvent interface. *J. Phys. Chem. B,* **2002**, *106*, 6617-6633.
[http://dx.doi.org/10.1021/jp020100m]
b) Russo, D.; Murarka, R.K.; Copley, J.R.; Head-Gordon, T. Molecular view of water dynamics near model peptides. *J. Phys. Chem. B,* **2005**, *109*(26), 12966-12975.
[http://dx.doi.org/10.1021/jp051137k] [PMID: 16852609]
c) Russo, D.; Hura, G.; Head-Gordon, T. Hydration dynamics near a model protein surface. *Biophys. J.,* **2004**, *86*(3), 1852-1862.
[http://dx.doi.org/10.1016/S0006-3495(04)74252-6] [PMID: 14990511]

[37] Mandumpal, J.B. *The molecular mechanism of solvent cryoprotection: Molecular Dynamics study of cryosolvents*; LAP Lambert Academic Publishing, **2012**, p. 237.

[38] a) Autenrieth, F.; Tajkhorshid, E.; Shulten, K.; Luthey-Shulten, Z.L. Role of water in transient cytochorme c2 docking. *J. Phys. Chem. B,* **2004**, *108*, 20376-20387.
[http://dx.doi.org/10.1021/jp047994q]
b) Royer, W.E., Jr; Pardanani, A.; Gibson, Q.H.; Peterson, E.S.; Friedman, J.M. Ordered water

molecules as key allosteric mediators in a cooperative dimeric hemoglobin. *Proc. Natl. Acad. Sci. USA,* **1996**, *93*(25), 14526-14531.
[http://dx.doi.org/10.1073/pnas.93.25.14526] [PMID: 8962085]

[39]   Ray, C.; Jain, R. Drinking Water Treatment Technology - Comparative Analysis. In: *Drinking Water Treatment, Focussing on appropriate Technology and sustainability*; Chittaranjan Ray, R.J., Ed.; Springer, **2011**; pp. 9-36.
[http://dx.doi.org/10.1007/978-94-007-1104-4_2]

[40]   Hester, J.F.; Mayes, A.M. Design and performance of foul-resistant poly(vinylidene fluoride) membranes prepared in a single step by surface segregation. *J. Membr. Sci.,* **2002**, *202*, 119-135.
[http://dx.doi.org/10.1016/S0376-7388(01)00735-9]

[41]   Raman, H.; Sunilkumar, B. Multivariate modelling of water resources time series using artificial neural networks. *Hydrol. Sci. J.,* **1995**, *40*, 145-163.
[http://dx.doi.org/10.1080/02626669509491401]

[42]   Nielsen, S.O.; Lopez, C.F.; Srinivas, G.; Klein, M.L. Coarse grain models and the computer simulation of soft materials. *J. Phys. Condens. Matter,* **2004**, *16*, R481-R512.
[http://dx.doi.org/10.1088/0953-8984/16/15/R03]

[43]   Karplus, M.; Petsko, G.A. Molecular dynamics simulations in biology. *Nature,* **1990**, *347*(6294), 631-639.
[http://dx.doi.org/10.1038/347631a0] [PMID: 2215695]

[44]   Castellan, G.W. *Physical Chemistry,* 3 ed; Addison–Wesley Publishing Company, **1983**.

[45]   Shik, J.M.; Eyring, H. Liquid Theory and the structure of water. *Annu. Rev. Phys. Chem.,* **1976**, *27*, 45-57.
[http://dx.doi.org/10.1146/annurev.pc.27.100176.000401]

[46]   Leach, A.R. *Molecular Modelling Principles and Applications,* 2nd ed; , **2001**, p. 773.

[47]   Stone, A.J. *The Theory of Intermolecular Forces*; Oxford University Press, **2013**, p. 339.
[http://dx.doi.org/10.1093/acprof:oso/9780199672394.001.0001]

[48]   de With, G. *Liquid-state Physical Chemistry*; Wiley-VCH: Amsterdam, **2013**, p. 526.
[http://dx.doi.org/10.1002/9783527676750]

[49]   Barker, J.A.; Henderson, D. Theories of liquids. *Annu. Rev. Phys. Chem.,* **1972**, *23*, 439-484.
[http://dx.doi.org/10.1146/annurev.pc.23.100172.002255]

[50]   Stillinger, F.H. A topographic view of supercooled liquids and glass formation. *Science,* **1995**, *267*(5206), 1935-1939.
[http://dx.doi.org/10.1126/science.267.5206.1935] [PMID: 17770102]

[51]   Ediger, M.D.; Angell, C.A.; Nagel, S.R. Supercooled liquids and glasses. *J. Phys. Chem.,* **1996**, *100*, 13200-13212.
[http://dx.doi.org/10.1021/jp953538d]

[52]   a) Kauzmann, W. The nature of the glassy state and the behaviour of liquids at low temperatures. *Chem. Rev.,* **1948**, *1*, 219-256.
[http://dx.doi.org/10.1021/cr60135a002]
b)Stillinger, F.H.; Debenedetti, P.G. Glass transition thermodynamics and kinetics. *Annu. Rev.*

Condens. Matter Phys., **2013**, *4*, 263-285.
[http://dx.doi.org/10.1146/annurev-conmatphys-030212-184329]

[53] Ediger, M.D. Spatially heterogeneous dynamics in supercooled liquids. *Annu. Rev. Phys. Chem.,* **2000**, *51*, 99-128.
[http://dx.doi.org/10.1146/annurev.physchem.51.1.99] [PMID: 11031277]

[54] Wowk, B. Thermodynamic aspects of vitrification. *Cryobiology,* **2010**, *60*(1), 11-22.
[http://dx.doi.org/10.1016/j.cryobiol.2009.05.007] [PMID: 19538955]

[55] Kivelson, S.; Tarjus, G. *Constraints on the theory of supercooled liquids as they become glassy*; Condensed Material-Statistical Mechanics, **2008**, pp. 1-4.

[56] Angell, C.A. Liquid fragility and the glass transition in water and aqueous solutions. *Chem. Rev.,* **2002**, *102*(8), 2627-2650.
[http://dx.doi.org/10.1021/cr000689q] [PMID: 12175262]

[57] Debenedetti, P.G.; Truskett, T.M.; Lewis, C.P.; Stillinger, F.H. Theory of supercooled liquids, and glasses: Energy landscape and statistical geometry perspectives. *Adv. Chem. Eng.,* **2001**, *28*, 21-79.
[http://dx.doi.org/10.1016/S0065-2377(01)28003-X]

[58] Alba-Simionesco, C.; Fan, J.; Angell, C.A. Thermodynamic aspects of the glass transition phenomenon. II. molecular liquids with variable interactions. *J. Chem. Phys.,* **1999**, *110*, 5262-5272.
[http://dx.doi.org/10.1063/1.478800]

[59] Adam, G.; Gibbs, J.H. On the temperature dependence of cooperative relaxation properties in glass-forming liquids. *J. Chem. Phys.,* **1965**, *43*, 139-146.
[http://dx.doi.org/10.1063/1.1696442]

[60] Angell, C.A.; Sichina, W. *Thermodynamics of the glass transition: empirical aspects*; The Glass Transition and the Nature of the Glassy State, **1976**, pp. 53-67.

[61] Tarjus, G.; Kivelson, S.A.; Nussinov, Z.; Viot, P. The frustration-based approach of supercooled liquids and the glass transition: a review and critical assessment. *J. Phys. Condens. Matter,* **2005**, *17*, R1143-R1182.
[http://dx.doi.org/10.1088/0953-8984/17/50/R01]

[62] Vilgis, T.A. Strong and fragile glasses: A powerful classification and its consequences. *Phys. Rev. B Condens. Matter,* **1993**, *47*(5), 2882-2885.
[http://dx.doi.org/10.1103/PhysRevB.47.2882] [PMID: 10006352]

[63] Angell, C.A. Entropy and Fragility in supercooled liquids. *J. Res. Natl. Inst. Stand. Technol.,* **1997**, *102*(2), 171-185.
[http://dx.doi.org/10.6028/jres.102.013] [PMID: 27805135]

[64] Binder, K.; Baschnagel, J.; Paul, W. Glass transition of polymer melts: test of theoretical concepts by computer simulation. *Prog. Polym. Sci.,* **2003**, *28*, 115-172.
[http://dx.doi.org/10.1016/S0079-6700(02)00030-8]

[65] Starr, F.W.; Angell, C.A.; Nave, E.L.; Sastry, S.; Scala, A.; Sciortino, F.; Stanley, H.E. Recent results on the connection between thermodynamics and dynamics in supercooled water. *Biophys. Chem.,* **2003**, *105*(2-3), 573-583.
[http://dx.doi.org/10.1016/S0301-4622(03)00067-X] [PMID: 14499919]

[66]   Ohimine, I.; Saito, S. Water dynamics: fluctuation, relaxation, and chemical reactions in hydrogen bond network arrangement. *Acc. Chem. Res.,* **1998**, *32*, 741-749.
[http://dx.doi.org/10.1021/ar970161g]

[67]   Tanaka, H. Two order parameter model of the liquid-glass transition. I. Relation between glass transition and crystallisation. *J. Non-Cryst. Solids,* **2005**, *351*, 3371-3384.
[http://dx.doi.org/10.1016/j.jnoncrysol.2005.09.008]

[68]   Diploma Program Chemistry Guide First Assessment 2016. In: *Baccalaureate, International*, **2014**.

[69]   Speedy, R.J.; Angell, C.A. Isothermal compressibility of supercooled water and evidence for a thermodynamic singularity at -45°C. *J. Chem. Phys.,* **1976**, *65*, 851-858.
[http://dx.doi.org/10.1063/1.433153]

[70]   a) Leetmaa, M.; Wikfeldt, K.T.; Ljungberg, M.P.; Odelius, M.; Swenson, J.; Nilsson, A.; Pettersson, L.G. Diffraction and IR/Raman data do not prove tetrahedral water. *J. Chem. Phys.,* **2008**, *129*(8), 084502.
[http://dx.doi.org/10.1063/1.2968550] [PMID: 19044830]
b) Head-Gordon, T.; Hura, G. Water structure from scattering experiments and simulation. *Chem. Rev.,* **2002**, *102*(8), 2651-2670.
[http://dx.doi.org/10.1021/cr0006831] [PMID: 12175263]

[71]   Bosio, L.; Johari, G.P.; Teixeira, J. X-ray study of high-density amorphous water. *Phys. Rev. Lett.,* **1986**, *56*(5), 460-463.
[http://dx.doi.org/10.1103/PhysRevLett.56.460] [PMID: 10033198]

[72]   Hobbs, P.V. *Ice Physics*; Oxford University Press: New York, **2010**, pp. 668-670.

[73]   a) Soper, A. K. The radial distribution functions of water and ice from 220 to 673 K and at pressures up to 400 MPa. *Chemical Physics,* **2000**, *258*, 121-137.
b) Postorino, P.; Tromp, R. H.; Ricci, M. A.; Soper, A. K.; Neilson, G. W. The interatomic structure of water at supercritical temperatures. *Nature,* **1993**, *366*, 668-670.

[74]   Liu, D.; Zhang, Y.; Chen, C.; Mou, C.; Poole, P.H.; Chen, S. Observation of the density minimum in deeply supercooled confined water. *Proc. Natl. Acad. Sci. USA,* **2007**, *104*, 9570-9574.
[http://dx.doi.org/10.1073/pnas.0701352104]

[75]   Bellisent ^'Funel, M.; Teixeira, J.; Bosio, L. Structure of high-density amorphous water. II. Neutron scattering study. *J. Chem. Phys.,* **1987**, *87*, 2231-2235.
[http://dx.doi.org/10.1063/1.453150]

[76]   Ricci, M.A.; Nardone, N.; Fontana, A.; Andreani, C.; Hahn, W. Light and neutron scattering studies of the OH stretching band in liquid and supercritical water. *J. Chem. Phys.,* **1998**, *108*, 450-454.
[http://dx.doi.org/10.1063/1.475407]

[77]   Lobban, C.; Finney, J.L.; Kuhs, W.F. The structure of a new phase of ice. *Nature,* **1998**, *391*, 268-270.
[http://dx.doi.org/10.1038/34622]

[78]   Teixeira, J.; Bellissent-Funel, M.; Chen, S.H.; Dianoux, A.J. Experimental determination of the nature of diffusive motions of water molecules at low temperatures. *Phys. Rev. A Gen. Phys.,* **1985**, *31*(3), 1913-1917.
[http://dx.doi.org/10.1103/PhysRevA.31.1913] [PMID: 9895699]

[79]   Beta, I.A.; Li, J.; Bellisent Funel, M. A quasi ^'elastic neutron scattering study of the dynamics of supercritical water. *Chem. Phys.,* **2003**, *292*, 229-234.
[http://dx.doi.org/10.1016/S0301-0104(03)00228-3]

[80]   a) Mezei, F.; Russina, M. Intermediate range order dynamics near the glass transition. *J. Phys. Condens. Matter,* **1999**, *11*, A341-A354.
[http://dx.doi.org/10.1088/0953-8984/11/10A/031]
b) Andreani, C.; Colognessi, D.; Degiorgi, E.; Ricci, M.A. Proton dynamics in supercritical water. *J. Chem. Phys.,* **2001**, *115*, 11243-11248.
[http://dx.doi.org/10.1063/1.1420751]
c) Tassaing, T.; Bellisent Funel, M. The dynamics of supercritical water: A quasielastic incoherent nuetron scattering study. *J. Chem. Phys.,* **2000**, *113*, 3332-3337.
[http://dx.doi.org/10.1063/1.1286599]

[81]   Sit, P.H.; Bellin, C.; Barbiellini, B.; Testemale, D.; Hazemann, J.L.; Buslaps, T.; Marzari, N.; Shukla, A. *Hydrogen bonding and coordination in normal and supercritical water from X-ray inelastic scattering*; Condesed Matter-Softmatter, **2008**, pp. 1-14.

[82]   Tulk, C.A.; Benmore, C.J.; Urquidi, J.; Klug, D.D.; Neuefeind, J.; Tomberli, B.; Egelstaff, P.A. Structural studies of several distinct metastable forms of amorphous ice. *Science,* **2002**, *297*(5585), 1320-1323.
[http://dx.doi.org/10.1126/science.1074178] [PMID: 12193779]

[83]   Ohmine, I.; Saito, S. Water dynamics: fluctuation, relaxation, and chemical reactions in hydrogen bond network rearrangement. *Acc. Chem. Res.,* **1999**, *32*, 741-749.
[http://dx.doi.org/10.1021/ar970161g]

[84]   Sellberg, J.A.; Huang, C.; McQueen, T.A.; Loh, N.D.; Laksmono, H.; Schlesinger, D.; Sierra, R.G.; Nordlund, D.; Hampton, C.Y.; Starodub, D.; DePonte, D.P.; Beye, M.; Chen, C.; Martin, A.V.; Barty, A.; Wikfeldt, K.T.; Weiss, T.M.; Caronna, C.; Feldkamp, J.; Skinner, L.B.; Seibert, M.M.; Messerschmidt, M.; Williams, G.J.; Boutet, S.; Pettersson, L.G.; Bogan, M.J.; Nilsson, A. Ultrafast X-ray probing of water structure below the homogeneous ice nucleation temperature. *Nature,* **2014**, *510*(7505), 381-384.
[http://dx.doi.org/10.1038/nature13266] [PMID: 24943953]

[85]   Wiedersich, J.; Blochowicz, T.; Benkhof, S.; Kudlik, A.; Surovtsev, N.V.; Tschirwitz, C.; Novikov, V.N. Rössler, E. Fast and slow relaxation processes in glasses. *J. Phys. Condens. Matter,* **1999**, *11*, A147-A156.
[http://dx.doi.org/10.1088/0953-8984/11/10A/010]

[86]   a) Stewart, G.W. X-Ray diffraction in liquids. *Rev. Mod. Phys.,* **1930**, *2*, 116-122.
[http://dx.doi.org/10.1103/RevModPhys.2.116]
b) Gorbaty, Y.E.; Kalinichev, A.G. Hydrogen bonding in supercritical water. 1. experimental results. *J. Phys. Chem.,* **1995**, *99*, 5336-5340.
[http://dx.doi.org/10.1021/j100015a016]

[87]   Perrin, C.L.; Nielson, J.B. Strong hydrogen bonds in chemistry and biology. *Annu. Rev. Phys. Chem.,* **1997**, *48*, 511-544.
[http://dx.doi.org/10.1146/annurev.physchem.48.1.511] [PMID: 9348662]

[88]   Souda, R. Effects of methanol on crystallization of water in the deeply supercooled region. *Phys. Rev. B,* **2007**, *75*, 184116.
[http://dx.doi.org/10.1103/PhysRevB.75.184116]

[89]   Angell, C.A.; Rodgers, V. Near infrared spectra and the disrupted network model of normal and supercooled water. *J. Chem. Phys.,* **1984**, *80*, 6245-6252.
[http://dx.doi.org/10.1063/1.446727]

[90]   Liu, K.; Cruzan, J.D.; Saykally, R.J. Water clusters. *Science,* **1996**, *271*, 929-933.
[http://dx.doi.org/10.1126/science.271.5251.929]

[91]   Brown, M.G.; Keutsch, F.N.; Saykally, R.J. The bifurcation rearrangement in cyclic water clusters: Breaking and making hydrogen bonds. *J. Chem. Phys.,* **1998**, *109*, 9645-9647.
[http://dx.doi.org/10.1063/1.477630]

[92]   Scheiner, S. *Ab initio* studies of hydrogen bonds: the water dimer paradigm. *Annu. Rev. Phys. Chem.,* **1994**, *45*, 23-56.
[http://dx.doi.org/10.1146/annurev.pc.45.100194.000323] [PMID: 7811354]

[93]   a) Nakahara, M. Structure, dynamics, and reactions of supercritical water studied by NMR and computer simulation. In: *14th international conference on the properties of water and steam, Kyoto* **2004**.
b) Matubayashi, N.; Nakao, N.; Nakahara, M. Structural study of supercritical water. III Rotational dynamics. *J. Chem. Phys.,* **2001**, *114*, 4107-4115.
[http://dx.doi.org/10.1063/1.1336571]

[94]   Matubayashi, N.; Wakai, C.; Nakahara, M. Structural study of supercritical water. 1. Nuclear magnetic resonance spectroscopy. *J. Chem. Phys.,* **1997**, *107*, 9133-9140.
[http://dx.doi.org/10.1063/1.475205]

[95]   Jonas, J.; DeFries, T.; Wilbur, D.J. Molecular motions in compressed liquid water. *J. Chem. Phys.,* **1976**, *65*, 582-588.
[http://dx.doi.org/10.1063/1.433113]

[96]   Price, W.S.; Ide, H.; Arata, Y. Self-Diffusion of Supercooled Water to 238 K Using PGSE NMR Diffusion Measurements. *J. Phys. Chem. A,* **1999**, *103*, 448-450.
[http://dx.doi.org/10.1021/jp9839044]

[97]   a) Angell, C.A. Supercooled water. *Annu. Rev. Phys. Chem.,* **1983**, *34*, 593-630.
[http://dx.doi.org/10.1146/annurev.pc.34.100183.003113]
b) Sun, Q.; Wang, Q.; Ding, D. Hydrogen bonded networks in supercritical water. *J. Phys. Chem. B,* **2014**, *118*(38), 11253-11258.
[http://dx.doi.org/10.1021/jp503474s] [PMID: 25187291]
c) Bernal, J.D.; Fowler, R.H. A theory of water and ionic solution, with particular referece to hydrogn and hydroxyl ions. *J. Chem. Phys.,* **1933**, *1*, 515-548.
[http://dx.doi.org/10.1063/1.1749327]

[98]   a) Behrends, R.; Fuchs, K.; Kaatze, U.; Hayashi, Y.; Feldman, Y. Dielectric properties of glycerol/water mixtures at temperatures between 10 and 50 A C. *J. Chem. Phys.,* **2006**, *124*(14), 144512.
[http://dx.doi.org/10.1063/1.2188391] [PMID: 16626219]

b) Sudo, S.; Yagihara, S. Universality of separation behavior of relaxation processes in supercooled aqueous solutions as revealed by broadband dielectric measurements. *J. Phys. Chem. B,* **2009**, *113*(33), 11448-11452.
[http://dx.doi.org/10.1021/jp901765a] [PMID: 19637896]

[99]   a) Cerveny, S.; Schwartz, G.A.; Bergman, R.; Swenson, J. Glass transition and relaxation processes in supercooled water. *Phys. Rev. Lett.,* **2004**, *93*(24), 245702.
[http://dx.doi.org/10.1103/PhysRevLett.93.245702] [PMID: 15697826]
b) Bergman, R.; Swenson, J. Dynamics of supercooled water in confined geometry. *Nature,* **2000**, *403*(6767), 283-286.
[http://dx.doi.org/10.1038/35002027] [PMID: 10659841]

[100]   Qvist, J.; Schober, H.; Halle, B. Structural dynamics of supercooled water from quasielastic neutron scattering and molecular simulations. *J. Chem. Phys.,* **2011**, *134*(14), 144508.
[http://dx.doi.org/10.1063/1.3578472] [PMID: 21495765]

[101]   Taschin, A.; Bartolini, P.; Eramo, R.; Righini, R.; Torre, R. Optical Kerr effect of liquid and supercooled water: the experimental and data analysis perspective. *J. Chem. Phys.,* **2014**, *141*(8), 084507.
[http://dx.doi.org/10.1063/1.4893557] [PMID: 25173021]

[102]   Torre, R.; Bartolini, P.; Righini, R. Structural relaxation in supercooled water by time-resolved spectroscopy. *Nature,* **2004**, *428*(6980), 296-299.
[http://dx.doi.org/10.1038/nature02409] [PMID: 15029190]

[103]   Kuhs, W.F.; Sippel, C.; Falenty, A.; Hansen, T.C. Extent and relevance of stacking disorder in ice I(c). *Proc. Natl. Acad. Sci. USA,* **2012**, *109*(52), 21259-21264.
[http://dx.doi.org/10.1073/pnas.1210331110] [PMID: 23236184]

[104]   Algara-Siller, G.; Lehtinen, O.; Wang, F.C.; Nair, R.R.; Kaiser, U.; Wu, H.A.; Geim, A.K.; Grigorieva, I.V. Square ice in graphene nanocapillaries. *Nature,* **2015**, *519*(7544), 443-445.
[http://dx.doi.org/10.1038/nature14295] [PMID: 25810206]

[105]   Simperler, A.; Kornherr, A.; Chopra, R.; Jones, W.; Motherwell, W.D.; Zifferer, G. The glass transition temperatures of amorphous trehalose-water mixtures and the mobility of water: an experimental and *in silico* study. *Carbohydr. Res.,* **2007**, *342*(11), 1470-1479.
[http://dx.doi.org/10.1016/j.carres.2007.04.011] [PMID: 17511976]

[106]   Hallbrucker, A.; Mayer, E.; Johari, G.P. Glass transition in Pressure-amorphized Hexagonal Ice. A comparison with amorphous forms made from the vapor and liquid. *J. Phys. Chem.,* **1989**, *93*, 7751-7752.
[http://dx.doi.org/10.1021/j100360a003]

[107]   Rasmussen, D.H.; MacKenzie, A.P. The glass transition in amorphous water. Application of the measurements to problems arising in cryobiology. *J. Phys. Chem.,* **1971**, *75*(7), 967-973.
[http://dx.doi.org/10.1021/j100677a022] [PMID: 5135344]

[108]   a) Speedy, R.J. Thermodynamic properties of supercooled water at 1 atm. *J. Phys. Chem.,* **1987**, *91*, 3354-3358.
[http://dx.doi.org/10.1021/j100296a049]
b) Hallett, J. The temperature dependence of the viscosity of supercooled water. *Proc. Phys. Soc.,*

1963, *82*, 1046-1050.
[http://dx.doi.org/10.1088/0370-1328/82/6/326]

[109] Geiger, P.; Dellago, C.; Macher, M.; Franchini, C.; Kresse, G.; Bernard, J.; Stern, J.N.; Loerting, T. Proton ordering of cubic ice Ic: Spectroscopy and computer simulations. *J Phys Chem C Nanomater Interfaces,* **2014**, *118*(20), 10989-10997.
[http://dx.doi.org/10.1021/jp500324x] [PMID: 24883169]

[110] Angell, C.A. Insights into phases of liquid water from study of its unusual glass-forming properties. *Science,* **2008**, *319*(5863), 582-587.
[http://dx.doi.org/10.1126/science.1131939] [PMID: 18239117]

[111] Allen, M.P.; Tildesley, D.J. *Computer Simulation of Liquids*; Clarendon Press Oxford, **1991**.

[112] Jensen, F. *Introduction to Computational Chemistry,* 2nd ed; John Wiley & sons, **2007**.

[113] Haile, J.M. *Molecular Dynamics Simulation Elementary Methods*; John Wiley & sons, **1992**.

[114] Chen, B.; Siepmann, J.I. A novel Monte Carlo Algorithm for simulating strongly associating fluids: applications to water, hydrogen fluoride, and acetic acid. *J. Phys. Chem. B,* **2000**, *104*, 8725-8734.
[http://dx.doi.org/10.1021/jp001952u]

[115] Guillot, B. A reappraisal of what we have learnt during three decades of computer simulations on water. *J. Mol. Liq.,* **2002**, *101*, 219-260.
[http://dx.doi.org/10.1016/S0167-7322(02)00094-6]

[116] MacBride, C.; Vega, C.; Noya, E.G.; Ramirez, R.; Sese, L.M. *Quantum contributions in the ice phases: the path to a new empirical model for water - TIP4PQ/2005*; Condensed Mater - Statistical Mechanics, **2009**, pp. 1-41.

[117] Starr, F.W.; Bellissent-Funel, M.C.; Stanley, H.E. Structure of supercooled and glassy water under pressure. *Phys. Rev. E Stat. Phys. Plasmas Fluids Relat. Interdiscip. Topics,* **1999**, *60*(1), 1084-1087.
[http://dx.doi.org/10.1103/PhysRevE.60.1084] [PMID: 11969860]

[118] Lewars, E.G. *Computational Chemistry. Introduction to the Theory and Applications of Molecular and Quantum Mechanics,* 2nd ed; Springer, **2011**, p. 681.

[119] Young, D.C. *Computational Chemistry A practical guide for Applying Techniques to Real-World problems*; John Wiley & sons, **2001**.

[120] Manby, F.R., Ed. *Accurate Condensed-Phase Quantum Chemistry*; Taylor and Francis Group, **2011**, p. 202.

[121] Finney, J.L. Water? Whats so special about it? *Philos. Trans. R. Soc. Lond. B Biol. Sci.,* **2004**, *359*(1448), 1145-1163.
[http://dx.doi.org/10.1098/rstb.2004.1495] [PMID: 15306373]

[122] Davis, C.M., Jr; Litovitz, T.A. Two-State theory of the structure of water. *J. Chem. Phys.,* **1965**, *42*, 2563-2576.
[http://dx.doi.org/10.1063/1.1696333]

[123] Umeyama, H.; Morkuma, K. The origin of Hydrogen Bonding. An Energy decomposition study. *J. Am. Chem. Soc.,* **1977**, *99*, 1316-1332.
[http://dx.doi.org/10.1021/ja00447a007]

[124] Ignatov, I.; Mosin, O. Methods for measurements of water spectrum. Differential Non-equilibrium Energy Spectrum Method (DNES). *J. Health Med. Nurs.,* **2014**, *6*, 50-72.

[125] Alkorta, I.; Rozas, I.; Elguero, J. Non-conventional Hydrogen Bonds. *Chem. Soc. Rev.,* **1998**, *27*, 163-170.
[http://dx.doi.org/10.1039/a827163z]

[126] Nibbering, E.T.; Elsaesser, T. Ultrafast vibrational dynamics of hydrogen bonds in the condensed phase. *Chem. Rev.,* **2004**, *104*(4), 1887-1914.
[http://dx.doi.org/10.1021/cr020694p] [PMID: 15080715]

[127] Chaplin, M.F. A proposal for the structuring of water. *Biophys. Chem.,* **2000**, *83*(3), 211-221.
[http://dx.doi.org/10.1016/S0301-4622(99)00142-8] [PMID: 10647851]

[128] Falk, M.; Ford, T.A. Infrared spectrum and structure of liquid water. *Can. J. Chem.,* **1966**, *44*, 1699-1707.
[http://dx.doi.org/10.1139/v66-255]

[129] Benson, S.W.; Siebert, E.D. A simple Two-structure model for liquid water. *J. Am. Chem. Soc.,* **1992**, *114*, 4269-4276.
[http://dx.doi.org/10.1021/ja00037a034]

[130] Head-Gordon, T.; Stillinger, F.H. An orientational peturbation theory for pure liquid water. *J. Chem. Phys.,* **1992**, *98*, 3313-3327.
[http://dx.doi.org/10.1063/1.464103]

[131] Kozack, R.E.; Jordan, P.C. Polarizability effects in water. *J. Chem. Phys.,* **1992**, *96*, 3120-3130.
[http://dx.doi.org/10.1063/1.461956]

[132] Rahman, A.; Stillinger, F.H.; Lemberg, H.L. Study of a central force model for liquid water by molecular dynamics. *J. Chem. Phys.,* **1975**, *63*, 5223-5230.
[http://dx.doi.org/10.1063/1.431307]

[133] Duh, D.; Perera, D.N.; Haymet, A.D. Structure and properties of CF1 central force model of water: Integral equation theory. *J. Chem. Phys.,* **1995**, *102*, 3736-3746.
[http://dx.doi.org/10.1063/1.468556]

[134] Svishchev, I.M.; Kusalik, P.G. Structure in liquid water: A study of spatial distribution functions. *J. Chem. Phys.,* **1993**, *99*, 3049-3058.
[http://dx.doi.org/10.1063/1.465158]

[135] Narten, A.H.; Levy, H.A. Observed diffraction pattern and proposed models of liquid water. *Science,* **1969**, *165*(3892), 447-454.
[http://dx.doi.org/10.1126/science.165.3892.447] [PMID: 17831028]

[136] Jedlovszky, P.; Brodholt, J.P.; Bruni, F.; Ricci, M.A.; Soper, A.K.; Vallauri, R. Analysis of the hydrogen-bonded structure of water from ambient to supercritical conditions. *J. Chem. Phys.,* **1998**, *108*, 8528-8540.
[http://dx.doi.org/10.1063/1.476282]

[137] Guo, J.H.; Luo, Y.; Augustsson, A.; Rubensson, J.E.; SA the, C.; Agren, H.; Siegbahn, H.; Nordgren, J. X-ray emission spectroscopy of hydrogen bonding and electronic structure of liquid water. *Phys. Rev. Lett.,* **2002**, *89*(13), 137402.

[http://dx.doi.org/10.1103/PhysRevLett.89.137402] [PMID: 12225062]

[138] a) Smith, J.D.; Cappa, C.D.; Wilson, K.R.; Messer, B.M.; Cohen, R.C.; Saykally, R.J. Energetics of hydrogen bond network rearrangements in liquid water. *Science,* **2004**, *306*(5697), 851-853.
[http://dx.doi.org/10.1126/science.1102560] [PMID: 15514152]
b) Stillinger, F.H. Water revisited. *Science,* **1980**, *209*(4455), 451-457.
[http://dx.doi.org/10.1126/science.209.4455.451] [PMID: 17831355]
c) Chandler, D. Hydrophobicity: two faces of water. *Nature,* **2002**, *417*(6888), 491.
[http://dx.doi.org/10.1038/417491a] [PMID: 12037545]
d) Stanley, H.E.; Teixeira, J. Interpretation of the unusual behaviour of H_2O and D_2O at low temperatures: Tests of a percolation model. *J. Chem. Phys.,* **1980**, *73*, 3404-3422.
[http://dx.doi.org/10.1063/1.440538]
e) Finney, J.L. Bernal and structure of water. *J. Phys. Conf. Ser.,* **2007**, *57*, 40-52.
[http://dx.doi.org/10.1088/1742-6596/57/1/004]

[139] Raiteri, P.; Laio, A.; Parrinello, M. Correlations among hydrogen bonds in liquid water. *Phys. Rev. Lett.,* **2004**, *93*(8), 087801.
[http://dx.doi.org/10.1103/PhysRevLett.93.087801] [PMID: 15447226]

[140] Geiger, A.; Stillinger, F.H.; Rahman, A. Aspects of the percolation process for Hydrogen-Bond networks in water. *J. Chem. Phys.,* **1979**, *70*, 4185-4193.
[http://dx.doi.org/10.1063/1.438042]

[141] Walrafen, G.E. Raman spectral studies of the effects of temperature on water structure. *J. Chem. Phys.,* **1967**, *47*, 114-126.
[http://dx.doi.org/10.1063/1.1711834]

[142] Sciortino, F.; Geiger, A.; Stanley, H.E. Effect of defects on molecular mobility in liquid water. *Nature,* **1991**, *354*, 218-221.
[http://dx.doi.org/10.1038/354218a0]

[143] Geiger, A.; Stanley, H.; Low-density, E. "Patches" in the Hydrogen-bonded network of liquid water: Evidence from Molecular Dynamics computer simulations. *Phys. Rev. Lett.,* **1982**, *49*, 1749-1752.
[http://dx.doi.org/10.1103/PhysRevLett.49.1749]

[144] Blumberg, R.L.; Stanley, H.E.; Geiger, A.; Mausbach, P. Connectivity of hydrogen bonds in liquid water. *J. Chem. Phys.,* **1984**, *80*, 5230-5241.
[http://dx.doi.org/10.1063/1.446593]

[145] a) Wernet, P.; Nordlund, D.; Bergmann, U.; Cavalleri, M.; Odelius, M.; Ogasawara, H.; Naslund, L.A.; Hirsch, T.K.; Ojamae, L.; Glatzel, P.; Pettersson, L.G.; Nilsson, A. The structure of the first coordination shell in liquid water. *Science,* **2004**, *304*(5673), 995-999.
[http://dx.doi.org/10.1126/science.1096205] [PMID: 15060287]
b) Myneni, S.; Luo, Y.; Naslund, L.A.; Cavalleri, M.; Ojamae, L.; Ogasawara, H.; Pelmenschikov, A.; Wernet, P.; Vaterlein, P.; Heske, C.; Hussain, Z.; Pettersson, L.G.; Nilsson, A. Spectroscopic probing of local Hydrogen bonding structures in liquid water. *J. Phys. Condens. Matter,* **2002**, *14*, L213-L219.
[http://dx.doi.org/10.1088/0953-8984/14/8/106]

[146] Huang, C.; Wikfeldt, K.T.; Tokushima, T.; Nordlund, D.; Harada, Y.; Bergmann, U.; Niebuhr, M.; Weiss, T.M.; Horikawa, Y.; Leetmaa, M.; Ljungberg, M.P.; Takahashi, O.; Lenz, A.; Ojamae, L.; Lyubartsev, A.P.; Shin, S.; Pettersson, L.G.; Nilsson, A. The inhomogeneous structure of water at

ambient conditions. *Proc. Natl. Acad. Sci. USA,* **2009**, *106*(36), 15214-15218.
[http://dx.doi.org/10.1073/pnas.0904743106] [PMID: 19706484]

[147] Ball, P. Water: wateran enduring mystery. *Nature,* **2008**, *452*(7185), 291-292.
[http://dx.doi.org/10.1038/452291a] [PMID: 18354466]

[148] Ludwig, R. Water: From Clusters to the bulk. *Angew. Chem. Int. Ed. Engl.,* **2001**, *40*(10), 1808-1827.
[http://dx.doi.org/10.1002/1521-3773(20010518)40:10<1808::AID-ANIE1808>3.0.CO;2-1] [PMID: 11385651]

[149] Petkov, V.; Ren, Y.; Suchomel, M. Molecular arrangement in water: random but not quite. *J. Phys. Condens. Matter,* **2012**, *24*(15), 155102.
[http://dx.doi.org/10.1088/0953-8984/24/15/155102] [PMID: 22418283]

[150] Steinel, T.; Asbury, J.B.; Corcelli, S.A.; Lawrence, C.P.; Skinner, J.L.; Fayer, M.D. Water Dynamics: dependence on local structure probed with vibrational echo spectroscopy. *Chem. Phys. Lett.,* **2004**, *386*, 295-300.
[http://dx.doi.org/10.1016/j.cplett.2004.01.042]

[151] Laage, D.; Hynes, J.T. A molecular jump mechanism of water reorientation. *Science,* **2006**, *311*(5762), 832-835.
[http://dx.doi.org/10.1126/science.1122154] [PMID: 16439623]

[152] Canpolat, M.; Starr, F.W.; Scala, A.; Sadr-Lahijany, M.R.; Mishima, O.; Halvin, S.; Stanley, H.E. Local structure heterogeneities in liquid water under pressure. *Chem. Phys. Lett.,* **1998**, *294*, 9-12.
[http://dx.doi.org/10.1016/S0009-2614(98)00828-8]

[153] Paolantoni, M.; Lago, N.F.; AlbertA-, M.; LaganA , A. Tetrahedral ordering in water: Raman profiles and their temperature dependence. *J. Phys. Chem. A,* **2009**, *113*(52), 15100-15105.
[http://dx.doi.org/10.1021/jp9052083] [PMID: 19894708]

[154] Hutchings, R. *Effects of supercooled water ingestion on engine performance*; University of Tennessee: Knoxville, **2011**.

[155] Franzese, G.; Stanley, H.E. The Widom line of supercooled water. *J. Phys. Condens. Matter,* **2007**, *19*, 1-16.
[http://dx.doi.org/10.1088/0953-8984/19/20/205126]

[156] Starr, F.W.; Sastry, S.; La Nave, E.; Scala, A.; Stanley, H.E.; Sciortino, F. Thermodynamic and structural aspects of the potential energy surface of simulated water. *Phys. Rev. E Stat. Nonlin. Soft Matter Phys.,* **2001**, *63*, 041201.
[http://dx.doi.org/10.1103/PhysRevE.63.041201] [PMID: 11308829]

[157] Sciortino, F.; Poole, P.H.; Stanley, H.E.; Havlin, S. Lifetime of the bond network and gel-like anomalies in supercooled water. *Phys. Rev. Lett.,* **1990**, *64*(14), 1686-1689.
[http://dx.doi.org/10.1103/PhysRevLett.64.1686] [PMID: 10041461]

[158] Giovambattista, N.; Buldyrev, S.V.; Stanley, H.E.; Starr, F.W. Clusters of mobile molecules in supercooled water. *Phys. Rev. E Stat. Nonlin. Soft Matter Phys.,* **2005**, *72*, 011202.
[http://dx.doi.org/10.1103/PhysRevE.72.011202] [PMID: 16089946]

[159] Zasetsky, A.Y.; Remorov, R.; Svishchev, I.M. Evidence of enhanced local order and clustering in supercooled water near liquid-vapor interface: Molecular Dynamics simulations. *Chem. Phys. Lett.,*

2007, *435*, 50-53.
[http://dx.doi.org/10.1016/j.cplett.2006.12.043]

[160]   Rasmussen, D.H.; MacKenzie, A.P. Clustering in supercooled water. *J. Chem. Phys.,* **1973**, *59*, 5003-5013.
[http://dx.doi.org/10.1063/1.1680718]

[161]   a) Paschek, D.; Ludwig, R. Advancing into waters no mans land: two liquid states? *Angew. Chem. Int. Ed. Engl.,* **2014**, *53*(44), 11699-11701.
[http://dx.doi.org/10.1002/anie.201408057] [PMID: 25252122]
b) Paschek, D. *How the liquid-liquid transition affects hydrophobic hydration in deeply supercooled water*; Condensed Mater-Statistical Mechanics, **2005**, pp. 1-4.

[162]   Mishima, O.; Stanley, H.E. Decompression-induced melting of ice IV and the liquid-liquid transition in water. *Nature,* **1998**, *392*, 164-168.
[http://dx.doi.org/10.1038/32386]

[163]   Harrington, S.; Zhang, R.; Poole, P.H.; Sciortino, F.; Stanley, H.E. Liquid ^'Liquid phase transition: evidence from simulations. *Phys. Rev. Lett.,* **1997**, *78*, 2409-2412.
[http://dx.doi.org/10.1103/PhysRevLett.78.2409]

[164]   Stanley, H.E.; Buldyrev, S.V.; Canpolat, M.; Mishima, O.; Sadr ^'Lahijany, M.R.; Scala, A.; Starr, F.W. The puzzling behaviour of water at low temperature. *Phys. Chem. Chem. Phys.,* **2000**, *2*, 1551-1558.
[http://dx.doi.org/10.1039/b000058m]

[165]   Paschek, D.; Geiger, A. *Characterising the stepwise transformation from a low density to a very high density form of supercooled liquid water*; Condensed Mater-Statistical Mechanics, **2005**, pp. 1-4.

[166]   Stanley, H.E.; Kumar, P.; Xu, L.; Yan, Z.; Mazza, M.G.; Buldyrev, S.V.; Chen, S.H.; Mallamace, F. The puzzling untold mystries of liquid water: Some recent progress. *Physica A,* **2007**, *386*, 729-743.
[http://dx.doi.org/10.1016/j.physa.2007.07.044]

[167]   Scala, A.; Starr, F.W.; Sciortino, F.; Stanley, H.E.; Stanley, H.E. Instantaneous normal mode analysis of supercooled water. *Phys. Rev. Lett.,* **2000**, *84*(20), 4605-4608.
[http://dx.doi.org/10.1103/PhysRevLett.84.4605] [PMID: 10990751]

[168]   Scala, A.; Starr, F.W.; Sciortino, F.; Stanley, H.E.; Stanley, H.E. Configurational entropy and diffusivity of supercooled water. *Nature,* **2000**, *406*(6792), 166-169.
[http://dx.doi.org/10.1038/35018034] [PMID: 10910351]

[169]   Giovambattista, N.; Mazza, M.G.; Buldyrev, S.V.; Starr, F.W.; Stanley, H.E. Dynamic hetergeneities in supercooled water. *J. Phys. Chem. B,* **2004**, *108*, 6655-6662.
[http://dx.doi.org/10.1021/jp037925w]

[170]   Starr, F. W.; Sastry, S.; Sciortino, F.; Stanley, H. E. Supercooled water: dynamics, structure, and thermodynamics. *Condens Mater Statistical Mech,* **2000**, 1-4.

[171]   Hawkes, L. Supercooled water. *Nature,* **1929**, *124*, 225-226.
[http://dx.doi.org/10.1038/124225b0]

[172]   Angell, C.A. Formation of glasses from liquids and biopolymers. *Science,* **1995**, *267*(5206), 1924-1935.

[http://dx.doi.org/10.1126/science.267.5206.1924] [PMID: 17770101]

[173] Debenedetti, P.G. Supercooled and glassy water. *J. Phys. Condens. Matter,* **2003**, *15*, R1669-R1726.
[http://dx.doi.org/10.1088/0953-8984/15/45/R01]

[174] Frank, F.C. Molecular strucutre of deeply supercooled water. *Nature,* **1946**, 267.
[http://dx.doi.org/10.1038/157267a0]

[175] Angell, A. Thermodynamics: highs and lows in the density of water. *Nat. Nanotechnol.,* **2007**, *2*(7), 396-398.
[http://dx.doi.org/10.1038/nnano.2007.201] [PMID: 18654321]

[176] Brovchenko, I.; Oleinikova, A. Four phases of amorphous water: Simulations *versus* experiment. *J. Chem. Phys.,* **2006**, *124*(16), 164505.
[http://dx.doi.org/10.1063/1.2194906] [PMID: 16674144]

[177] Mishima, O.; Stanley, H.E. The relationship between liquid, supercooled and glassy water. *Nature,* **1998**, *396*, 329-335.
[http://dx.doi.org/10.1038/24540]

[178] Stanley, H.E.; Kumar, P.; Franzese, G.; Xu, L.; Yan, Z.; Mazza, M.G.; Buldyrev, S.V.; Chen, S.H.; Mallamace, F. Liquid Polymorphism: Possible relation to the anomalous behaviour of water. *Eur. Phys. J. Spec. Top.,* **2008**, *161*, 1-17.
[http://dx.doi.org/10.1140/epjst/e2008-00746-3]

[179] Poole, P.H.; Essmann, U.; Sciortino, F.; Stanley, H.E. Phase diagram for amorphous solid water. *Phys. Rev. E Stat. Phys. Plasmas Fluids Relat. Interdiscip. Topics,* **1993**, *48*(6), 4605-4610.
[http://dx.doi.org/10.1103/PhysRevE.48.4605] [PMID: 9961142]

[180] Sellberg, J.A. *X-ray scattering and spectroscopy of supercooled water and ice*; Stockholm University, **2014**.

[181] Mishima, O.; Suzuki, Y. Propagation of the polyamorphic transition of ice and the liquid-liquid critical point. *Nature,* **2002**, *419*(6907), 599-603.
[http://dx.doi.org/10.1038/nature01106] [PMID: 12374974]

[182] Stanley, H.E.; Buldyrev, S.V.; Franzese, G.; Giovambattista, N.; Starr, F.W. Static and dynamic heterogeneities in water. *Philos Trans A Math Phys Eng Sci,* **2005**, *363*(1827), 509-523.
[http://dx.doi.org/10.1098/rsta.2004.1505] [PMID: 15664896]

[183] Hallbrucker, A.; Mayer, E.; Johari, G.P. Glass ^'liquid transition and the enthalpy of devitrification of annealed vapor ^'deposited amorphous solid water. A comparison with hyperquenched glassy water. *J. Phys. Chem.,* **1989**, *93*, 4986-4990.
[http://dx.doi.org/10.1021/j100349a061]

[184] Brovchenko, I.; Geiger, A.; Oleinikova, A. Liquid-liquid phase transitions in supercooled water studied by computer simulations of various water models. *J. Chem. Phys.,* **2005**, *123*(4), 044515.
[http://dx.doi.org/10.1063/1.1992481] [PMID: 16095377]

[185] Giovambattista, N.; Angell, C.A.; Sciortino, F.; Stanley, H.E. Glass-transition temperature of water: a simulation study. *Phys. Rev. Lett.,* **2004**, *93*(4), 047801.
[http://dx.doi.org/10.1103/PhysRevLett.93.047801] [PMID: 15323794]

[186] Giovambattista, N.; Loerting, T.; Lukanov, B. R.; Starr, F. W. Interplay of the Glass transition and the liquid-liquid phase transition in water. *Scienitific reports,* **2012**, *390*(2), 1-8.

[187] Bohmer, R.; Ngai, K.L.; Angell, C.A.; Plazek, D.J. Nonexponential relaxations in strong and fragile glass formers. *J. Chem. Phys.,* **1993**, *99*, 4201-4209.
[http://dx.doi.org/10.1063/1.466117]

[188] Angell, C.A. Glass formation and glass transition in supercooled liquids, with insights from study of related phenomena in crystals. *J. Non-Cryst. Solids,* **2008**, *354*, 4703-4712.
[http://dx.doi.org/10.1016/j.jnoncrysol.2008.05.054]

[189] a) Murphy, D.M.; Koop, T. Review of the vapour pressures of ice and supercooled water for atmospheric applications. *Quat. J. Royal Meteorolog. Soci.,* **2005**, *131*, 1539-1565.
[http://dx.doi.org/10.1256/qj.04.94]
b) Durham, W.B.; Stern, L.A. Rheological properties of water ice-applications to satellites of the outer planets. *Annu. Rev. Earth Planet. Sci.,* **2001**, *29*, 295-330.
[http://dx.doi.org/10.1146/annurev.earth.29.1.295]

[190] Bartels-Rausch, T. Chemistry: Ten things we need to know about ice and snow. *Nature,* **2013**, *494*(7435), 27-29.
[http://dx.doi.org/10.1038/494027a] [PMID: 23389527]

[191] Egolf, P.W.; Kauffeld, M. From physical properties of ice slurries to industrial ice slurry applications. *Int. J. Refrig.,* **2005**, *28*, 4-12.
[http://dx.doi.org/10.1016/j.ijrefrig.2004.07.014]

[192] Petzold, G.; Aguilera, J.M. Ice morphology: Fundamentals and technological applications in foods. *Food Biophys.,* **2009**, *4*, 378-396.
[http://dx.doi.org/10.1007/s11483-009-9136-5]

[193] Vali, G. Supercooling of water and Nucleation of ice. *Am. J. Phys.,* **1971**, *39*, 1125-1128.

[194] Sastry, S. Water: ins and outs of ice nucleation. *Nature,* **2005**, *438*(7069), 746-747.
[http://dx.doi.org/10.1038/438746a] [PMID: 16340997]

[195] Zachariassen, K.E.; Kristiansen, E. Ice nucleation and antinucleation in nature. *Cryobiology,* **2000**, *41*(4), 257-279.
[http://dx.doi.org/10.1006/cryo.2000.2289] [PMID: 11222024]

[196] Sanz, E.; Vega, C.; Espinosa, J.R.; Caballero-Bernal, R.; Abascal, J.L.; Valeriani, C. Homogenous ice nucleation at moderate supercooling from molecular simulation. *Condensed Material. Soft Matter,* **2013**, 1-13.

[197] Noya, E.G.; Menduina, C.; Aragones, J.L.; Vega, C. Equation of state, Thermal expansion coefficient, and Isothermal compressibility for ices Ih, II, III, V and VI, as obtained from computer simulation. *J. Phys. Chem. C,* **2007**, *111*, 15877-15888.
[http://dx.doi.org/10.1021/jp0743121]

[198] Fuentes-Landete, V.; Mitterdorfer, C.; Handle, P.H.; Rutz, G.N.; Bernard, J.; Bogdan, A.; Seidl, M.; Amman-Winkel, A.; Stern, J.; Fuhrmann, S.; Loerting, T. Crystalline and amorphous ices. *Proceedings of the international school of physics "Enrico Fermi",* **2015**.

[199] Buch, V.; Sandler, P.; Sadlej, J. Simulations of H₂O solid, liquid, and clusters with an emphasis on ferroelectric ordering transition in hexagonel ice. *J. Phys. Chem. B,* **1998**, *102*, 8641-8653.
[http://dx.doi.org/10.1021/jp980866f]

[200] Chiu, J.; Starr, F.W.; Giovambattista, N. Pressure-induced transformations in computer simulations of glassy water. *J. Chem. Phys.,* **2013**, *139*(18), 184504.
[http://dx.doi.org/10.1063/1.4829276] [PMID: 24320281]

[201] Kryachko, E.S. On the red shift of OH stretching region vibrations in ice and water. *Int. J. Quantum Chem.,* **1986**, *30*, 495-508.
[http://dx.doi.org/10.1002/qua.560300405]

[202] Kuo, J.L.; Kuhs, W.F. A first principles study on the structure of ice-VI: static distortion, molecular geometry, and proton ordering. *J. Phys. Chem. B,* **2006**, *110*(8), 3697-3703.
[http://dx.doi.org/10.1021/jp055260n] [PMID: 16494426]

[203] Podeszwa, R.; Buch, V. Structure and dynamcis of orientational defects in ice. *Phys. Rev. Lett.,* **1999**, *83*, 4570-4573.
[http://dx.doi.org/10.1103/PhysRevLett.83.4570]

[204] Lindberg, G.E.; Wang, F. Efficient sampling of ice structures by electrostatic switching. *J. Phys. Chem. B,* **2008**, *112*(20), 6436-6441.
[http://dx.doi.org/10.1021/jp800736t] [PMID: 18438999]

[205] a) Sciortino, F.; Corongiu, G. Structure and dynamics in hexagonal ice: A molecular dynamics simulation with an *ab initio* polarizable and flexible potential. *J. Chem. Phys.,* **1992**, *98*, 5694-5700.
[http://dx.doi.org/10.1063/1.464884]
b) Herrero, C.P.; Ramirez, R. Path-integral simulation of ice Ih: the effect of pressure. *Condensed Matter-Mater. Sci.,* **2012**, 1-12.

[206] Ikeda-Fukazawa, T.; Kawamura, K. Molecular-dynamics studies of surface of ice Ih. *J. Chem. Phys.,* **2004**, *120*(3), 1395-1401.
[http://dx.doi.org/10.1063/1.1634250] [PMID: 15268265]

[207] Herrero, C.P.; Ramirez, R. Isotope effects in ice Ih: A path-integral simulation. *Condensed Mater-Mater. Sci.,* **2011**.

[208] Rick, S.W. Simulations of proton order and disorder in ice Ih. *J. Chem. Phys.,* **2005**, *122*(9), 094504.
[http://dx.doi.org/10.1063/1.1853351] [PMID: 15836147]

[209] Tse, J.S.; Klein, M.L.; McDonald, I.R. Molecular Dynamics studies of Ice Ic and the structure I Clathrate Hydrate of methane. *J. Chem. Phys.,* **1983**, *87*, 4198-4203.
[http://dx.doi.org/10.1021/j100244a044]

[210] Borzsak, I.; Cummings, P.T. Molecular dynamics simulation of Ice XII. *Chem. Phys. Lett.,* **1999**, *300*, 359-363.
[http://dx.doi.org/10.1016/S0009-2614(98)01387-6]

[211] Martin-Conde, M.; MacDowell, L.G.; Vega, C. Computer simulation of two new solid phases of water: Ice XIII and ice XIV. *J. Chem. Phys.,* **2006**, *125*(11), 116101.
[http://dx.doi.org/10.1063/1.2354150] [PMID: 16999507]

[212] Ramirez, R.; Neuerburg, N.; Herrero, C.P. *The phase diagram of ice Ih, II, and III: a quasi-harmonic study*; PhysicsChemistry, **2012**, pp. 1-14.

[213] Donadio, D.; Raiteri, P.; Parinello, M. Topological defects and bulk melting of hexagonal ice. *Cond. Matt. Stat. Mech.,* **2005**, 1-5.

[214] Lee, C.; Vanderbilt, D.; Laasonen, K.; Car, R.; Parrinello, M. *Ab initio* studies on the structural and dynamical properties of ice. *Phys. Rev. B Condens. Matter,* **1993**, *47*(9), 4863-4872. [http://dx.doi.org/10.1103/PhysRevB.47.4863] [PMID: 10006644]

[215] Itoh, H.; Kawamura, K.; Hondoh, T.; Mae, S. Polarized librational spectra of proton-ordered ice XI by molecular dynamics simulations. *J. Chem. Phys.,* **1998**, *109*, 4894-4899. [http://dx.doi.org/10.1063/1.477100]

[216] Abascal, J.L.; Sanz, E.; GarcA-a FernA ndez, R.; Vega, C. A potential model for the study of ices and amorphous water: TIP4P/Ice. *J. Chem. Phys.,* **2005**, *122*(23), 234511. [http://dx.doi.org/10.1063/1.1931662] [PMID: 16008466]

[217] Vega, C.; Sanz, E.; Abascal, J.L. The melting temperature of the most common models of water. *J. Chem. Phys.,* **2005**, *122*(11), 114507. [http://dx.doi.org/10.1063/1.1862245] [PMID: 15836229]

[218] Fernandez, R.G.; Abascal, J.L.F.; Vega, C. The melting point of ice Ih for common water models calculated from direct coexistence of the solid−liquid interface. *J. Chem. Phys.,* **2006**, *124*, 144506--
-11. [http://dx.doi.org/10.1063/1.2183308] [PMID: 16626213]

[219] Rick, S.W. Simulations of ice and liquid water over a range of temperatures. *J. Chem. Phys.,* **2001**, *114*, 2276-2283. [http://dx.doi.org/10.1063/1.1336805]

[220] Vega, C.; Abascal, J.L.; Conde, M.M.; Aragones, J.L. What ice can teach us about water interactions: a critical comparison of the performance of different water models. *Faraday Discuss.,* **2009**, *141*, 251-276. [http://dx.doi.org/10.1039/B805531A] [PMID: 19227361]

[221] Yamada, M.; Mossa, S.; Stanley, H.E.; Sciortino, F. Interplay between time-temperature transformation and the liquid-liquid phase transition in water. *Phys. Rev. Lett.,* **2002**, *88*(19), 195701. [http://dx.doi.org/10.1103/PhysRevLett.88.195701] [PMID: 12005645]

[222] Chaichama, C.; Charters, W.W.; Aye, L. An ice thermal storage computer model. *Appl. Therm. Eng.,* **2001**, *21*, 1769-1778. [http://dx.doi.org/10.1016/S1359-4311(01)00046-1]

[223] Brock, T.D. Life at high temperatures. Evolutionary, ecological, and biochemical significance of organisms living in hot springs is discussed. *Science,* **1967**, *158*(3804), 1012-1019. [http://dx.doi.org/10.1126/science.158.3804.1012] [PMID: 4861476]

[224] Broll, D.; Kaul, C.; Kraemer, A.; Krammer, P.; Richter, T.; Jung, M.; Vogel, H.; Zehner, P. Chemistry in supercritical water. *Angew. Chem. Int. Ed. Engl.,* **1999**, *38*(20), 2998-3014. [http://dx.doi.org/10.1002/(SICI)1521-3773(19991018)38:20<2998::AID-ANIE2998>3.0.CO;2-L] [PMID: 10540405]

[225] Narayan, R.; Antal, M.J. Kinetic elucidation of the acid-catalysed mechanism of 1-propanol dehydration in supercritical water. In: *Supercritical Fluid Science and Technology*; Johnston, K.P.; Penninger, J.M., Eds.; American Chemical Society: Washington, D.C., **1989**; pp. 226-241.
[http://dx.doi.org/10.1021/bk-1989-0406.ch015]

[226] Peterson, A.A.; Vogel, F.; Lachance, R.P.; Froling, M.; Antal, M.J., Jr; Tester, J.W. Thermochemical biofuel production in hydrothermal media: A review of sub- and supercritical water technologies. *Energy Environ. Sci.,* **2008**, *1*, 32-65.
[http://dx.doi.org/10.1039/b810100k]

[227] a) Modell, M.; Reid, R. C.; Amin, S. O. *Gasification process.,* **1978**.
b) Modell, M.; Gaudet, G.G.; Simson, M.; Hong, G.T.; Bieman, K. *Supercritical water*; Solid Wastes Management, **1982**, pp. 26-30.

[228] Antal, M. J., Jr; Brittain, A.; DeAlmeida, C.; Ramayya, S. Heterolysis and homolysis in Supercritical Water. In: ACS Symposium Series, 1987; vol. 329, pp.77-86.
[http://dx.doi.org/10.1021/bk-1987-0329.ch007]

[229] Akiya, N.; Savage, P.E. Roles of water for chemical reactions in high-temperature water. *Chem. Rev.,* **2002**, *102*(8), 2725-2750.
[http://dx.doi.org/10.1021/cr000668w] [PMID: 12175266]

[230] An, P.N.; Halstead, S.; Zhang, S. Classical simulation of acid and base dissociation constants in supercritical water at constant density. *J. Supercrit. Fluids,* **2014**, *86*, 145-149.
[http://dx.doi.org/10.1016/j.supflu.2013.12.015]

[231] Franco, A.; Diaz, A. R. The future challenges for "clean coal technologies": joining efficiency increase and pollutant emission control. Water, 2009, 34, 348-354.

[232] Connolly, J.F. Solubility of hydrocarbons in water near the critical solution temperatures. *J. Chem. Eng. Data,* **1966**, *11*, 13-16.
[http://dx.doi.org/10.1021/je60028a003]

[233] Thomson, W.H.; Snyder, J.R. Mutual solubilities of benzene and water Equilibria in the two phase liquid-liquid region. *J. Chem. Eng. Data,* **1964**, *9*, 516-520.
[http://dx.doi.org/10.1021/je60023a013]

[234] Botti, A.; Bruni, F.; Ricci, M.A.; Soper, A.K. Neutron Diffraction Study of high density supercritical water. *J. Chem. Phys.,* **1998**, *109*, 3180-3184.
[http://dx.doi.org/10.1063/1.476909]

[235] Tassaing, T.; Bellisent-Funel, M.; Guillot, B.; Guissani, Y. The partial pair correlation functions of dense supercritical water. *Europhys. Lett.,* **1998**, *42*, 265-270.
[http://dx.doi.org/10.1209/epl/i1998-00240-x]

[236] a) Yamanaka, K.; Yamaguchi, T.; Wakita, H. Structure of water in the liquid and supercritical states by rapid X-ray diffractometry using an imaging plate detector. *J. Chem. Phys.,* **1994**, 9830-9836.
[http://dx.doi.org/10.1063/1.467948]
b) Guissani, Y.; Guillot, B. A computer simulation study of the liquid-vapor coexistence curve of water. *J. Chem. Phys.,* **1993**, *98*, 8221-8235.
[http://dx.doi.org/10.1063/1.464527]

[237] Jin, Y.; Ikawa, S. Near Infrared spectroscopic study of water at high temperatures and pressures. *J. Chem. Phys.,* **2003**, *119*, 12432-12438.
       [http://dx.doi.org/10.1063/1.1628667]

[238] a) Soper, A.K. *Water and ice structure in the range 220 - 365 K from radiation total scattering experiments*; Condensed Matter Discussions, **2014**, pp. 1-14.
       b) Touba, H.; Mansoori, G.A.; Matteoli, E. Subcritical and supercritical water radial distribution function. *Int. J. Thermophys.,* **1998**, *19*, 1447-1471.
       [http://dx.doi.org/10.1023/A:1021939720336]

[239] Kalinichev, A.G.; Bass, J.D. Hydrogen bonding in supercritical water 2. Computer Simulations. *J. Phys. Chem. A,* **1997**, *101*, 9720-9727.
       [http://dx.doi.org/10.1021/jp971218j]

[240] Kalinichev, A.G.; Churakov, S.V. Size and topology of molecular clusters in supercritical water: a molecular dynamics simulation. *Chem. Phys. Lett.,* **1999**, *302*, 411-417.
       [http://dx.doi.org/10.1016/S0009-2614(99)00174-8]

[241] Fois, F.S.; Sprik, M.; Parinello, M. Properties of supercriitcal water: an *ab-initio* simulation. *Chem. Phys. Lett.,* **1994**, *223*, 411-415.
       [http://dx.doi.org/10.1016/0009-2614(94)00494-3]

[242] di Dio, P.J. Thermal stability of water up to super-critical states: Application of the singular value decomposition and grund functions. *J. Mol. Liq.,* **2013**, *187*, 206-217.
       [http://dx.doi.org/10.1016/j.molliq.2013.07.013]

[243] Yoshida, K.; Matubayasi, N.; Uosaki, Y.; Nakahara, M. Density effect on infrared spectrum for supercritical water in the low- and medium-density region studied by molecular dynamics simulation. *J. Chem. Phys.,* **2012**, *137*(19), 194506.
       [http://dx.doi.org/10.1063/1.4767352] [PMID: 23181325]

[244] Yoshida, K.; Matubayasi, N.; Nakahara, M. Self-diffusion of supercritical water in extremely low-density region. *J. Chem. Phys.,* **2006**, *125*(7), 074307.
       [http://dx.doi.org/10.1063/1.2333511] [PMID: 16942339]

[245] a) Liu, K.; Brown, M.G.; Cruzan, J.D.; Saykally, R.J. Vibration-Rotaion tunneling spectra of the water pentamer: structure and dynamics. *Science,* **1996**, *271*, 62-64.
       [http://dx.doi.org/10.1126/science.271.5245.62]
       b) Pugliano, N.; Saykally, R.J. Measurement of quantum tunneling between chiral isomers of the cyclic water trimer. *Science,* **1992**, *257*(5078), 1937-1940.
       [http://dx.doi.org/10.1126/science.1411509] [PMID: 1411509]

[246] Keutsch, F.N.; Fellers, R.S.; Brown, M.G.; Viant, M.R.; Petersen, P.B.; Saykally, R.J. Hydrogen bond breaking dynamics of the water trimer in the translational and librational band region of liquid water. *J. Am. Chem. Soc.,* **2001**, *123*(25), 5938-5941.
       [http://dx.doi.org/10.1021/ja003683r] [PMID: 11414826]

[247] Kalinichev, A.G.; Churakov, S.V. Thermodynamics and structure of molecular clusters in supercritical water. *Fluid Phase Equilib.,* **2001**, *183*, 271-278.
       [http://dx.doi.org/10.1016/S0378-3812(01)00438-1]

[248] Harker, H.A.; Viant, M.R.; Keutsch, F.N.; Michael, E.A.; McLaughlin, R.P.; Saykally, R.J. Water pentamer: characterization of the torsional-puckering manifold by terahertz VRT spectroscopy. *J. Phys. Chem. A,* **2005**, *109*(29), 6483-6497.
[http://dx.doi.org/10.1021/jp051504s] [PMID: 16833993]

[249] a) Ugalde, J.M.; Alkorta, I.; Elguero, J. Water clusters: Towards an understanding based on first principles of their static and dynamic properties. *Angew. Chem. Int. Ed. Engl.,* **2000**, *39*(4), 717-721.
[http://dx.doi.org/10.1002/(SICI)1521-3773(20000218)39:4<717::AID-ANIE717>3.0.CO;2-E]
[PMID: 10760847]
b) Hodges, M.P.; Stone, A.J.; Xantheas, S.S. Contribution of Many-body terms to the energy for small water clusters: A comparison of *ab initio* calculations and accurate model potentials. *J. Phys. Chem. A,* **1997**, *101*, 9163-9168.
[http://dx.doi.org/10.1021/jp9716851]

[250] Wang, B.; Xin, M.; Dai, X.; Song, R.; Meng, Y.; Han, J.; Jiang, W.; Wang, Z.; Zhang, R. Electronic delocalization in small water rings. *Phys. Chem. Chem. Phys.,* **2015**, *17*(5), 2987-2990.
[http://dx.doi.org/10.1039/C4CP05129G] [PMID: 25485752]

[251] Xantheas, S.S. Cooperativity and hydrogen bonding network in water clusters. *Chem. Phys.,* **2000**, *258*, 225-231.
[http://dx.doi.org/10.1016/S0301-0104(00)00189-0]

[252] Kazimirski, J.K.; Buch, V. Search for Low energy structures of Water Clusters (H₂O)ₙ, n=20 - 22, 48, 123, and 293. *J. Phys. Chem. A,* **2003**, *107*, 9762-9775.
[http://dx.doi.org/10.1021/jp0305436]

[253] Cruzan, J.D.; Viant, M.R.; Brown, M.G.; Saykally, R. Tetrahertz Laser Vibration-Rotation Tunneling Spectroscopy of the water tetramer. *J. Phys. Chem. A,* **1997**, *101*, 9022-9031.
[http://dx.doi.org/10.1021/jp970782r]

[254] Su, J.T.; Xu, X.; Godard, W.A., III Accurate energies and structures for large water clusters using the X3LYP hybrid density functional. *J. Phys. Chem. A,* **2004**, *108*, 10518-10526.
[http://dx.doi.org/10.1021/jp047502+]

[255] Saykally, R.J.; Wales, D.J. Chemistry. Pinning down the water hexamer. *Science,* **2012**, *336*(6083), 814-815.
[http://dx.doi.org/10.1126/science.1222007] [PMID: 22605742]

[256] Kirchner, B. Cooperative *versus* dispersion effects: what is more important in an associated liquid such as water? *J. Chem. Phys.,* **2005**, *123*(20), 204116.
[http://dx.doi.org/10.1063/1.2126977] [PMID: 16351249]

[257] Maheshwary, S.; Patel, N.; Sathyamurthy, N.; Kulkarni, A.D.; Gadre, S.R. Structure and stability of water clusters (H2O)n, n = 8-20: An *Ab-initio* investigation. *J. Phys. Chem. A,* **2001**, *105*, 10525-10537.
[http://dx.doi.org/10.1021/jp013141h]

[258] Kabrede, H.; Hentschke, R. Global minima of water clusters (H₂O)ₙ, N < 25, described by three empirical potentials. *J. Phys. Chem. B,* **2003**, *107*, 3914-3920.
[http://dx.doi.org/10.1021/jp027783q]

[259] Liu, K.; Brown, M.G.; Carter, C.; Saykally, R.J.; Gregory, J.K.; Clary, D.C. Characterisation of a cage form of the water hexamer. *Nature,* **1996**, *381*, 501-503.
[http://dx.doi.org/10.1038/381501a0]

[260] Liu, K.; Loeser, J.G.; Elrod, M.J.; Host, B.C.; Rzepiela, J.A.; Pugliano, N.; Saykally, R.J. Dynamics of structural rearragements in the water trimer. *J. Am. Chem. Soc.,* **1994**, *116*, 3507-3512.
[http://dx.doi.org/10.1021/ja00087a042]

[261] Liu, K.; Brown, M.G.; Saikally, R.J. Tetrahertz Laser Vibration-Rotation Tunneling spectroscopy and dipole moment of a cage form of the water Hexamer. *J. Phys. Chem. A,* **1997**, *101*, 8995-9010.
[http://dx.doi.org/10.1021/jp9707807]

[262] Keutsch, F.N.; Saykally, R.J. Water clusters: untangling the mysteries of the liquid, one molecule at a time. *Proc. Natl. Acad. Sci. USA,* **2001**, *98*(19), 10533-10540.
[http://dx.doi.org/10.1073/pnas.191266498] [PMID: 11535820]

[263] Viant, M.A.; Cruzan, J.D.; Lucas, D.D.; Brown, M.G.; Liu, K.; Saykally, R.J. Pseudorotation in Water trimer Isotopomers using Tetrahertz Laser spectroscopy. *J. Phys. Chem. A,* **1997**, *101*, 9032-9041.
[http://dx.doi.org/10.1021/jp970783j]

[264] Keutsch, F.N.; Cruzan, J.D.; Saykally, R.J. The water trimer. *Chem. Rev.,* **2003**, *103*(7), 2533-2577.
[http://dx.doi.org/10.1021/cr980125a] [PMID: 12848579]

[265] Lin, W.; Han, J.X.; Takahashi, L.K.; Harker, H.A.; Keutsch, F.N.; Saykally, R.J. Terahertz vibration-rotation-tunneling spectroscopy of the water tetramer-d8: combined analysis of vibrational bands at 4.1 and 2.0 THz. *J. Chem. Phys.,* **2008**, *128*(9), 094302.
[http://dx.doi.org/10.1063/1.2837466] [PMID: 18331088]

[266] Bosma, W.B.; Fried, L.E.; Mukamel, S. Simulation of the intemolecular vibration spectra of liquid water and water clusters. *J. Chem. Phys.,* **1992**, *98*, 4413-4421.
[http://dx.doi.org/10.1063/1.465001]

[267] Liu, K.; Brown, M.G.; Cruzan, J.D.; Saykally, R.J. Tetrahertz laser spectroscopy for the water pentamer: structure and hydrogen bond rearrangement dynamics. *J. Phys. Chem. A,* **1997**, *101*, 9011-9021.
[http://dx.doi.org/10.1021/jp970781z]

[268] Cruzan, J.D.; Braly, L.B.; Liu, K.; Brown, M.G.; Loeser, J.G.; Saykally, R.J. Quantifying hydrogen bond cooperativity in water: VRT spectroscopy of the water tetramer. *Science,* **1996**, *271*(5245), 59-62.
[http://dx.doi.org/10.1126/science.271.5245.59] [PMID: 11536731]

[269] Gregory, J.K.; Clary, D.C.; Liu, K.; Brown, M.G.; Saykally, R.J. The water dipole moment in water clusters. *Science,* **1997**, *275*(5301), 814-817.
[http://dx.doi.org/10.1126/science.275.5301.814] [PMID: 9012344]

[270] Burnham, C.J.; Li, J.; Xantheas, S.S.; Leslie, M. The parametirization of a Thole-type all-atom polarizable water model from first principles and its application to the study of water clusters (n= 2-21) and photon spectrum of Ih. *J. Chem. Phys.,* **1999**, *110*, 4566-4581.
[http://dx.doi.org/10.1063/1.478797]

[271] Dill, K.A.; Truskett, T.M.; Vlachy, V.; Hribar-Lee, B. Modeling water, the hydrophobic effect, and ion

solvation. *Annu. Rev. Biophys. Biomol. Struct.,* **2005**, *34*, 173-199.
[http://dx.doi.org/10.1146/annurev.biophys.34.040204.144517] [PMID: 15869376]

[272] Meadley, S.L.; Angell, C.A. Water and its relatives: the stable, supercooled and particularly the stretched, regimes. *Proceedings of the International school of physics "Enrico Fermi" course CLXXXVII,* **2014**.

[273] Poole, P.H.; Sciortino, F.; Grande, T.; Stanley, H.E.; Angell, C.A. Effect of hydrogen bonds on the thermodynamic behavior of liquid water. *Phys. Rev. Lett.,* **1994**, *73*(12), 1632-1635.
[http://dx.doi.org/10.1103/PhysRevLett.73.1632] [PMID: 10056844]

[274] Stanley, H.E.; Barboza, M.C.; Mossa, S.; Netz, P.A.; Sciortino, F.; Starr, F.W.; Yamada, M. *Water at positive and negative pressures*; Condensed Matter, **2002**, pp. 1-10.

[275] a) Vega, C.; Abascal, J.L. Relation between the melting temperature and the temperature of maximum density for the most common models of water. *J. Chem. Phys.,* **2005**, *123*(14), 144504.
[http://dx.doi.org/10.1063/1.2056539] [PMID: 16238404]
b) Pi, H.L.; Aragones, J.L.; Vega, C.; Noya, E.G.; Abascal, J.L.; Gonsalez, M.A.; McBride, C. *Anomalies in water as obtained from computer simulations of the TIP4P/2005 model: density maxima, and density, isothermal compressibility and heat capacity minima*; Condensed Matter. Statistical Mechanics, **2009**.

[276] Roemer, F.; Lervik, A.; Bresme, F. Non equilibrium molecular dynamics simulations of the thermal conductivity of water: a systematic investigation of the SPC/E and TIP4P/2005 models. *J. Chem. Phys.,* **2012**, *137*, 074503.

[277] Kumar, P. *Anomalies of bulk, nano confined and protein-hydration water*; Boston University: Boston, **2008**.

[278] Prielmeier, F.X.; Lang, E.W.; Speedy, R.J.; LA1/4demann, H. Diffusion in supercooled water to 300 MPa. *Phys. Rev. Lett.,* **1987**, *59*(10), 1128-1131.
[http://dx.doi.org/10.1103/PhysRevLett.59.1128] [PMID: 10035147]

[279] Starr, F.W.; Harrington, S.; Sciortino, F.; Stanley, H.E. *Slow dynamics of water under pressure*; Condensed Matter, **1999**, pp. 1-5.

[280] Netz, P.A.; Starr, F.A.; Barbosa, M.A.; Stanley, H.E. Relation between structural and dynamical anomalies in supercooled water. *Physica A,* **2002**, *314*, 470-476.
[http://dx.doi.org/10.1016/S0378-4371(02)01083-X]

[281] Kumar, P.; Buldyrev, S.V.; Becker, S.R.; Poole, P.H.; Starr, F.W.; Stanley, H.E. Relation between the Widom line and the breakdown of the Stokes-Einstein relation in Supercooled Water. *Proc. Natl. Acad. Sci. USA,* **2007**, *104*, 9575-9579.

[282] Stanley, H.E.; Cruz, L.; Harrington, S.T.; Poole, P.H.; Sastry, S.; Sciortino, F.; Starr, F.W.; Zhang, R. Cooperative molecular motions in water: The liquid-liquid critical point hypothesis. *Physica A,* **1997**, *236*, 19-37.
[http://dx.doi.org/10.1016/S0378-4371(96)00429-3]

[283] Sastry, S.; Debenedetti, P.G.; Sciortino, F.; Stanley, H.E. Singularity-free interpretation of the thermodynamics of supercooled water. *Phys. Rev. E Stat. Phys. Plasmas Fluids Relat. Interdiscip. Topics,* **1996**, *53*(6), 6144-6154.

References   *A Journey Through Water*   **273**

[http://dx.doi.org/10.1103/PhysRevE.53.6144] [PMID: 9964976]

[284] Angell, C.A. Supercooled water: Two phases? *Nat. Mater.,* **2014**, *13*(7), 673-675.
[http://dx.doi.org/10.1038/nmat4022] [PMID: 24947781]

[285] Errington, J.R.; Debenedetti, P.G. Relationship between structural order and the anomalies of liquid water. *Nature,* **2001**, *409*(6818), 318-321.
[http://dx.doi.org/10.1038/35053024] [PMID: 11201735]

[286] Nemethy, G. Hydrophobic interactions. *Angew. Chem. Int. Ed. Engl.,* **1967**, *6*(3), 195-206.
[http://dx.doi.org/10.1002/anie.196701951] [PMID: 4962948]

[287] Wilhelm, E.; Battino, R.; Wilcock, R.J. Low-pressure solubility of gases in liquid water. *Chem. Rev.,* **1977**, *77*, 219-262.

[288] Hadlington, S. Molecular surgery stiches up water. *Chemistry world,* **2016**, 29.

[289] Gunther, M. Alkali metal explosion explained. *Chemistry world,* **2015**, 25.

[290] Saito, S.; Ohmine, I. Dynamics and relaxation of an intermediate size water cluster $(H_2O)_{108}$. *J. Chem. Phys.,* **1994**, *101*, 6063-6075.
[http://dx.doi.org/10.1063/1.467321]

[291] Stanley, H.E.; Buldyrev, S.V.; Canpolat, M.; Meyer, M.; Mishima, O.; Sadr-Lahijany, M.R.; Scala, A.; Starr, F.W. The puzzling statistical physics of liquid water. *Physica A,* **1998**, *257*, 213-232.
[http://dx.doi.org/10.1016/S0378-4371(98)00264-7]

[292] Alvarez, S. What we mean when we talk about bonds. *Chemistry world,* **2015**, 36-37.

[293] Ball, P. Scientists report 'new state of water'. *Chemistry world,* **2016**, 28.

[294] The nano4water cluster. Available from: https://nano4water.vito.be/Pages/home.aspx

[295] Ahmed, E.M. Hydrogel: Preparation, characterization, and applications: A review. *J. Adv. Res.,* **2015**, *6*(2), 105-121.
[http://dx.doi.org/10.1016/j.jare.2013.07.006] [PMID: 25750745]

Glossary

Amplifier: An electronic equipment that increases input signal to the desired level as output

Auto ionisation: A process by which a molecule ionises itself into constituent cations and anions (in the case of water, hydronium ions and hydroxyl ions)

Basis set: A set of functions that represent molecular orbitals by the linear combination of atomic orbitals

Bifurcated Hydrogen Bond (BHB): The bonding pattern by which a hydrogen atom can participate in two hydrogen bonds, rather than the conventional one

Binding energy: The energy required to crush a molecule/atom into its constituents

Boiling point: The temperature at which a substance converted to its vapour phase

Bond cleavage: The splitting or breaking of a chemical bond

Born−Oppenheimer approximation: The approximation that the movement of nucleus is negligible with respect to electrons so that their motions can be separated

Bragg's equation: The mathematical condition that connects the wavelength, inter planar spacing and angle of incident wave to the surface

Catalytic activity: The action of catalysts for enhancing the rate of reaction by reducing activation energy

Chemical Vapour Decomposition (CVD): The process by which a chemical substrate is subject to vapours in order to make the surface coated with the material contained in the vapour

Climate change: The change in weather pattern

Clusters: Association of molecules constituted in a regular fashion

Coefficient of compressibility: The measure of variation in volume with respect to pressure at constant temperature

Coefficient of thermal expansion: the measure of variation of volume of a material with respect to temperature at constant pressure

Coordination number: The number of neighbours of a given atom

Critical Pressure: The minimum pressure that must be applied to bring about the liquefaction at the critical temperature (see critical temperature)

Critical Temperature (T_c): The temperature of a substance above which its gas form cannot be made to liquefy regardless of the pressure applied. Critical temperature is the highest temperature at which a substance can exist as a liquid

Critical Solution Temperature (CST): The temperature above which water and solute are completely miscible

Crystallisation: A process by which a solid (crystal) is formed

D-defect: A defect in hydrogen bond caused by the presence of two protons in it

Density: The ratio of mass to volume

Deterministic: A system that evolves without randomness

Dimer: a chemical structure formed as a result of the union of two identical sub units, without needing to have bond with each other

Electric field: Electric force per unit charge

Electromagnetic wave: The wave resulted when an electric wave combines with a magnetic wave

Electronic correlation: The interactions between electrons in a system

Electrostatic interactions: The interactions exerted by charged species, positively charged and negatively charged

Entropy: The measure of a system's disorder, which stands for the unavailable thermal energy in the system which cannot be converted to mechanical work.

Ergodic hypothesis: The theory posits that every possible micro states of a surface of constant energy have equal probability to be visited by the system

Forcefield: The mathematical form and parameter sets required to calculate the potential energy of a system

Foulants: The matter absorbed onto the membranes (filers) used in the water

Filtration process: The process of the separation of suspended solid particles from a liquid by passing it through a filter

Fragile Liquids: The liquids whose dynamics do not slow down linearly towards glass transition temperature

Free energy: The maximum energy that can be converted to work

Freezing point: The temperature at which a liquid freezes

Gaussian function: This is a function that takes a bell shaped curve. Three constants in the function indicate the height, width and the centre of the curve

Gaussian type orbitals: One of the types of orbitals that are used for generating molecular orbitals

Glass transition: The transition from liquid state to highly viscous amorphous state

Greenhouse gas: A gas that can absorb Infra−Red (IR) rays, thereby increasing the temperature of the atmosphere

Hamiltonian: An operator that represents the total energy of the system

Hartree−Fock calculation: The fundamental theoretical method for the determination of energy and wave function of a system

HDA (High Density Amorphous): A high density amorphous form of water

Hexamer: A chemical structure formed as a result of the union of six identical sub units, without needing to have bond with each other

Hole theory: An interpretation to account of for the properties of liquids, accounting for vacant spaces between particles

Hydrolysis: The process by which a chemical bond is broken by the addition of water

Hydrophobic hydration: Hydration of non−soluble compounds

Hydrophilic: A substance that has strong affinity towards water

Hyperquenching: Very high cooling of materials

Intermolecular forces: Forces of attraction between molecules

Infra-Red: An electromagnetic radiation that has higher wavelength than the visible light

Kauzmann temperature: The temperature at which the entropy of liquid would be lower than that of a solid

Kinetic energy: The energy acquired by an object due to motion.

LDA: Low Density Amorphous form of water (observed at lower temperatures)

Lattice theory: A proposition designed to understand the behaviour of liquids

L Defect: A defect that leaves a hydrogen bond with no proton in it

Ligands: It is a group or molecule that binds to central atom to form a coordination compound

Lipids: A group of naturally occurring organic compounds

London force: The weakest intermolecular force

Long range forces: Types of forces are even experienced at long distance

Magnetisation: The degree to which a material magnetised

Means square displacement: Measure of deviation over time between the position of a particle and some reference particle

MENA: Abbreviation for Middle East and North Africa, the countries within this region

Metastable: The state of pseudo stability, which gives way to more stable state upon the action of external forces

Molecular Dynamics: A computational algorithm that is used to investigate the time dependent properties of molecules

Monomer: The smallest unit (one molecule) in a polymer

Moor's law: It states that the number of transistors per square inch of Integrated Circuits double every year

NMR: A technique that is used to characterise the internal structure of a material, exploiting magnetic properties of the material

Octamer: A chemical structure formed as a result of the union of eight identical sub units, without needing to have bond with each other

Orbital: Holder of electrons

Pentamer: A chemical structure formed as a result of the union of five identical sub units, without needing to have bond with each other

Peptide bond: A covalent chemical bond between two amino acid molecules

Phase: A region of uniform chemical composition and unique physical properties

Phonon: A unit of vibrational energy arising due to the vibration of oscillating atoms within a crystal

Polytype: A crystal being in more than one form

Proton Hopping: The jumping of one proton in one molecule to the other

Puckering: The distortion of certain atoms in a compound from the molecular plane

Radial distribution function: A function that describes the variance of density with respect to the distance

Raman scattering: A type of inelastic scattering of a photon upon interaction with matter

Rayleigh scattering: Elastic scattering of light by particles having much lower wavelength

Schrödinger equation: A differential equation which shows the dependence of a state of system with time

Self Consistent Field theory: A numerical approach used to employ for solving many particle (many electron) problems in chemistry and physics

Short range forces: A type of force that is confined to very short distances

Slater type orbitals: One of the types of orbitals that are used for generating molecular orbitals

Spectroscopy: A group of techniques that are employed for the characterisation of molecules

Spinodal line: The locus of points that refers to the limit of local stability with respect to small fluctuations

Stochastic event: An event that is unpredictable due to randomness

Stokes line: The radiation of particular wavelength associated with Raman Scattering

Strong Liquids: The liquids that show a systematic variation in certain physical properties (for example relaxation times) upon variation of temperature

Supercooled: A state achieved by lowering temperature of a liquid, without being converted to a solid

Superheated: A state achieved by increasing temperature of a liquid, without being converted to a vapour

Thermoelectric effect: Conversion of the temperature gradient to the differences in voltage

Thermodynamics: A branch of physical science concerning heat and temperature

Thermoluminescence: Emission of light that is not resulted from heat

Tetramers: A chemical structure formed as a result of the union of four identical sub units, without needing to have bond with each other

Trimer: A chemical structure formed as a result of the union of three identical sub units, without needing to have bond with each other

Tyndal flowers: Small cavities (often in hexagonal shape) that are appeared in ice crystals

Van der Waal force: A type of attractive or repulsive force exerted between neighbouring molecules

Vogel–Fulcher–Tamman (VFT) equation: An equation described for the description of temperature dependence of viscosity

Walrafen pentamers: A water cluster consists of five water molecules wherein four water molecules occupying the corners of regular tetrahedron linked to a water molecule at the centre

Water crisis: A situation that affects most part of the world due to the availability of clean water

Widom line: A line emanating from critical points in phase diagram, on which certain physical quantities show maximum

SUBJECT INDEX

A

Ab-initio 185, 269, 270
Absorption 11, 20, 50, 112, 116, 182, 192, 193, 248
Acceptor 17, 36, 104, 105, 110, 111, 114, 116, 120, 121, 161, 197, 199, 200, 202, 204
Adhesion 155
ADP 15
Aeroplane 127
Africa 13, 246, 247, 250, 277
Aggregates 138, 195, 207
Ahlriches 93
Aldol 185, 187
Alpha 27, 33, 50
AMBER 82
Amino acids 14, 17
Ammonia 110, 162, 188
Ammonium fluoride 162
Ammonium hydroxide 162, 163
Amorphization 160
Amorphous 40, 52, 60, 124, 136, 152, 160, 166, 167, 171, 181, 184, 242, 255, 256, 258, 264, 265, 267, 276, 277
Applequist 34
Aquatic 12, 14, 182
Aquifers 6, 11, 12, 19
Argentina 9
Aromatic 41
Arsenic 12, 19
Asia Pacific 8
Asymmetric 62, 103, 104
Atactic 27, 41
Atomic spectrum 61
ATP 15
Australian tube worm 14

B

Bacteria 12, 19, 186
Bahrain 6
Barnacle 14
Basin 4, 8, 12, 49, 50, 246
Basis Sets 75, 77, 79, 96, 97
Belgium 7, 18
Benzamide 185, 187
Benzene 12, 188, 268
Bernal 87, 110, 161, 162, 167, 184, 243, 257, 261, 265
Berthelot 82
Beta 27, 50, 256
Bifurcation Tunnelling 200, 201
Blood 14, 17
Blue shift 67
Bonding 35, 36, 59, 63, 64, 67, 78, 81, 92, 102, 112, 114, 121, 122, 124, 129, 130, 161, 171, 173, 177, 185, 196, 197, 210, 213, 218, 241, 243, 245, 256, 269, 270, 274
Born Mayer 83
Butanediol 187
Butene 151, 152

C

CAMBA 8
Cannizzaro 185, 187
Carbohydrates 104
Carbonates 41
Catalytic 16, 17, 154, 187, 243, 252, 274
Cell 15, 16, 38, 56, 57, 60, 65, 167, 173, 252
Chaplin 114, 125, 260
Charge cloud 97
Chemical shift 64

www.ingramcontent.com/pod-product-compliance
Lightning Source LLC
Chambersburg PA
CBHW041725210326
41598CB00008B/783